Titles in This Series

For a complete list of titles in this series, visit the
AMS Bookstore at **www.ams.org/bookstore/.**

A Course in
Approximation
Theory

A Course in Approximation Theory

Ward Cheney
Will Light

Graduate Studies
in Mathematics

Volume 101

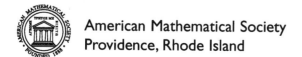

American Mathematical Society
Providence, Rhode Island

2000 *Mathematics Subject Classification.* Primary 41–01.

For additional information and updates on this book, visit
www.ams.org/bookpages/gsm-101

Library of Congress Cataloging-in-Publication Data

Cheney, E. W. (Elliott Ward), 1929–
  A course in approximation theory / Ward Cheney, Will Light.
    p. cm. — (Graduate studies in mathematics ; v. 101)
  Originally published: Pacific Grove : Brooks/Cole Pub. Co., c2000.
  Includes bibliographical references and index.
  ISBN 978-0-8218-4798-5 (alk. paper)
  1. Approximation theory—Textbooks. I. Light, W. A. (William Allan), 1950– II. Title.

QA221.C44 2009
511′.4—dc22
                                                                    2008047417

To our wives, Victoria and Anita.

Preface

This book offers a graduate-level exposition of selected topics in modern approximation theory. A large portion of the book focuses on multivariate approximation theory, where much recent research is concentrated. Although our own interests have influenced the choice of topics, the text cuts a wide swath through modern approximation theory, as can be seen from the table of contents. We believe the book will be found suitable as a text for courses, seminars, and even solo study. Although the book is at the graduate level, it does not presuppose that the reader already has taken a course in approximation theory.

Topics of This Book

A central theme of the book is the problem of interpolating data by smooth multivariable functions. Several chapters investigate interesting families of functions that can be employed in this task; among them are the polynomials, the positive definite functions, and the radial basis functions. Whether these same families can be used, in general, for approximating functions to arbitrary precision is a natural question that follows; it is addressed in further chapters.

The book then moves on to the consideration of methods for concocting approximations, such as by convolutions, by neural nets, or by interpolation at more and more points. Here there are questions of limiting behavior of sequences of operators, just as there are questions about interpolating on larger and larger sets of nodes.

A major departure from our theme of multivariate approximation is found in the two chapters on univariate wavelets, which comprise a significant fraction of the book. In our opinion wavelet theory is so important a development in recent times—and is so mathematically appealing—that we had to devote some space to expounding its basic principles.

The Style of This Book

In style, we have tried to make the exposition as simple and clear as possible, electing to furnish proofs that are complete and relatively easy to read without the reader needing to resort to pencil and paper. Any reader who finds this style too prolix can proceed quickly over arguments and calculations that are routine. To paraphrase Shaw: We have done our best to avoid conciseness! We have also made considerable efforts to find simple ways to introduce and explain each topic. We hope that in doing so, we encourage readers to delve deeper into some areas. It should be borne in mind that further exploration of some topics may require more mathematical sophistication than is demanded by our treatment.

Organization of the Book

A word about the general plan of the book: we start with relatively elementary matters in a series of about ten short chapters that do not, in general, require more of the reader than undergraduate mathematics (in the American university system). From that point on, the gradient gradually increases and the text becomes more demanding, although still largely self-contained. Perhaps the most significant demands made on the technical knowledge of the reader fall in the areas of measure theory and the Fourier transform. We have freely made use of the Lebesgue function spaces, which bring into play such measure-theoretic results as the Fubini Theorem. Other results such as the Riesz Representation Theorem for bounded linear functionals on a space of continuous functions and the Plancherel Theorem for Fourier transforms also are employed without compunction; but we have been careful to indicate explicitly how these ideas come into play. Consequently, the reader can simply accept the claims about such matters as they arise. Since these theorems form a vital part of the equipment of any applied analyst, we are confident that readers will want to understand for themselves the essentials of these areas of mathematics. We recommend Rudin's *Real and Complex Analysis* (McGraw-Hill, 1974) as a suitable source for acquiring the necessary measure theoretic ideas, and the book *Functional Analysis* (McGraw-Hill, 1973) by the same author as a good introduction to the circle of ideas connected with the Fourier transform.

Additional Reading

We call the reader's attention to the list of books on approximation theory that immediately precedes the main section of references in the bibliography. These books, in general, are concerned with what we may term the "classical" portion of approximation theory—understood to mean the parts of the subject that already were in place when the authors were students. As there are very few textbooks covering recent theory, our book should help to fill that "much needed gap," as some wag phrased it years ago. This list of books emphasizes only the systematic textbooks for the subject as a whole, not the specialized texts and monographs.

Acknowledgments

It is a pleasure to have this opportunity of thanking three agencies that supported our research over the years when this book was being written: the Division of Scientific Affairs of NATO, the Deutsche Forschungsgemeinschaft, and the Science and Engineering Research Council of Great Britain. For their helpful reviews of our manuscript, we thank Robert Schaback, University of Göttingen, and Larry Schumaker, Vanderbilt University. We acknowledge also the contribution made by many students, who patiently listened to us expound the material contained in this book and who raised incisive questions. Students and colleagues in Austin, Leicester, Würzburg, Singapore, and Canterbury (NZ) all deserve our thanks. Professor S. L. Lee was especially helpful.

A special word of thanks goes to Ms. Margaret Combs of the University of Texas Mathematics Department. She is a superb technical typesetter in the modern sense of the word, that is, an expert in TEX. She patiently created the TEX files for lecture notes, starting about six years ago, and kept up with the constant editing of these notes, which were to become the backbone of the book.

The staff of Brooks/Cole Publishing has been most helpful and professional in guiding this book to its publication. In particular, we thank Gary Ostedt, sponsoring editor; Ragu Raghavan, marketing representative; and Janet Hill, production editor, for their personal contact with us during this project.

How to Reach Us

Readers are encouraged to bring errors and suggestions to our attention. E-mail is excellent for this purpose: our addresses are cheney@math.utexas.edu and pwl@mcs.le.ac.uk. A web site for the book is maintained at http:www.math.utexas.edu/user/cheney/ATBOOK.

Ward Cheney

Will Light

Contents

A Course in
Approximation
Theory

1

Introductory Discussion of Interpolation

We shall be concerned with real-valued functions defined on a domain X, which need not be specified at this moment. (It will often be a subset of $\mathbb{R}$, $\mathbb{R}^2$, ..., but can be more general.) In the domain X a set of n distinct points is given:

$$\mathcal{N} = \{x_1, x_2, \ldots, x_n\}$$

These points are called **nodes**, and $\mathcal{N}$ is the **node set**. For each node x_i an **ordinate** $\lambda_i \in \mathbb{R}$ is given. (Each λ_i is a real number.) The problem of **interpolation** is to find a suitable function $F : X \to \mathbb{R}$ that takes these prescribed n values. That is, we want

$$F(x_i) = \lambda_i \qquad (1 \le i \le n)$$

When this occurs, we say that F **interpolates** the given data $\{(x_i, \lambda_i)\}_{i=1}^{n}$. Usually F must be chosen from a preassigned family of functions on X.

A wide variety of functions F may be suitable. Figures 1.1 and 1.2 show 12 different interpolation functions for a single data set. The nodes are 5 real numbers. They and the specified ordinates are given in this table:

x	1.2	3.6	4.3	6.1	7.8
y	1.2	3.1	-1.3	2.7	1.4

In Figure 1.1a, the raw data points are shown. In Figures 1.1b to 1.1f, F has the form $F(x) = \sum_1^5 c_j u(x - x_j)$, in which u is a function of our choosing. First we took a B-spline of degree 0. To avoid the discontinuous nature of this example, we then took u to be a B-spline of degree 1, as shown in Figure 1.1c. To avoid discontinuities in the first two derivatives, we then let u be a cubic B-spline, as in Figure 1.1d. In Figure 1.1e we show the interpolant when $u(x) = |x|$, and in Figure 1.1f we used $u(x) = |x|^{1/2}$.

1

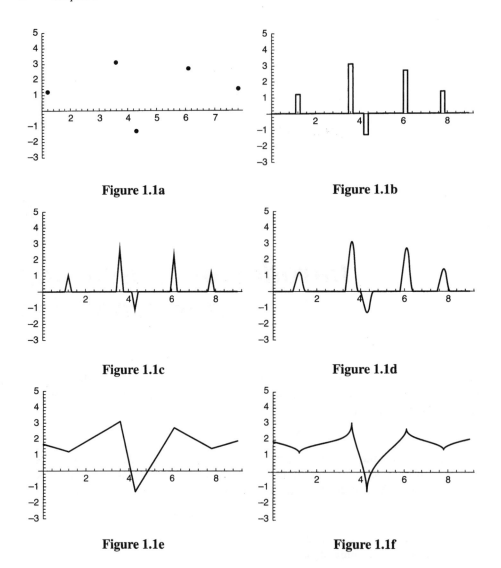

Figure 1.1a

Figure 1.1b

Figure 1.1c

Figure 1.1d

Figure 1.1e

Figure 1.1f

Further examples are shown in Figure 1.2. Here we have used the same data as in Figure 1.1, but a different choice of interpolating functions. Specifically, 1.2a employs a fourth-degree polynomial; 1.2b employs a natural cubic spline; 1.2c is given by the Interpolation command in Mathematica and is also a cubic spline. In 1.2d, we used a cubic B-spline, B^3, determined by integer knots, and interpolated with $\sum_1^5 c_i B^3 (x - x_i)$. In 1.2e, we used $\sum c_i e^{-(x-x_i)^2}$, and in 1.2f we used, in the same manner, a 0-degree B-spline. Some variations in scaling are noticeable in the figures.

The examples in Figures 1.1 and 1.2 suggest the great diversity among different types of interpolating functions. The selection of an appropriate type of interpolant must be made according to further criteria, above and beyond the basic requirement of inter-

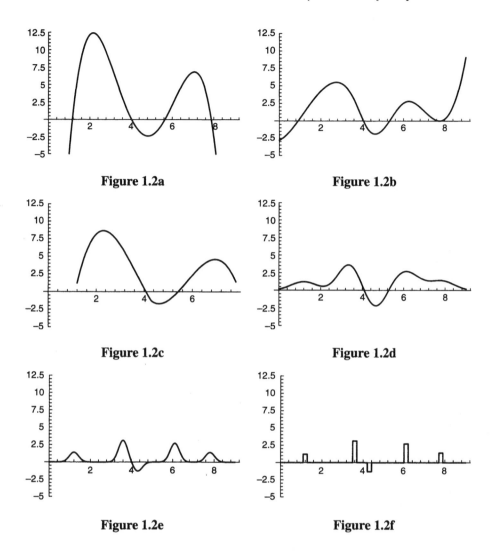

Figure 1.2a

Figure 1.2b

Figure 1.2c

Figure 1.2d

Figure 1.2e

Figure 1.2f

polation. For example, in a specific application we may want the interpolating function to have a continuous first derivative. (That requirement would disqualify most of the functions in Figure 1.1.)

The **linear interpolation problem** is a special case that arises when F is to be chosen from a prescribed n-dimensional vector space of functions on X. Suppose that U is this vector space and that a basis for U is $\{u_1, u_2, \ldots, u_n\}$. The function F that we seek must have the form

$$F = \sum_{j=1}^{n} c_j u_j$$

When the interpolation conditions are imposed on F, we obtain

$$\lambda_i = F(x_i) = \sum_{j=1}^{n} c_j u_j(x_i) \qquad (1 \leq i \leq n)$$

This is a system of n linear equations in n unknowns. It can be written in matrix form as $Ac = y$, or in detail as

$$\begin{bmatrix} u_1(x_1) & u_2(x_1) & \cdots & u_n(x_1) \\ u_1(x_2) & u_2(x_2) & \cdots & u_n(x_2) \\ \cdots & \cdots & \cdots & \cdots \\ u_1(x_n) & u_2(x_n) & \cdots & u_n(x_n) \end{bmatrix} \begin{bmatrix} c_1 \\ c_2 \\ \cdots \\ c_n \end{bmatrix} = \begin{bmatrix} \lambda_1 \\ \lambda_2 \\ \cdots \\ \lambda_n \end{bmatrix}$$

The $n \times n$ matrix A appearing here is called the **interpolation matrix.** In order that our problem be solvable for any choice of ordinates λ_i it is necessary and sufficient that the interpolation matrix be nonsingular. The ideal situation is that this matrix be nonsingular for all choices of n distinct nodes.

THEOREM 1. *Let U be an n-dimensional linear space of functions on X. Let $x_1, x_2, \ldots, x_n$ be n distinct nodes in X. In order that U be capable of interpolating arbitrary data at the nodes it is necessary and sufficient that zero data be interpolated only by the zero-element in U.*

Proof. The space U can furnish an interpolant for arbitrary data if and only if the interpolation matrix A is nonsingular. An equivalent condition on the matrix A is that the equation $Ac = 0$ can be true only if $c = 0$.   ∎

Example. Let $X = \mathbb{R}$ and let $u_j(x) = x^{j-1}$, for $j = 1, 2, \ldots, n$. An $n \times n$ interpolation matrix in this special case is called a **Vandermonde matrix.** It looks like this:

$$V = \begin{bmatrix} 1 & x_1 & x_1^2 & \cdots & x_1^{n-1} \\ 1 & x_2 & x_2^2 & \cdots & x_2^{n-1} \\ \cdots & \cdots & \cdots & \cdots & \cdots \\ 1 & x_n & x_n^2 & \cdots & x_n^{n-1} \end{bmatrix}$$

The determinant of V is given by the formula

$$\det V = \prod_{1 \leq j < i \leq n} (x_i - x_j)$$

This is obviously nonzero if and only if the nodes are distinct. Hence the interpolation problem has a unique solution for any choice of distinct nodes. We can also use Theorem 1 to see that V is nonsingular. Thus, we consider the "homogeneous" linear problem, in which we attempt to interpolate zero data. The solution will be a polynomial of degree at most $n - 1$ that takes the value 0 at each of the n nodes. Since a nonzero polynomial of degree at most $n - 1$ can have at most $n - 1$ zeros, we conclude that the zero polynomial is the only possible solution. ∎

The Vandermonde matrix occurs often in mathematics. Refer to Rushanan [Rush], Grosof and Taiani [GT], Cheney [C1], for example. It is ill-conditioned for numerical work. See Gautschi, [Gau1, Gau2].

An n-dimensional vector space U of functions on a domain X is said to be a **Haar space** if the only element of U which has more than $n - 1$ roots in X is the zero element. The next theorem provides some properties equivalent to the Haar property. In the theorem, we refer to **point-evaluation functionals.** If V is a vector space of functions on a set X, and if x is a point of X, then the point-evaluation functional corresponding to x is denoted by x^* and is defined on V by

$$x^*(f) = f(x) \qquad (f \in V)$$

Obviously x^* is linear, because

$$x^*(\alpha f + \beta g) = (\alpha f + \beta g)(x) = \alpha f(x) + \beta g(x) = \alpha x^*(f) + \beta x^*(g)$$

THEOREM 2. *Let U have the basis $\{u_1, u_2, \ldots, u_n\}$. These properties are equivalent:*

  a. *U is a Haar space*
  b. *$\det\left(u_j(x_i)\right) \neq 0$ for any set of distinct points $x_1, x_2, \ldots, x_n$ in X*
  c. *For any distinct points $x_1, x_2, \ldots, x_n$ in X, the set of point-evaluation functionals $x_1^*, x_2^*, \ldots, x_n^*$ spans the algebraic dual space U^**
  d. *If $x_1, x_2, \ldots, x_m$ are distinct in X and if $\sum_{i=1}^m \lambda_i u_j(x_i) = 0$ for $j = 1, 2, \ldots, n$ then either at least $n + 1$ of the coefficients λ_i are nonzero, or $\sum_{i=1}^m |\lambda_i| = 0$*

Proof. To show that **a** implies **b**, suppose **b** false. Since the determinant of $\left(u_j(x_i)\right)$ is zero, the matrix is singular, and there exists a nonzero vector $(c_1, c_2, \ldots, c_n)$ such that $\sum_{j=1}^n c_j u_j(x_i) = 0$, $(1 \leq i \leq n)$. Put $u = \sum_{j=1}^n c_j u_j$. Since $\{u_1, u_2, \ldots, u_n\}$ is linearly independent, $u \neq 0$. But $u(x_i) = 0$ for $1 \leq i \leq n$. Hence **a** is false.

To show that **b** implies **c**, suppose **b** true. Then the set $\{x_1^*, x_2^*, \ldots, x_n^*\}$ is linearly independent when these functionals are restricted to U. Indeed, if $\sum_{i=1}^n a_i x_i^* | U = 0$, then $\sum_{i=1}^n a_i x_i^*(u_j) = 0$ for $1 \leq j \leq n$, and by **b**, $\sum_{i=1}^n |a_i| = 0$. Since U^* is of dimension n, the functionals span U^*.

To show that **c** implies **d**, assume **c**. Let $x_1, \ldots, x_m$ be distinct points that satisfy $\sum_{i=1}^m \lambda_i u_j(x_i) = 0$ for $1 \leq j \leq n$. If $m \leq n$, then by **c** we can take additional points and obtain a basis $\{x_1^*, \ldots, x_n^*\}$ for U^*. Then the subset $\{x_1^*, \ldots, x_m^*\}$ is linearly independent on U and all λ_i are zero. If $m > n$ and $\sum_{i=1}^m |\lambda_i| \neq 0$ then at least $n + 1$ of the λ_i are nonzero, for otherwise we will have a nontrivial linear combination of n (or fewer) x_i^* that vanishes on U, contrary to **c**.

To prove that **d** implies **a**, assume **d** and take $m = n$. Then the equation $\sum_{i=1}^n \lambda_i u_j(x_i) = 0$ for $1 \leq j \leq n$ implies $\sum_{i=1}^n |\lambda_i| = 0$. Hence the matrix $\left(u_j(x_i)\right)$ is nonsingular. Thus if $(c_1, c_2, \ldots, c_n) \neq 0$, we cannot have $\sum_{j=1}^n c_j u_j(x_i) = 0$. In other words, a nonzero member of U cannot have n zeros. ∎

Any basis for a Haar space is called a **Chebyshev system.** Here are some examples of Chebyshev systems on $\mathbb{R}$:

1. $1, x, x^2, \ldots, x^n$
2. $e^{\lambda_1 x}, e^{\lambda_2 x}, \ldots, e^{\lambda_n x}$ $(\lambda_1 < \lambda_2 < \cdots < \lambda_n)$
3. $1, \cosh x, \sinh x, \ldots, \cosh nx, \sinh nx$

Here are some Chebyshev systems on $(0, \infty)$:

4. $x^{\lambda_1}, x^{\lambda_2}, \ldots, x^{\lambda_n}$ $(\lambda_1 < \lambda_2 < \cdots < \lambda_n)$
5. $(x + \lambda_1)^{-1}, \ldots, (x + \lambda_n)^{-1}$ $(0 \le \lambda_1 < \lambda_2 < \cdots < \lambda_n)$

Here is a Chebyshev system on the circle $\mathbb{R}/2\pi$:

6. $1, \cos\theta, \sin\theta, \ldots, \cos n\theta, \sin n\theta$

The notation $\mathbb{R}/2\pi$ denotes the set of reals with an equivalence relation: $x \equiv y$ if $x - y$ is an integer multiple of 2π.

Are there any Chebyshev systems of continuous functions on $\mathbb{R}^2$ and on the higher-dimensional Euclidean spaces? No, there is an immediate and absolute barrier:

THEOREM 3. *On $\mathbb{R}^2, \mathbb{R}^3, \ldots$ there are no Haar subspaces of continuous functions except one-dimensional ones.*

Proof. Suppose that $\{u_1, u_2, \ldots, u_n\}$ is a Chebyshev system of continuous functions on $\mathbb{R}^s$, where $s \ge 2$ and $n \ge 2$. By Theorem 2, $\det(u_j(x_i)) \ne 0$ for any set of distinct nodes $x_1, x_2, \ldots, x_n$ in $\mathbb{R}^s$. Select a closed path in $\mathbb{R}^s$ containing x_1 and x_2 but no other nodes. By moving x_1 and x_2 in the same direction continuously along this path, we can made x_1 and x_2 exchange positions without allowing them to coincide at any stage in the process. In the determinant above, rows 1 and 2 will exchange positions, and the determinant will change sign. Since the determinant is a continuous function of x_1 and x_2, it will assume the value 0 during this process, contrary to Theorem 2.   ∎

Even on domains X that are subsets of $\mathbb{R}^s (s \ge 2)$ it may be impossible to have Haar subspaces (of continuous functions) with dimension 2 or higher. Suppose that X is, or contains, a subset homeomorphic to the letter Y. (For example, in $\mathbb{R}^2$ consider the case when X is the union of two nonparallel lines.) Then there can exist no continuous Haar space of dimension 2 or more on X. The argument is as before. By a continuous movement of nodes x_1 and x_2 along the Y-shaped figure, their positions can interchange without being coincident at any stage. See the diagram.

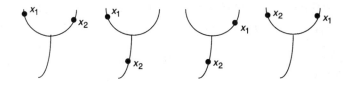

The general theorem along these lines is due to Mairhuber [Mai]. Later elucidations and extensions are by Curtis [Cu], Sieklucki [Si], Lutts [Lut], McCullough and Wulbert [McW], Schoenberg [S10] and Schoenberg and Yang [SY]. The result is as follows.

THEOREM 4. *Let X be a compact Hausdorff space, and suppose that C(X) contains a Haar subspace of dimension 2 or more. Then X is homeomorphic to a subset of the circumference of a circle.*

Linear interpolation is closely connected to the notion of linear independence of a set of functions. Suppose that $\{u_1, u_2, \ldots, u_m\}$ is a set of m real-valued functions defined on a set X. The set of functions is said to be **linearly independent on** X if this implication is valid:

(1)
$$\sum_{j=1}^{m} c_j u_j(x) = 0 \text{ for all } x \in X \implies \sum_{j=1}^{m} |c_j| = 0$$

If D is a subset of X, we will say that $\{u_1, u_2, \ldots, u_m\}$ is **linearly independent on** D if the following implication is valid:

(2)
$$\sum_{j=1}^{m} c_j u_j(x) = 0 \text{ for all } x \in D \implies \sum_{j=1}^{m} |c_j| = 0$$

Observe that the property in Implication (2) is stronger than the one in (1).

To connect these notions to the interpolation problem, let us consider a finite set of nodes in X:

(3)
$$\mathcal{N} = \{x_1, x_2, \ldots, x_n\}$$

Interpolating arbitrary data on $\mathcal{N}$ by the functions u_j requires the solution of the system

(4)
$$\sum_{j=1}^{m} c_j u_j(x_i) = \lambda_i \qquad (1 \leq i \leq n)$$

THEOREM 5. *Equation (4) is solvable for arbitrary $\lambda_1, \ldots, \lambda_n$ if and only if the $n \times m$ matrix $A_{ij} = u_j(x_i)$ has rank n.*

Proof. Write Equation (4) in the expanded form

(5)
$$c_1 \begin{bmatrix} u_1(x_1) \\ \vdots \\ u_1(x_n) \end{bmatrix} + \cdots + c_m \begin{bmatrix} u_m(x_1) \\ \vdots \\ u_m(x_n) \end{bmatrix} = \begin{bmatrix} \lambda_1 \\ \vdots \\ \lambda_n \end{bmatrix}$$

This system is solvable if and only if the vector $\lambda = (\lambda_1, \lambda_2, \ldots, \lambda_n)^T$ is in the column space of A. Equation (5) is solvable for all λ if and only if the column space of A contains $\mathbb{R}^n$. This occurs if and only if the column rank (and the rank) of A is n. (The rank cannot exceed n.) ∎

THEOREM 6. *If $m \leq n$ and if Equation (4) is solvable for arbitrary $\lambda_1, \ldots, \lambda_n$, then the set of functions $u_1, \ldots, u_m$ is linearly independent on the set of nodes $x_1, \ldots, x_n$.*

Proof. If $\{u_1, \ldots, u_m\}$ is linearly dependent on the node set, then for suitable c, not zero, we have $\sum_{j=1}^{m} c_j u_j(x_i) = 0$ for $1 \leq i \leq n$. The set of columns displayed in Equation (5)

is linearly dependent and has at most n elements. Hence, the column space of A is a proper subspace of $\mathbb{R}^n$, and System (5) fails to be solvable for some vectors λ.   ∎

Problems

One must learn by doing the thing;
for though you think you know it,
you have no certainty until you try.

—Sophocles

1. Prove the formula given for a Vandermonde determinant.

2. Prove that the functions $e^{\lambda_j x}$ ($1 \leq j \leq n$) form a Chebyshev system on $\mathbb{R}$.

3. Under what conditions will the functions $\cosh \lambda_j x$ ($1 \leq j \leq n$) form a Chebyshev system on $\mathbb{R}$? What about $(0, \infty)$?

4. Consider the space $C(X)$, where X is a compact Hausdorff space. Use the norm $\|f\| = \sup_{x \in X} |f(x)|$ in this space. Prove that each point-evaluation functional has norm 1.

5. In the space $C[0, 1]$ use the norm $\|f\| = \int_0^1 |f(x)|\, dx$. What is the norm of a point-evaluation functional?

6. Prove that if X is a compact Hausdorff space and if $C(X)$ contains an n-dimensional Haar subspace with $n \geq 2$, then X is homeomorphic to a subset of $\mathbb{R}^n$. Hint: Let $\{u_1, u_2, \ldots, u_n\}$ be a basis for the Haar subspace, and define a map $f : X \to \mathbb{R}^n$ by writing $f(x) = [u_1(x), u_2(x), \ldots, u_n(x)]$. What is the correct theorem when $n = 1$?

7. Prove that Examples 4 and 5 given in the text are indeed Chebyshev systems on $(0, \infty)$.

8. Let $\{u_1, u_2, \ldots, u_n\}$ be a set of real-valued functions on a set X. Prove that the set of functions is linearly independent if and only if there exist n distinct points $x_1, x_2, \ldots, x_n$ in X such that $\det(u_j(x_i)) \neq 0$.

9. Let $\{u_1, u_2, \ldots, u_n\}$ be a linearly independent set of functions from a set X to the reals. Prove that there is a subset Y of X on which $\{u_1, u_2, \ldots, u_n\}$ is a Chebyshev system. Does there necessarily exist a maximal such Y? Illustrate with $n = 2$, $u_1(x) = x$, $u_2(x) = x^2$ on $\mathbb{R}$.

10. In the space Π_n consisting of all polynomials of degree at most n, let $\|p\|_\infty = \max_{-1 \leq x \leq 1} |p(x)|$. Find the norm of the functional x^* when $x = 2$. That is, compute

$$\sup_{\|p\|_\infty \leq 1} |p(2)|$$

You may need the theory of Chebyshev polynomials.

11. Let $S = \{u_1, u_2, \ldots, u_n\}$, where each u_i is a continuous function on a domain X. Let $Y \subset X$. Prove these assertions: (a) If S is linearly independent on Y then it is linearly

independent on X. (b) If S is a Chebyshev system on X then it is a Chebyshev system on Y, provided that Y has at least n elements. Show by examples that in assertions (a) and (b) we cannot interchange X and Y.

12. Is a subset of a Chebyshev system necessarily a Chebyshev system? For each $n = 1, 2, 3, \ldots$ give an example of a Chebyshev system having n elements such that each subset is also a Chebyshev system.

13. Let U be an n-dimensional subspace in $C^{(n)}[a, b]$. Let $D = d/dt$. Thus $D^k(U) = \{u^{(k)} : u \in U\}$. Prove that if $\dim D^k(U) = n - k$ then $\dim D^i(U) = n - i$ for all $i \in \{0, 1, \ldots, k\}$. Under the same hypothesis, prove that $\Pi_{k-1} \subset U$. Prove that if $D^k(U)$ is an $(n - k)$-dimensional Haar space, then $D^i(U)$ is an $(n - i)$-dimensional Haar space for each $i \in \{0, 1, \ldots, k\}$.

14. In the discussion of Mairhuber's Theorem, sets homeomorphic to the letter Y were employed. These sets are called "triods." Prove that any set of disjoint triods in $\mathbb{R}^2$ must be countable. *References:* Moore [RLM] and Problem 6598 proposed by W. Rudin in *American Math. Monthly* 98 (1991), 70–71.

15. Let U be an m-dimensional space of functions defined on a set X. Let $\mathcal{N}$ be a set of n points ("nodes") in X. Prove the equivalence of these properties:
    a. for any function f on $\mathcal{N}$ there is a unique element u in U such that $f(x) = u(x)$ on $\mathcal{N}$.
    b. $m = n$, and no element $u \in U$ (other than $u = 0$) vanishes on $\mathcal{N}$.

16. There exist discontinuous Haar spaces on all the spaces $\mathbb{R}^s$. Prove that if X is any infinite set of cardinality at most c (the cardinal number of $\mathbb{R}$), then for each n there is an n-dimensional Haar space of functions on X. (A suitable reference is Zielke's book [Zi].)

17. This chapter emphasizes linear interpolation problems. Investigate the question of whether the function $F(x) = a(1 + bx)^{-1}$ can be used to interpolate arbitrary data at two points.

18. (Continuation of Problem 17.) Investigate whether the function $F(x) = ae^{bx} + ce^{dx}$ can be used to interpolate arbitrary data at four points.

19. Find a set of three functions (defined on $\mathbb{R}^2$) such that interpolation of arbitrary data at any two points in $\mathbb{R}^2$ is possible by a linear combination of the three functions. Explain why this does not contradict Theorem 3. References: Wulbert [Wul1] and Shekhtman [Shek1].

20. Explain why point-evaluation functionals cannot be defined on $L^p[0, 1]$, $(1 \le p \le \infty)$.

21. Prove or disprove the converse of Theorem 6.

22. A concept stronger than that of a Chebyshev system is that of a Markov system. A sequence of continuous functions $u_0, u_1, u_2, \ldots$ defined on $\mathbb{R}$ is a **Markov system** if, for each n, $\{u_0, u_1, \ldots, u_n\}$ is a Chebyshev system. For example, the functions $u_n(x) = x^n$ form a Markov system. Notice that if the order of this sequence is changed, the Markov property is lost. If possible, make Markov systems from the examples 1–6 given on page 6.

23. Prove that one base for $\Pi_n(\mathbb{R})$ is the set of functions $u_k(x) = x^k(1-x)^{n-k}$, where $0 \le k \le n$.

References

[C1]   Cheney, E. W. *Introduction to Approximation Theory*. McGraw-Hill, New York, 1966. 2nd ed., Chelsea, New York, 1982. Amer. Math. Soc. 1998. Japanese edition, Kyoritsu Shuppan, Tokyo, 1978. Chinese edition, Beijing, 1981.

[Cu]   Curtis, P. C. "N-Parameter families and best approximation." *Pacific J. Math.* 9 (1959), 1013–1027.

[Gau1]   Gautschi, W. "The condition of Vandermonde-like matrices involving orthogonal polynomials." *Linear Alg. Appl.* 52 (1983), 293–300.

[Gau2]   Gautschi, W. "Lower bounds on the condition number of Vandermonde matrices." *Numer. Math.* 52 (1988), 241–250.

[GT]   Groshof, M. S., and G. Taiani. "Vandermonde strikes again." *Amer. Math. Monthly* 100 (1993), 575–578.

[KS]   Karlin, S., and W. J. Studden. *Tchebycheff Systems: with Applications in Analysis and Statistics*. Interscience, New York, 1966.

[Lut]   Lutts, J. A. "Topological spaces which admit unisolvent systems." *Trans. Amer. Math. Soc.* 111 (1964), 440–448.

[Mai]   Mairhuber, J. C. "On Haar's theorem concerning Chebyshev approximation problems having unique solution." *Proc. Amer. Math. Soc.* 7 (1956), 609–615.

[McW]   McCullough, S., and D. E. Wulbert. "The topological spaces that support Haar systems." *Proc. Amer. Math. Soc.* 94 (1985), 687–692.

[RLM]   Moore, R. L. "Concerning triods in the plane and the junction points of plane continua." *Proc. Nat. Acad. Science* 14 (1928), 85–88.

[Rush]   Rushanan, J. J. "On the Vandermonde matrix." *Amer. Math. Monthly* 96 (1989), 921–924.

[S10]   Schoenberg, I. J. "On the question of unicity in the theory of best approximation." *New York Acad. Sci. Annals* 86 (1960), 682–692.

[SY]   Schoenberg, I. J., and C. T. Yang. "On the unicity of solutions of problems of best approximation." *Annali di Mathematica Pura ed Applicata* 54 (1961), 1–12.

[Shek1]   Shekhtman, B. "Interpolating subspaces in R^n." in *Approximation Theory, Wavelets and Applications* ed. by S. P. Singh, Kluwer, Amsterdam, 1995, 465–471.

[Si]   Sieklucki, K. "Topological properties of sets admitting Tchebycheff systems." *Bull. Acad. Polon. Sci. Ser. Sci. Math.* 6 (1958), 603–606.

[Wul1]   Wulbert, D. "Interpolation at a few points." *J. Approx. Theory* 96 (1999), 139–148.

[Zi]   Zielke, R. *Discontinuous Čebyšev Systems*. Springer-Verlag Lecture Notes in Math., vol. 707, 1979.

2
Linear Interpolation Operators

We continue with notation and concepts from Chapter 1. Thus we have a set X, which is the domain of the functions being considered, and a set of n points (called "nodes") $x_1, x_2, \ldots, x_n$ in X. Also prescribed is an n-dimensional vector space U of functions that we intend to use for interpolation. Let $\{u_1, u_2, \ldots, u_n\}$ be any basis for U. We assume that the matrix A given by $A_{ij} = u_j(x_i)$ is nonsingular. This ensures that for any ordinates $\lambda_1, \lambda_2, \ldots, \lambda_n$ there is a unique function u in U that interpolates these values at the nodes, that is

$$u(x_i) = \lambda_i \qquad (1 \leq i \leq n)$$

One method of obtaining u is to solve the $n \times n$ system of equations $Ac = \lambda$ for the coefficient vector c and to set $u = \sum_{j=1}^{n} c_j u_j$. This procedure can be used for any basis of U. The coefficient matrix A will change, of course, if we change the basis. This remark raises the question of whether some bases for U may be more favorable than others for the problem at hand. In solving for c, one must "invert" the linear system, usually *not* by actually inverting the matrix, but by using Gaussian elimination on the system of equations. From this point of view, the ideal situation is that A be the identity matrix. There *is* a basis for U that makes $A = I$. All we need to do is to determine the basis $\{v_1, v_2, \ldots, v_n\}$ by solving the n interpolation problems

$$v_j(x_i) = \delta_{ij} \qquad (1 \leq i \leq n, \, 1 \leq j \leq n)$$

In this equation, δ_{ij} is the **Kronecker** δ, defined by $\delta_{ij} = 0$ if $i \neq j$ and $\delta_{ij} = 1$ if $i = j$. Our assumptions guarantee that the functions v_j exist in U. Such a basis is called the **cardinal basis** for U (with respect to the given nodes).

Example 1. Let $X = \mathbb{R}$ and let U be the vector space Π_{n-1} consisting of all polynomials of degree less than n. For any set of distinct nodes $\{x_1, x_2, \ldots, x_n\}$ define

$$v_j(x) = \prod_{\substack{i=1 \\ i \neq j}}^{n} \frac{x - x_i}{x_j - x_i} \qquad (1 \leq j \leq n)$$

Since the product contains $n - 1$ linear factors, $v_j \in \Pi_{n-1}$. It is readily verified that $v_j(x_i) = \delta_{ij}$ for all i and j. The interpolant to data $\lambda_1, \ldots, \lambda_n$ will be

$$F(x) = \sum_{j=1}^{n} \lambda_j v_j(x)$$

This is the classical **Lagrange interpolation formula.** It produces a polynomial of degree at most $n - 1$ that interpolates arbitrary data at n arbitrary nodes.   ∎

Returning to the general theory, suppose that $\{v_1, v_2, \ldots, v_n\}$ is the basis for U such that $v_j(x_i) = \delta_{ij}$. If f is any function on X, we can find a function in U that interpolates f at the nodes by writing

$$Lf = \sum_{j=1}^{n} f(x_j) v_j$$

It is trivial to verify that $(Lf)(x_i) = f(x_i)$ for $1 \leq i \leq n$. The operator L is linear and idempotent. That is,

(1) $$L(\alpha f + \beta g) = \alpha Lf + \beta Lg$$

(2) $$L^2 = L$$

The proof of (2) is as follows. There is a unique element u in U that interpolates f, and $u = Lf$. Likewise, there is a unique element in U that interpolates u and it is $Lu = L^2 f$. Since $u \in U$ and u interpolates u, $Lu = u$ and $L^2 f = Lf$.

For a linear transformation, T, acting between normed linear spaces, the **norm** of T is defined by the equation

$$\|T\| = \sup \{\|Tf\| : \|f\| \leq 1\}$$

If $\|T\| < \infty$, we say that T is **bounded.** This property is equivalent to continuity. The equation $\|Tf\| \leq \|T\| \, \|f\|$ is always valid.

The **distance** of a point f to a subspace U in a normed linear space is defined by the equation

$$\text{dist}(f, U) = \inf \{\|f - u\| : u \in U\}$$

A linear map P acting on a linear space is called a **projection** if $P^2 = P$. Projections are frequently used to provide approximations. Some basic facts about projections are summarized in the next theorem. See also Chapter 6, which is devoted to projections.

THEOREM. *Let E be a normed linear space, and let P be a bounded linear projection, $P : E \rightarrow E$. Let V be the range of P. Then $Pv = v$ for each $v \in V$. Furthermore,*

   a. *V is a closed subspace of E*
   b. *For each $f \in E$, $\|f - Pf\| \leq \|I - P\| \, \text{dist}(f, V) \leq (1 + \|P\|) \, \text{dist}(f, V)$*

Proof. If $v \in V$, then $v = Pf$ for some f. Hence $Pv = P^2 f = Pf = v$. In order to establish **a** it will be enough to prove that $V = \ker(I - P)$, because the kernel of a continuous linear operator is necessarily a closed linear subspace. If $v \in V$, then $v = Pf$ for some f, whence

$$Pv = P^2 f = Pf = v$$

showing that $v \in \ker (I - P)$. If $f \in \ker (I - P)$ then $f = Pf$, showing that $f \in V$.

For the inequality in **b**, let v be any element of V. Since $Pv = v$, we have

$$\| f - Pf \| = \| (f - v) - P(f - v) \| = \| (I - P)(f - v) \|$$
$$\leq \| I - P \| \| f - v \|$$

Now take the infimum as v ranges over V.   ∎

Now let us assume that X is a compact Hausdorff space, for example, any closed and bounded set in $\mathbb{R}^s$. Let $C(X)$ denote the Banach space of all continuous real-valued functions on X, normed by defining

$$\| f \|_\infty = \sup_{x \in X} |f(x)|$$

It is a consequence of compactness that $|f|$ attains its supremum at some point of X, and therefore $\sup_x |f(x)| < \infty$. Let U be an n-dimensional linear subspace in $C(X)$, and let $x_1, x_2, \ldots, x_n$ be points in X at which interpolation is possible, using elements in U. From Chapter 1, we know that this occurs if and only if the conditions $u \in U$ and $\sum_{i=1}^n |u(x_i)| = 0$ imply that $u = 0$. We can find a basis (called a cardinal basis) $\{v_1, v_2, \ldots, v_n\}$ for U such that $v_j(x_i) = \delta_{ij}$. The operator L described previously now can be considered as a map of $C(X)$ into U. Since $L^2 = L$, L maps $C(X)$ onto U. The **Lebesgue function** of L is defined by

$$\Lambda(x) = \sum_{j=1}^n |v_j(x)|$$

LEMMA. *The operator norm* $\| L \|$ *of the map just defined is given by* $\| L \| = \| \Lambda \|_\infty$.

Proof. If $\| f \|_\infty \leq 1$, then

$$\| Lf \|_\infty = \sup_{x \in X} |(Lf)(x)| = \sup_x \left| \sum_{j=1}^n f(x_j) v_j(x) \right|$$
$$\leq \sup_x \sum_{j=1}^x |f(x_j)| |v_j(x)| \leq \sup_x \sum_{j=1}^n |v_j(x)| = \| \Lambda \|_\infty$$

By the definition of the norm of an operator, this implies that $\| L \| \leq \| \Lambda \|_\infty$. To prove the reverse inequality, select $\xi \in X$ such that $\Lambda(\xi) = \| \Lambda \|_\infty$. Then find a function $f \in C(X)$ such that $\| f \|_\infty = 1$ and such that $f(x_j) = \operatorname{sgn} v_j(\xi)$. For this function we have

$$\| L \| \geq \| Lf \|_\infty \geq (Lf)(\xi) = \sum_{j=1}^n f(x_j) v_j(\xi) = \sum_{j=1}^n |v_j(\xi)| = \Lambda(\xi) = \| \Lambda \|_\infty$$

In order to find such a function f, one can invoke the Tietze Extension Theorem. It asserts that if Y is a closed set in a normal topological space X, then each continuous function from Y to $[-1, 1]$ has a continuous extension from X to $[-1, 1]$. (See Kelley's "General Topology" [Ke], page 242.) ∎

Example 2. Let $X = [a, b]$ and $U = \Pi_{n-1}$. Then the cardinal basis $\{v_j\}$ is as described in Example 1. Since $1 \in U$, we have $L1 = 1$, and hence

$$\sum_{j=1}^{n} v_j(x) = 1$$

Consequently $\Lambda(x) = \sum_{j=1}^{n} |v_j(x)| \geq 1$. Also $\Lambda(x_i) = 1$ for each value of i. The graph of Λ will have the following appearance:

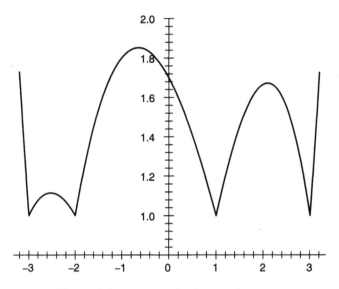

Figure 2.1 *A Typical Lebesgue Function*

In the classical (polynomial) theory of interpolation, there is another noteworthy procedure for computing interpolants. This one has Newton's name attached to it, and in our general way of looking at things, it corresponds to a special choice of basis functions. It leads to an interpolation matrix that is unit lower triangular; this makes the interpolation equations easy to solve. What sort of basis leads to a lower unit triangular matrix? Let us answer this in the general (nonpolynomial) case. The interpolation matrix is $A_{ij} = u_j(x_i)$. The condition of lower triangularity is that $u_j(x_i) = 0$ for $j > i$. Thus u_1 should be an element of U that takes the value 1 at x_1. The existence of such an element follows from our interpolation theory. In a similar way we can find u_2 that vanishes at x_1 and takes the value 1 at x_2, and so on. For general k, we find u_k so that $u_k(x_k) = 1$ and

$u_k(x_i) = 0$ for $i < k$. (Notice that these conditions do not specify the basis uniquely.) In some particular spaces, it is easy to construct such a basis, as we now illustrate. ∎

Example 3. Let $x_1, x_2, \ldots, x_n$ be distinct points of $\mathbb{R}$, and let $U = \Pi_{n-1}$. Define

$$u_1(x) = 1$$

$$u_2(x) = \frac{x - x_1}{x_2 - x_1}$$

$$u_3(x) = \frac{(x - x_1)(x - x_2)}{(x_3 - x_1)(x_3 - x_2)}$$

$$\ldots$$

$$u_n(x) = \frac{(x - x_1)(x - x_2) \cdots (x - x_{n-1})}{(x_n - x_1)(x_n - x_2) \cdots (x_n - x_{n-1})}$$

Notice that this basis has the additional property $u_j \in \Pi_{j-1}$ for $j = 1, 2, \ldots, n$. ∎

What are the advantages of such a "Newtonian" basis for interpolation problems? The principal advantage is that if we have constructed an element u in U that interpolates data $\lambda_1, \lambda_2, \ldots, \lambda_{k-1}$ at $x_1, x_2, \ldots, x_{k-1}$ (where $1 \le k \le n$) then an interpolant for data $\lambda_1, \lambda_2, \ldots, \lambda_k$ at $x_1, x_2, \ldots, x_k$ can be constructed of the form

$$F = u + cu_k$$

Notice that $u_k(x_i) = 0$ for $1 \le i < k$, and that therefore $F(x_i) = u(x_i) = \lambda_i$ for $1 \le i < k$. On the other hand, a proper choice for c leads to $F(x_k) = \lambda_k$ since $u_k(x_k) = 1$. Since the interpolant can be built up step by step in this way, we obtain in the end $F = \sum_{j=1}^{n} c_j u_j$, and each partial sum $\sum_{j=1}^{k} c_j u_j$ interpolates the "partial" data $\lambda_1, \lambda_2, \ldots, \lambda_k$ at $x_1, x_2, \ldots, x_k$.

Problems

> *I hear, I forget;*
> *I see, I remember;*
> *I do, I understand!*
>
> —Chinese proverb

1. For the functions v_j in Examples 1 and 2, show that each is a polynomial of exact degree $n - 1$. Then write $v_j(x) = \sum_{i=0}^{n-1} a_{ij} x^i$ for $1 \le j \le n$, and prove that $\sum_{j=1}^{n} a_{ij} = 0$ for $1 \le i \le n - 1$.

2. Suppose that for any f, the function $\sum_{j=1}^{n} f(x_j) v_j$ interpolates f at distinct nodes $x_1, x_2, \ldots, x_n$. Prove that the same is true of $\sum_{j=1}^{n} f(x_j) v_j^m$ for any exponent $m \in \mathbb{N}$. Don't neglect to *prove* that $v_j(x_i) = \delta_{ij}$.

3. Prove that if $f \in C[a, b]$ and $f \ge 0$, then for any set of nodes $x_1, x_2, \ldots, x_n$ there is a polynomial p such that $p \ge 0$ and $p(x_i) = f(x_i)$ for $1 \le i \le n$.

4. Prove that the Lebesgue function has this property: If $u \in U$, then $|u(x)| \leq \Lambda(x) \max_i |u(x_i)|$. (This applies to the general theory, not only to polynomials.) From this inequality prove that

$$\|u\|_\infty \leq \|\Lambda\|_\infty \max_i |u(x_i)|$$

5. In the general theory let $v_1, v_2, \ldots, v_n$ be the cardinal functions for a set of nodes $\{x_1, x_2, \ldots, x_n\}$. Put $u_k = \sum_{j=k}^n v_j$. Show that $\{u_1, u_2, \ldots, u_n\}$ is a Newtonian basis for the interpolation problem. Prove that if u interpolates data (x_i, λ_i) for $1 \leq i < k$, then $u + [\lambda_k - u(x_k)]u_k$ interpolates the data for $1 \leq i \leq k$.

6. Prove that the cardinal basis $\{v_1, v_2, \ldots, v_n\}$ has the property of the Newtonian basis, namely that the partial sum $\sum_{j=1}^k f(x_j)v_j$ interpolates data $f(x_i)$ for $1 \leq i \leq k$. (This is to be established for $k = 1, 2, \ldots, n$.)

7. Prove a theorem like the Tietze Theorem: Any function defined on a finite subset of $\mathbb{R}$ has an extension in $C^\infty(\mathbb{R})$. Then prove your result when the original and the extended function map $\mathbb{R}$ into the interval $[-1, 1]$.

8. Let P be a projection of a normed linear space E onto a subspace V. Prove that $I - P$ is a projection and that $E = \ker(P) \oplus \ker(I - P)$.

9. Prove that if U is a subspace of a linear space E and if $L : E \twoheadrightarrow U$ is a surjective linear map, then these properties of L are equivalent: (a) $L|U = I|U$, (b) $L^2 = L$. Here $L|U$ denotes the restriction of the operator L to the set U.

10. Let $x_1, x_2, \ldots, x_n$ be distinct points in $\mathbb{R}$, and let V be the corresponding Vandermonde matrix, whose generic element is x_i^{j-1}. Define the polynomials v_j as in Example 1. Write $v_j(x) = \sum_{i=1}^n a_{ij} x^{i-1}$ and prove that $A \equiv (a_{ij})$ is the inverse of V.

11. How large can the coefficients be in a polynomial p of degree at most n, if p satisfies the inequality $|p(x)| \leq A$ for all x in an interval $[a, b]$?

12. Let V be an n-dimensional linear space of real-valued functions on $\mathbb{R}^s$. Let $\mathcal{N}$ be a set of n points in $\mathbb{R}^s$ such that interpolation of arbitrary data on $\mathcal{N}$ is always possible by elements of V. Discuss the continuity of the interpolant, first as a function of the data given on $\mathcal{N}$, and second as a function of the points in $\mathcal{N}$.

13. Prove that the operator L discussed in the lemma has the property $\|I - L\| = 1 + \|L\|$. For any operator T, if the equation $\|I + T\| = 1 + \|T\|$ is true, we say that T has the **Daugavet property.** Much is known about this, as can be seen by searching *Mathematical Reviews* for papers whose titles contain the word "Daugavet." We cite only three references, namely [Holb], [Wer], and [Woj].

References

[C3]   Cheney, E. W. "Projection operators in approximation theory." In *Studies in Functional Analysis*, ed. by R. G. Bartle. Math. Assoc. of America, Washington, D.C. 1980, pp. 50–80.

[C1]   Cheney, E. W. *Introduction to Approximation Theory*. McGraw-Hill, New York, 1966. 2nd ed., Chelsea, New York, 1982. Amer. Math. Soc. 1998. Japanese edition, Kyoritsu Shuppan, Tokyo, 1978. Chinese edition, Beijing, 1981.

[CP]   Cheney, E. W., and K. H. Price. "Minimal projections." In *Approximation Theory*, ed. by A. Talbot. Academic Press, New York, 1970, pp. 261–289.

[Da2]   Davis, P. J. *Interpolation and Approximation.* Blaisdell, New York, 1963. Reprint, Dover Publ., New York.

[Holb]   Holub, J. R. "Daugavet's equation and ideals of operators on $C[0, 1]$." *Math. Nachr.* 141 (1989), 177–181.

[Ke]   Kelley, J. L. *General Topology.* Springer-Verlag, Graduate Texts in Mathematics 27, New York, 1955.

[LS]   Lancaster, P., and K. Salkauskas. *Curve and Surface Fitting.* Academic Press, New York, 1986.

[Riv3]   Rivlin, T. J. *An Introduction to the Approximation of Functions.* Dover, New York, 1981.

[Wer]   Werner, D. "The Daugavet equation for operators on function spaces." *J. Funct. Anal.* 143 (1997), 117–128.

[Woj]   Wojtaszczyk, P. "Some remarks on the Daugavet equation." *Proc. Amer. Math. Soc.* 115 (1992), 1047–1052.

3
Optimization of the Lagrange Operator

This chapter is concerned (almost exclusively) with polynomial interpolation in one variable. The standard interval $[-1, 1]$ is adopted, and n is the number of interpolation nodes. Thus interpolation is by members of the polynomial space Π_{n-1}, whose dimension is n. For nodes $x_1, x_2, \ldots, x_n$, arranged so that $-1 \leq x_1 < x_2 < \cdots < x_n \leq 1$, we define

$$(1) \qquad \ell_j(x) = \prod_{\substack{i=1 \\ i \neq j}}^{n} \left[(x - x_i)/(x_j - x_i) \right] \qquad (1 \leq j \leq n)$$

$$(2) \qquad Lf = \sum_{j=1}^{n} f(x_j) \ell_j$$

$$(3) \qquad \Lambda = \sum_{j=1}^{n} |\ell_j|$$

Some of the principal theorems in this subject will be quoted here without proofs. References are given, however, for all of them.

THEOREM 1. (Kharshiladze–Lozinski). *Every linear projection* $P : C[-1, 1] \to \Pi_{n-1}$ *satisfies the inequality*

$$(4) \qquad \|P\| \geq \frac{2}{\pi^2} \log (n - 1) - \frac{1}{2}$$

18

The proof of this and further references can be found in [C1] and Natanson [Nata]. When this theorem is combined with the Uniform Boundedness Principle, one can conclude that for any preassigned system of nodes

(5) $$-1 \le x_1^{(n)} < x_2^{(n)} < \cdots < x_n^{(n)} \le 1 \qquad (n = 1, 2, 3, \ldots)$$

there will exist a function $f \in C[-1, 1]$ whose polynomial interpolants, p_n, diverge, in the sense that $\lim \sup_{n \to \infty} \|p_n\| = \infty$. Here p_n is the polynomial of degree $n - 1$ that interpolates f at nodes $x_i^{(n)}$.

The Uniform Boundedness Principle asserts that if $\{A_\alpha\}$ is a collection of continuous linear transformations from a Banach space E into a normed linear space, and if $\sup_\alpha \|A_\alpha\| = \infty$, then there exists an f in E such that $\sup_\alpha \|A_\alpha f\| = \infty$. (A refinement of this result, called the Banach-Steinhaus Theorem, draws the stronger conclusion that the set of f as described is a set of the second Baire category.)

Since the Lagrange operator L, defined earlier, is a projection, it obeys Inequality (4). However, a stronger result is available.

THEOREM 2. (Erdős, Brutman). *The Lagrange operator L in Equation (2) obeys the inequality*

(6) $$\|L\| > \frac{2}{\pi} \log n + 0.5212$$

The original article by Erdős [Er] gave the correct factor $2/\pi$ but not a very good constant. Brutman [Bru] proved the Inequality (6).

The Chebyshev polynomial is defined by $T_n(x) = \cos(n \cos^{-1} x)$. An easy calculation shows that its zeros are the points $x_j = \cos\left[(2j - 1)/2n\right]\pi$, for $1 \le j \le n$. They are represented graphically by taking the equally spaced angles $\theta_j = \left[(2j - 1)/2n\right]\pi$ and projecting points on a unit semicircle down on the horizontal axis, as shown in Figure 3.1. These points can be used for interpolation and are often referred to as the **Chebyshev nodes.**

In the 1930s, Bernstein already knew [Be2] that interpolation at the Chebyshev nodes achieves the order of growth $(2/\pi) \log n$ in its Lebesgue function. It was also

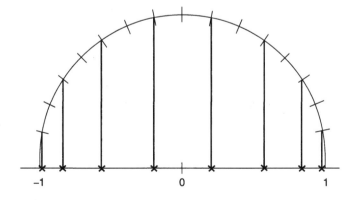

Figure 3.1   *Chebyshev Nodes for $n = 8$*

known that this special interpolation operator does *not* have minimal norm. Rivlin [Riv1] proved the following:

THEOREM 3. *If L is the Lagrange interpolation operator for the Chebyshev nodes then*

(7)
$$0.9625 < \|L\| - \frac{2}{\pi} \log n < 1$$

As indicated in Chapter 2, the Lebesgue function Λ has a relative maximum on every interval between adjacent nodes. There are two other local maxima at -1 and $+1$ if these points are not nodes. Brutman [Bru] proved the next theorem.

THEOREM 4. *Let the nodes x_j be the roots of T_n, and let $\lambda_0, \lambda_1, \ldots, \lambda_n$ be the local maxima of Λ on the intervals $[-1, x_1], [x_1, x_2], \ldots, [x_n, 1]$. Then*

(8)
$$\max \lambda_i - \min \lambda_i < 1/2$$

In all these quoted results there is not much to be inferred about the *optimal* nodes—that is, the nodes that lead to a minimal value. For each n, let σ_n denote the minimum of $\|L\|$, when we consider all choices of n nodes in $[-1, 1]$. Theorems 2 and 3 tell us that

(9)
$$0.5212 < \sigma_n - \frac{2}{\pi} \log n < 1$$

In 1932, Serge Bernstein published an important paper on this subject, [Be2], in which he conjectured (roughly) the following theorem: If the interpolation problem is standardized so that the endpoints of the interval are nodes, then the minimum norm of the interpolation operator is achieved when and only when the local maxima (λ_i) in the Lebesgue function are all equal. This conjecture resisted all efforts to prove it for 44 years. In 1976, Theodore Kilgore succeeded in establishing Bernstein's conjecture and announced his result (together with the lemmas leading up to it) in [Ki1]. The methods contained in Kilgore's work became the basis for much work to follow. A fuller account of his work appeared in [Ki2]. In 1978, Carl de Boor and Allan Pinkus proved in [BP1] a related conjecture of Erdős [Er2] that for any set of nodes, the inequality

(10)
$$\min_i \lambda_i \leq \sigma_n \leq \max_i \lambda_i$$

must hold.

A numerical determination of the optimal nodes (for $n \leq 100$) has been undertaken by Angelos, Kaufman, Henry, and Lenker in [AKHL].

Another specific set of nodes termed the "expanded Chebyshev nodes" recommends itself. These are the nodes $x_j = \alpha \cos \left[(2j - 1)\pi/2n \right]$ where α is chosen to make $x_n = -1$ and $x_1 = +1$. For these nodes Brutman proved

(11)
$$\|L\| - \sigma_n < 0.201$$

Hence, for all practical purposes the expanded Chebyshev nodes are sufficiently close to the optimal ones. In Figure 3.2, the Lebesgue function for this choice of nodes is shown, with $n = 6$.

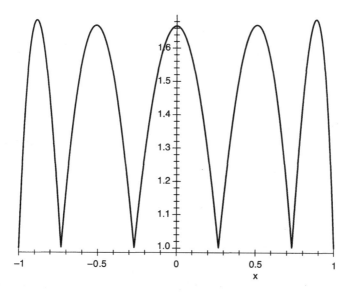

Figure 3.2   *Lebesgue Function for Expanded Chebyshev Nodes*

The following questions are still open:

1. Is there a set of relatively simple functions f_n such that the roots of f_n are the optimal nodes for Lagrange interpolation?

2. Is the Erdös conjecture true for arbitrary Haar subspaces in $C[-1, 1]$?

3. What is the correct asymptotic relationship of the form

$$\sigma_n = \frac{2}{\pi} \log n + c + \varepsilon_n$$

in which $\varepsilon_n \to 0$?

4. What is the projection constant of Π_n in $C[-1, 1]$? What are the minimal projections in this setting? Here we require the infimum of $\|P\|$ when P ranges over all projections of $C[-1, 1]$ onto Π_n.

A theorem related to question 2 is the following one by Kilgore and Cheney, [KC].

THEOREM 5. *Let G be an n-dimensional Haar subspace in $C[a, b]$. Assume that G contains the constant functions. Then there exists a set of nodes x_i such that the corresponding Lebesgue function has equal local maxima in the intervals $[a, x_1]$, $[x_1, x_2]$, ..., $[x_n, b]$.*

Question 4 has resisted all efforts to obtain a general solution. Characterization theorems for minimal projections were given by Cheney and Morris in [CM]. Chalmers and Metcalf have given the minimal projection onto Π_2 in [CMe].

An efficient arrangement of the proofs for Theorems 2, 3, 4, and Inequality (10) would be to start with Rivlin's result (Theorem 3), followed by the proofs of Inequality

(10) and Theorem 4. Theorem 2 in a slightly weakened form will then follow trivially: Use superscript t to signify that Chebyshev nodes are being used, and write

$$\sigma_n \geq \min_i \lambda_i^t \geq \max_i \lambda_i^t - \frac{1}{2} = \|L^t\| - \frac{1}{2} \geq \frac{2}{\pi} \log n - 0.4625$$

It has probably occurred to the reader that the case of equally spaced nodes has been omitted from the preceding discussion of polynomial interpolation. The reason is that uniformly distributed nodes lead to very unfavorable Lebesgue constants. Turetskii [Tur1] has proved that if n nodes are chosen in $[a, b]$ according to the prescription

$$x_i = a + (i - 1)\frac{b - a}{n} \qquad (1 \leq i \leq n)$$

then the corresponding Lebesgue function Λ satisfies

$$\max_{a \leq x \leq b} \Lambda(x) \sim \frac{2^n}{en \log n}$$

This rate of growth compares very badly with the $\log n$ growth for optimal nodes in Inequality (9). The reader is referred to Brutman's article [Bru2], which provides considerable detail and has 71 references.

Problems

If it doesn't kill me, it makes me stronger.

—Nietzsche

1. Use the notation

$$\Lambda(x_1, \ldots, x_n; x) = \sum_{j=1}^{n} \prod_{\substack{i=1 \\ i \neq j}}^{n} |x - x_i| / |x_j - x_i|$$

Prove that if $t = cx + d$ and $t_i = cx_i + d$ then

$$\Lambda(t_1, \ldots, t_n; t) = \Lambda(x_1, \ldots, x_n; x)$$

2. Use the Banach–Steinhaus Theorem with Theorem 3 in this chapter to prove that if L_n is the Lagrange operator for nodes at the roots of T_n, then the set of functions f in $C[-1, 1]$ for which $\limsup \|L_n f\| = \infty$ is a set of second category, in the sense of Baire.

3. Prove that Λ in Equation (3) has a single local maximum in each of the intervals $[-1, x_1], [x_1, x_2], \ldots, [x_n, 1]$. If you find this difficult, work Problems 4–8.

4. Assume that $x_1 < \cdots < x_n$ and that $x_k < x < x_{k+1}$. Prove that $x_j > x$ if and only if $j \geq k + 1$. Prove that $x_j < x$ if and only if $j \leq k$.

5. Prove that

$$\text{sgn } \ell_i(x) = \text{sgn}\left[\prod_{j=1}^{i-1} (x - x_j)\right] \text{sgn}\left[\prod_{j=i+1}^{n} (x_j - x)\right]$$

6. Prove that under the assumptions of Problem 4, sgn $\ell_i(x)$ is $(-1)^{k-i}$ when $k \geq i$, but it is $(-1)^{i-1-k}$ when $k \leq i - 1$.

7. Let $p(x) = \sum_{i=1}^{k}(-1)^{k-i}\ell_i(x) - \sum_{i=k+1}^{n}(-1)^{k-i}\ell_i(x)$. Prove that for $x \in (x_k, x_{k+1})$ we have $\Lambda(x) = p(x)$.

8. Count the zeros of p and p', where p is defined in Problem 7.

9. Prove that for n nodes equally spaced in $[a, b]$, including endpoints, the norm of the interpolating operator is $\mathcal{O}(2^n)$.

References

[AKHL] Angelos, J. R., E. H. Kaufman, Jr., M. S. Henry, and T. D. Lenker. "Optimal nodes for polynomial interpolation." In *Approximation Theory VI,* ed. by C. K. Chui, L. L. Schumaker and J. D. Ward, Academic Press, New York, 1989. pp. 17–20 in vol. I.

[Be2] Bernstein, S. N. "Sur la limitation des valeurs d'un polynôme $P_n(x)$ de degré n sur tout un segment par ses valeurs en $n + 1$ points du segment." *Bull. Acad. Sci. URSS, Leningrad* (1931), 1025–1050. *Zentralblatt* 4 (1932), page 6.

[BooP1] de Boor, C., and A. Pinkus. "Proof of the conjectures of Bernstein and Erdős concerning the optimal nodes for polynomial interpolation." *J. Approx. Theory* 24 (1978), 289–303.

[Bru] Brutman, L. "On the Lebesgue function for polynomial interpolation." *SIAM J. Numerical Analysis* 15 (1978), 694–704.

[Bru2] Brutman, L. "Lebesgue functions for polynomial interpolation: a survey." *Ann. Numer. Math.* 4 (1997) 111–127.

[CMe] Chalmers, B. L., and F. T. Metcalf. "Determination of a minimal projection from $C[-1, 1]$ onto the quadratics." *Numer. Func. Anal. Optim.* 11 (1990), 1–10.

[C1] Cheney, E. W. *Introduction to Approximation Theory.* McGraw-Hill, New York, 1966. 2nd ed., Chelsea, New York, 1982. Amer. Math. Soc, 1998. Japanese edition, Kyoritsu Shuppan, Tokyo, 1978. Chinese edition, Beijing, 1981.

[CM] Cheney, E. W., and P. D. Morris. "On the existence and characterization of minimal projections." *J. Reine Angew. Math.* 270 (1974), 215–227.

[Er] Erdős, P. "Problems and results on the theory of interpolation, II." *Acta Math. Acad. Sci. Hungar.* 12 (1961), 235–244.

[Er2] Erdős, P. "Some remarks on polynomials." *Bull. Amer. Math. Soc.* 53 (1947), 1169–1176.

[Ki1] Kilgore, T. A. "Optimization of the norm of the Lagrange interpolation operator." *Bull. Amer. Math. Soc.* 83 (1977), 1069–1071.

[Ki2] Kilgore, T. A. "A characterization of the Lagrange interpolating projection with minimal Tchebycheff norm." *J. Approx. Theory* 24 (1978), 273–288.

[KC] Kilgore, T. A., and E. W. Cheney. "A theorem on interpolation in Haar subspaces." *Aequationes Math.* 14 (1976), 391–400.

[Nata] Natanson, I. P. *Constructive Theory of Functions.* U.S. Atomic Energy Commission, Office of Technical Information, AEC–tr–4503, 1961. Original Russian Text, Moscow, 1949. German translation, "Konstructive Funktionentheorie," Akademie-Verlag, Berlin, 1955.

[Riv1]   Rivlin, T. J. *Chebyshev Polynomials*. Wiley, New York, 1974. 2nd Edition, 1990.

[Scho2]   Schönhage, A. "Fehlerfortpflanzung bei Interpolation." *Numer. Math.* 3 (1961), 62–71.

[SzV]   Szabados, J. and P. Vértesi. *Interpolation of Functions*. World Scientific Publishers, Singapore, 1990.

[Tur1]   Turetskii, A. H. "The bounding of polynomials prescribed at equally distributed points." Proc. Pedag. Inst. Vitebsk 3 (1940), 117–127. (In Russian)

[Tur2]   Turetskii, A. H. *Theory of Interpolation in Problem Form*. Minsk, 1968. (In Russian)

4
Multivariate Polynomials

In order to deal efficiently with polynomials in many variables, we use multi-index notation. Let us consider functions of s real variables. The variables can be denoted by $\xi_1, \xi_2, \ldots, \xi_s$, and can be put into a vector called x. Thus $x = (\xi_1, \xi_2, \ldots, \xi_s)$, and $x \in \mathbb{R}^s$. A real-valued function of s real variables is then a map $f : \mathbb{R}^s \to \mathbb{R}$. A polynomial in s variables is such a map.

As is usual, let $\mathbb{Z}$ denote the set of all integers (positive, negative, and 0). The set of non-negative integers is denoted by $\mathbb{Z}_+$. The notation $\mathbb{Z}_+^s$ then means the set of all s-tuples $\alpha = (\alpha_1, \alpha_2, \ldots, \alpha_s)$, where $\alpha_i \in \mathbb{Z}_+$ for $1 \leq i \leq s$. Such s-tuples are called **multi-indices**. (The integer s remains fixed throughout.) We shall reserve the Greek letters α and β for multi-indices, that is, elements of $\mathbb{Z}_+^s$.

For a multi-index α, we write

$$|\alpha| = \alpha_1 + \alpha_2 + \cdots + \alpha_s$$

(Thus, unfortunately, the symbol $|\ |$ changes meaning according to its context.) If $x = (\xi_1, \xi_2, \ldots, \xi_s)$ and $\alpha = (\alpha_1, \alpha_2, \ldots, \alpha_s)$ then, by definition,

$$x^\alpha := \prod_{i=1}^s \xi_i^{\alpha_i} = \xi_1^{\alpha_1} \cdot \xi_2^{\alpha_2} \cdots \xi_s^{\alpha_s}$$

The function $x \mapsto x^\alpha$ is a **monomial**. For $s = 3$, here are seven typical monomials:

$$\xi_1 \xi_2^7 \xi_3^2, \quad \xi_1^4 \xi_3^5, \quad \xi_2^3, \quad 1, \quad \xi_1, \quad \xi_2, \quad \xi_3$$

These are the building blocks for **polynomials**. The **degree** of a monomial x^α is defined to be $|\alpha|$. Thus, in the examples, the degrees are 10, 9, 3, 0, 1, 1, and 1. A polynomial in s variables is a function

$$p(x) = \sum_\alpha c_\alpha x^\alpha \qquad (x \in \mathbb{R}^s)$$

in which the sum is finite, the c_α are real numbers, and $\alpha \in \mathbb{Z}_+^s$. The **degree** of p is

$$\max \{|\alpha| : c_\alpha \neq 0\}$$

25

If all c_α are 0, then $p(x) = 0$, and we assign the degree $-\infty$ in this case. A polynomial of degree 0 is a constant function. Here are some examples, again with $s = 3$:

$$p_1(x) = 3 + 2\xi_1 - 7\xi_2^2\xi_3 + 2\xi_1\xi_2^4\xi_3^7$$

$$p_2(x) = \sqrt{2}\xi_1\xi_3 - \pi\xi_1^3\xi_2^5\xi_3$$

These have degrees 12 and 9, respectively.

The completely general polynomial of degree at most k can be written as

$$\sum_{|\alpha|\le k} c_\alpha x^\alpha$$

This sum has $\binom{k+s}{s}$ terms, as established later in Theorem 2.

Multi-indices are also useful in defining differential operators. It is customary to set $D = \left(\dfrac{\partial}{\partial\xi_1}, \dfrac{\partial}{\partial\xi_2}, \cdots, \dfrac{\partial}{\partial\xi_s}\right)$. Then in a natural way, we define

$$D^\alpha = \prod_{i=1}^{s}\frac{\partial^{\alpha_i}}{\partial\xi_i^{\alpha_i}} = \frac{\partial^{|\alpha|}}{\partial\xi_1^{\alpha_1}\,\partial\xi_2^{\alpha_2}\cdots\partial\xi_s^{\alpha_s}}$$

Within the set of multi-indices, $\mathbb{Z}_+^s$, we can define addition in the natural way:

$$(\alpha + \beta)_i = \alpha_i + \beta_i \qquad (1 \le i \le s)$$

Also there is a natural partial ordering, namely,

$$\alpha \le \beta \quad \text{means} \quad \alpha_i \le \beta_i \quad \text{for} \quad \text{all } i$$

If $\alpha \le \beta$, then $\beta - \alpha$ is a multi-index having components $\beta_i - \alpha_i$. Further definitions are

$$\alpha! = \alpha_1!\alpha_2!\cdots\alpha_s! = \prod_{i=1}^{s}\alpha_i!$$

$$\binom{\alpha}{\beta} = \begin{cases} \dfrac{\alpha!}{\beta!(\alpha-\beta)!} & \text{if} \quad \beta \le \alpha \\[2mm] 0 & \text{in all other cases} \end{cases}$$

BINOMIAL THEOREM. *For all* $x, y \in \mathbb{R}^s$ *and* $\alpha \in \mathbb{Z}_+^s$,

$$(x + y)^\alpha = \sum_{0\le\beta\le\alpha}\binom{\alpha}{\beta}x^\beta y^{\alpha-\beta}$$

Proof.

$$(x + y)^\alpha = \prod_{i=1}^{s}(\xi_i + \eta_i)^{\alpha_i} = \prod_{i=1}^{s}\sum_{\beta_i=0}^{\alpha_i}\binom{\alpha_i}{\beta_i}\xi_i^{\beta_i}\eta_i^{\alpha_i-\beta_i}$$

$$= \left[\sum_{\beta_1=0}^{\alpha_1}\binom{\alpha_1}{\beta_1}\xi_1^{\beta_1}\eta_1^{\alpha_1-\beta_1}\right]\cdots\left[\sum_{\beta_s=0}^{\alpha_s}\binom{\alpha_s}{\beta_s}\xi_s^{\beta_s}\eta_s^{\alpha_s-\beta_s}\right]$$

$$= \sum_{\beta_1=0}^{\alpha_1}\cdots\sum_{\beta_s=0}^{\alpha_s}\prod_{i=1}^{s}\binom{\alpha_i}{\beta_i}\xi_i^{\beta_i}\eta_i^{\alpha_i-\beta_i} = \sum_{0\le\beta\le\alpha}\prod_{i=1}^{s}\binom{\alpha_i}{\beta_i}\prod_{j=1}^{s}\xi_j^{\beta_j}\eta_j^{\alpha_j-\beta_j}$$

$$= \sum_{0\le\beta\le\alpha}\binom{\alpha}{\beta}x^\beta y^{\alpha-\beta}$$

In this calculation we used Problem 4 and the rule

$$\left[\sum_{j=1}^{n} a_j\right]\left[\sum_{k=1}^{m} b_k\right] = \sum_{j=1}^{n}\sum_{k=1}^{m} a_j b_k \qquad \blacksquare$$

We will usually abbreviate the inner product $\langle x, y\rangle$ of two vectors $x, y \in \mathbb{R}^s$ by the simpler notation xy.

MULTINOMIAL THEOREM. *Let $x, y \in \mathbb{R}^s$ and $n \in \mathbb{N}$. Then*

$$(xy)^n = \sum_{|\alpha|=n} \frac{n!}{\alpha!} x^\alpha y^\alpha$$

Proof. It suffices to consider only the special case $y = (1, 1, \ldots, 1)$ because

$$xy = \langle (\xi_1\eta_1, \ldots, \xi_s\eta_s), (1, \ldots, 1)\rangle$$

For this special case we proceed by induction on s. The case $s = 1$ is trivially true, and the case $s = 2$ is the usual binomial formula:

$$(\xi_1 + \xi_2)^n = \sum_{\alpha_1+\alpha_2=n} \frac{n!}{\alpha_1!\alpha_2!} \xi_1^{\alpha_1} \xi_2^{\alpha_2} = \sum_{j=0}^{n} \frac{n!}{j!(n-j)!} \xi_1^j \xi_2^{n-j}$$

Suppose that the multinomial formula is true for a particular value of s. The proof for the next case goes as follows. Let $x = (\xi_1, \ldots, \xi_s)$, $w = (\xi_1, \ldots, \xi_{s+1})$, $\alpha = (\alpha_1, \ldots, \alpha_s)$ and $\beta = (\alpha_1, \ldots, \alpha_{s+1})$. Then

$$(\xi_1 + \cdots + \xi_{s+1})^n = [(\xi_1 + \cdots + \xi_s) + \xi_{s+1}]^n$$

$$= \sum_{j=0}^{n} \binom{n}{j} (\xi_1 + \cdots + \xi_s)^j \xi_{s+1}^{n-j}$$

$$= \sum_{j=0}^{n} \binom{n}{j} \sum_{|\alpha|=j} \frac{j!}{\alpha!} x^\alpha \xi_{s+1}^{n-j}$$

$$= \sum_{j=0}^{n} \sum_{|\alpha|=j} \frac{n!}{j!(n-j)!} \frac{j!}{\alpha!} x^\alpha \xi_{s+1}^{n-j}$$

$$= \sum_{|\beta|=n} \frac{n!}{(n-j)!\alpha!} w^\beta = \sum_{|\beta|=n} \frac{n!}{\beta!} w^\beta$$

In this calculation, we let $\beta = (\alpha_1, \ldots, \alpha_s, n-j)$, where $|\alpha| = j$. $\qquad \blacksquare$

The linear space of all polynomials of degree at most n in s real variables is denoted by $\Pi_n(\mathbb{R}^s)$. Thus each element of this space can be written as

$$p(x) = \sum_{|\alpha|\le n} c_\alpha x^\alpha$$

Consequently the set of monomials

$$\{x \longmapsto x^\alpha : |\alpha| \le n\}$$

spans $\Pi_n(\mathbb{R}^s)$. Is this set in fact a basis?

THEOREM 1. *The set of monomials $x \mapsto x^\alpha$ on $\mathbb{R}^s$ is linearly independent.*

Proof. If $s = 1$, the monomials are the elementary functions $\xi_1 \mapsto \xi_1^j$ for $j = 0, 1, 2, \ldots$. They form a linearly independent set since a nontrivial linear combination $\sum_{j=0}^n c_j \xi_1^j$ cannot vanish as a function. (Indeed, it can have at most n zeros.)

Suppose now that our assertion has been proved for dimension $s - 1$. Let $x = (\xi_1, \ldots, \xi_s)$ and $\alpha = (\alpha_1, \ldots, \alpha_s)$. Suppose that $\sum_{\alpha \in J} c_\alpha x^\alpha = 0$, where the sum is over a finite set $J \subset \mathbb{Z}_+^s$. Put

$$J_k = \{\alpha \in J : \alpha_1 = k\}$$

Then for some n, $J = J_0 \cup \cdots \cup J_n$, and we can write

$$0 = \sum_{k=0}^n \sum_{\alpha \in J_k} c_\alpha \xi_1^{\alpha_1} \cdots \xi_s^{\alpha_s} = \sum_{k=0}^n \xi_1^k \sum_{\alpha \in J_k} c_\alpha \xi_2^{\alpha_2} \cdots \xi_s^{\alpha_s}$$

By the one-variable case, we infer that for $k = 0, \ldots, n$

$$\sum_{\alpha \in J_k} c_\alpha \xi_2^{\alpha_2} \cdots \xi_s^{\alpha_s} = 0$$

By the induction hypothesis, we then infer that, for all $\alpha \in J_k$, $c_\alpha = 0$. Since k runs from 0 to n, all c_α are 0. As α runs over J_k the multi-indices $(\alpha_2, \ldots, \alpha_s)$ are distinct. ∎

We want to calculate the dimension of $\Pi_n(\mathbb{R}^s)$. The following lemma is needed before this can be done.

LEMMA 1. *For $s = 1, 2, 3, \ldots$ and $n = 0, 1, 2, \ldots$ we have*

$$\sum_{k=0}^n \binom{k+s}{s} = \binom{n+1+s}{s+1}$$

Proof. We use induction on n. When $n = 0$, the assertion is that

$$\binom{s}{s} = \binom{1+s}{s+1}$$

and this is true. Assuming the correctness of the formula for a particular n, we prove that it is true for $n + 1$:

$$\sum_{k=0}^{n+1} \binom{k+s}{s} = \sum_{k=0}^n \binom{k+s}{s} + \binom{n+1+s}{s} = \binom{n+1+s}{s+1} + \binom{n+1+s}{s}$$

$$= \frac{(n+1+s)!}{(s+1)!n!} + \frac{(n+1+s)!}{(n+1)!s!}$$

$$= \frac{(n+1+s)!}{(s+1)!(n+1)!}[(n+1) + (s+1)]$$

$$= \frac{(n+2+s)!}{(s+1)!(n+1)!} = \binom{n+2+s}{s+1}$$ ∎

THEOREM 2. *The dimension of* $\Pi_n(\mathbb{R}^s)$ *is* $\binom{n+s}{s}$.

Proof. The preceding theorem asserts that a basis for $\Pi_n(\mathbb{R}^s)$ is $\{x \mapsto x^\alpha : |\alpha| \le n\}$. Here $x \in \mathbb{R}^s$. Using $\#$ to denote the number of elements in a set, we only have to prove

$$\#\{\alpha \in \mathbb{Z}_+^s : |\alpha| \le n\} = \binom{n+s}{s}$$

We use induction on s. For $s = 1$, the formula is correct, since

$$\#\{\alpha \in \mathbb{Z}_+ : \alpha \le n\} = n + 1 = \binom{n+1}{1}$$

Assume that the formula is correct for a particular s. For the next case we write

$$\# \{\alpha \in \mathbb{Z}_+^{s+1} : |\alpha| \le n\} = \# \bigcup_{k=0}^n \Big\{\alpha \in \mathbb{Z}_+^{s+1} : \alpha_{s+1} = k, \sum_{i=1}^s \alpha_i \le n - k\Big\}$$

$$= \sum_{k=0}^n \#\{\alpha \in \mathbb{Z}_+^s : |\alpha| \le n - k\} = \sum_{k=0}^n \binom{n-k+s}{s}$$

$$= \sum_{k=0}^n \binom{k+s}{s} = \binom{n+s+1}{s+1}$$

In the last step, we applied Lemma 1. ∎

From Theorem 3 in Chapter 1, page 6, we know that there exist no Haar subspaces of dimension greater than 1 in any of the spaces $\mathbb{R}^2$, $\mathbb{R}^3$, ... It follows that interpolation by $\Pi_n(\mathbb{R}^s)$ is not possible on arbitrary sets of m nodes, where $m = \dim[\Pi_n(\mathbb{R}^s)]$. A simple example will illustrate this phenomenon.

Example 1. Consider the case $s = 2$ and $n = 1$. The monomials that form a basis for $\Pi_1(\mathbb{R}^2)$ are $1, \xi_1, \xi_2$. Let three points be given as nodes:

$$x_i = (\xi_{i1}, \xi_{i2}) \qquad (1 \le i \le 3)$$

The *homogeneous* interpolation problem is to find coefficients c_1, c_2, c_3 so that

$$c_1 + c_2\xi_{i1} + c_3\xi_{i2} = 0 \qquad (1 \le i \le 3)$$

This equation will have a nontrivial solution if and only if the three points x_1, x_2, and x_3 are colinear. Thus, if the nodes are on a line, the interpolation matrix will be singular, and interpolation of arbitrary data is not possible. ∎

Example 2. On $\mathbb{R}^2$ there are no 2-dimensional Haar spaces of continuous functions, by Theorem 3 of Chapter 1 (page 6). Thus, in order to interpolate at any two points in $\mathbb{R}^2$, we shall require a linear space of dimension at least 3. Consider $\Pi_1(\mathbb{R}^2)$. (It has dimension 3.) Given any two nodes, $x_1 = (\xi_1^{(1)}, \xi_2^{(1)})$ and $x_2 = (\xi_1^{(2)}, \xi_2^{(2)})$, can we interpolate arbitrary data by a function in $\Pi_1(\mathbb{R}^2)$? The interpolation conditions are

$$\begin{pmatrix} 1 & \xi_1^{(1)} & \xi_2^{(1)} \\ 1 & \xi_1^{(2)} & \xi_2^{(2)} \end{pmatrix} \begin{pmatrix} c_1 \\ c_2 \\ c_3 \end{pmatrix} = \begin{pmatrix} \lambda_1 \\ \lambda_2 \end{pmatrix}$$

This equation is always solvable for $(c_1, c_2, c_3)^T$ because the coefficient matrix has rank 2. In order to verify that assertion, suppose that row 2 is C times row 1. Then $C = 1$, $\xi_1^{(1)} = \xi_1^{(2)}$, and $\xi_2^{(1)} = \xi_2^{(2)}$. (Recall that the row rank, the column rank, and the dimension of the range of a matrix are all equal.)   ∎

For more information about interpolation when the number of nodes is not equal to the dimension of the space of interpolating functions, see recent work in [Shel1] and [Wul1]. Some recent papers on polynomial interpolation are [Blo1], [Blo2], [Bos1], [Bos2], [ChuLa], [BR1], and [BR2].

Problems

1. Prove that interpolation by a function

$$p(x) = a + b\xi_1 + c\xi_2 + d\xi_1\xi_2$$

   is possible at the four corners of any rectangle whose sides are parallel to the two coordinate axes. Functions having the form of $p(x)$ in this problem are said to be **bilinear.** They are useful in the numerical solution of partial differential equations.

2. Prove that interpolation at six points in $\mathbb{R}^2$ is possible using $\Pi_2(\mathbb{R}^2)$ if and only if the nodes do not lie on a conic section or on a pair of lines.

3. Determine whether interpolation by the function in Problem 1 is possible at the vertices of a nondegenerate triangle and at one interior point of the triangle.

4. Prove that

$$\binom{\alpha}{\beta} = \prod_{i=1}^{s} \binom{\alpha_i}{\beta_i}$$

5. Prove that

$$\prod_{i=1}^{n} \sum_{j=1}^{m} A_{ij} = \sum_{\substack{\alpha \in \mathbb{N}^n \\ \|\alpha\|_\infty \leq m}} \prod_{i=1}^{n} A_{i\alpha_i}$$

   Here $\|a\|_\infty = \max(\alpha_1, \alpha_2, \dots, \alpha_n)$.

6. Prove that, formally,

$$\prod_{i=1}^{n} \sum_{j=1}^{\infty} A_{ij} = \sum_{k=1}^{\infty} \sum_{|\alpha|=k} \prod_{i=1}^{n} A_{i\alpha_i}$$

7. A polynomial $p(x) = \sum_{|\alpha| \leq k} c_\alpha x^\alpha$ is of degree k if $\sum_{|\alpha|=k} |c_\alpha| \neq 0$. In other words, there exists at least one nonzero term $c_\alpha x^\alpha$ in which $|\alpha| = k$. Prove that the product of a polynomial of degree k with one of degree n is a polynomial of degree $n + k$.

8. Prove that $x^\alpha x^\beta = x^{\alpha+\beta}$. Prove that if $\beta \leq \alpha$ then $x^\alpha / x^\beta = x^{\alpha-\beta}$. Prove that $(x^\alpha)^k = x^{k\alpha}$ for $k \in \mathbb{Z}_+$.

9. Prove that if $|\alpha| = k$ then $D^\alpha (vx)^k = k! v^\alpha$. (Here v and x are in $\mathbb{R}^s$.)

10. Fix a multi-index $\beta \in \mathbb{Z}_+^s$ and compute the dimension of the polynomial space spanned by $\{x \mapsto x^\alpha : \alpha \leq \beta\}$.

11. How many monomials of exact degree k are there in $\Pi_k (\mathbb{R}^s)$? What is the recurrence relation by which we can compute dim $\left(\Pi_k (\mathbb{R}^s) \right)$ in terms of dim $\left(\Pi_{k-1} (\mathbb{R}^s) \right)$?

12. Repeat Problem 11 for computing the dimension of $\Pi_k (\mathbb{R}^s)$ in terms of the dimension of $\Pi_k (\mathbb{R}^{s-1})$.

13. Any set of monomials x^α is linearly independent on any open subset of $\mathbb{R}^s$.

14. Refer to Problem 1, and prove that the determinant of the interpolation matrix is the square of the area of the rectangle.

References

[AG]   Arcangeli, R. and J. L. Gout. "Sur l'interpolation de Lagrange dans R^n." *Comptes Rendues Acad. Sci. Paris* 281 Ser. A, (1975), 357–359.

[Blo1]   Bloom, T. "The Lebesgue constant for Lagrange interpolation in the simplex." *J. Approx. Theory* 54 (1988), 338–353.

[Blo2]   Bloom, T. "On the convergence of multivariable Lagrange interpolants." *Constr. Approx.* 5 (1989), 415-435.

[BR2]   de Boor, C., and A. Ron. "Computational aspects of polynomial interpolation in several variables." *Math. Comp.* 58 (1992), 705–727.

[Bo1]   de Boor, C. "Polynomial interpolation in several variables." In *Proceedings of a Conference Honoring S. D. Conte*. Ed. by R. DeMillo and J. R. Rice, Plenum Press, New York, 1994.

[Bos3]   Bos, L. "Bounding the Lebesgue function for Lagrange interpolation in a simplex." *J. Approx. Theory* 38 (1983), 43–59.

[Bos6]   Bos, L. "On Lagrange interpolation at points in a disk." In *Approximation Theory V*. Ed. by C. K. Chui, L. L. Schumaker, and J. D. Ward, Academic Press, Boston, 1986. pp. 275–278.

[Bos1]   Bos, L. "A characteristic of points in R^2 having Lebesgue function of polynomial growth." *J. Approx. Theory* 56 (1989), 316–329.

[Bos2]   Bos, L. "Some remarks on the Fejér problem for Lagrange interpolation in several variables." *J. Approx. Theory* 60 (1990), 133–140.

[ChuLa]   Chui, Charles, and M. J. Lai. "Vandermonde determinant and Lagrange interpolation in R^s." In *Nonlinear and Convex Analysis*. Ed. by B. L. Lin and L. Simons. Marcel Dekker, New York, 1988. 23–26.

[Shek1]   Shekhtman, B. "Interpolating subspaces in R^n." In *Approximation Theory, Wavelets and Applications*. Ed. by S. P. Singh. Kluwer, Amsterdam, 1995, 465–471.

[Wul1]   Wulbert, D. "Interpolation at a few points." *J. Approx Theory* 96 (1999), 139–148.

5
Moving the Nodes

In an interpolation problem, we often want to know the effect of moving the nodes. We begin with some elementary facts.

LEMMA 1. *Let* $\{f_1, \ldots, f_n\}$ *be a linearly independent set of functions from a set X into some linear space. Let* σ *be a map from some set Y onto X. Then* $\{f_1 \circ \sigma, \ldots, f_n \circ \sigma\}$ *is also linearly independent.*

Proof. If $\sum_{i=1}^n c_i f_i(\sigma(y)) = 0$ for all $y \in Y$, then $\sum_{i=1}^n c_i f_i(x) = 0$ for all $x \in X$. Hence $\sum_{i=1}^n |c_i| = 0$. ∎

Example 1. Let $X = Y = \mathbb{R}$, $f_i(x) = x^{i-1}$, and $\sigma(x) = x^3$. We conclude that the functions $x \mapsto 1, x^3, x^6, \ldots$ form a linearly independent set. Of course, this conclusion follows from the observation that this set is a subset of the set of all monomials. ∎

Example 2. Let $X = Y = [0, \infty)$, $f_i(x) = x^{i-1}$, and $\sigma(x) = \sqrt{x}$. We conclude that the functions $x \mapsto 1, x^{1/2}, x^{2/2}, x^{3/2}, \ldots$ form a linearly independent set on $[0, \infty)$. ∎

Example 3. The functions $x \mapsto (\tan x)^n$, $n \in \mathbb{N}$, form a linearly independent set on $(-\pi/2, \pi/2)$. ∎

Recall that the notation $f|A$ signifies the restriction of a function f to a domain A. In the same way, if V is a set of functions, we write $V|A$ for $\{f|A : f \in V\}$.

LEMMA 2. *Let V be a linear space of functions from a set X to some linear space. Let* $\sigma : X \twoheadrightarrow X$ *be a bijection such that* $v \circ \sigma^{-1} \in V$ *whenever* $v \in V$. *Then for any set* $\mathcal{N} \subset X$, $\dim [V|\sigma(\mathcal{N})] \geq \dim [V|\mathcal{N}]$.

Proof. If n is a natural number such that $n \leq \dim \left[V | \mathcal{N} \right]$, then there exist $v_1, \ldots, v_n \in V$ for which the set $\{v_1 | \mathcal{N}, \ldots, v_n | \mathcal{N}\}$ is linearly independent. By hypothesis, $v_i \circ \sigma^{-1} \in V$. Furthermore, these functions form a linearly independent set on $\sigma(\mathcal{N})$. Indeed, if

$$\sum_{i=1}^{n} c_i v_i \circ \sigma^{-1}(\sigma(x)) = 0 \qquad (x \in \mathcal{N})$$

then $\sum |c_i| = 0$ by the linear independence of $\{v_i | \mathcal{N}\}$. We conclude that $\dim \left[V | \sigma(\mathcal{N}) \right]$ is at least n, because $v_i \circ \sigma^{-1} \in V$. If $V | \mathcal{N}$ is finite dimensional, we can let $n = \dim (V | \mathcal{N})$ to complete the proof. If $V | \mathcal{N}$ is infinite dimensional, we can take n to be any natural number and conclude that $\dim \left[V | \sigma(\mathcal{N}) \right] = \infty$.   ∎

Example 4. Let $X = \mathbb{R}^2$. Let σ be a translation: $\sigma(x) = x + a$, where a is fixed in $\mathbb{R}^2$. Let $V = \Pi_k(\mathbb{R}^2)$ and $n = \dim (V) = \binom{k+2}{2} = \frac{1}{2}(k+1)(k+2)$. Let $\mathcal{N}$ be a set of n nodes on which interpolation by V is possible. This means that $\dim (V | \mathcal{N}) = n$. Since $v \circ \sigma^{-1} \in V$ whenever $v \in V$, we conclude that interpolation by V is possible on $\sigma(\mathcal{N})$. Lemma 4, below, gives the general case.   ∎

Example 5. Adopt the setting of Example 4, except that we allow σ to be a nonsingular linear transformation:

$$\sigma(\xi_1, \xi_2) = (a_{11}\xi_1 + a_{12}\xi_2, \, a_{21}\xi_1 + a_{22}\xi_2)$$

We shall prove in the following lemma that $v \circ \sigma^{-1} \in V$ whenever $v \in V$. Hence interpolation by V is possible on $\sigma(\mathcal{N})$ in this case.   ∎

LEMMA 3. *Let σ be a linear transformation from $\mathbb{R}^s$ into $\mathbb{R}^s$. The polynomial space $\Pi_k(\mathbb{R}^s)$ is invariant under the map $p \mapsto p \circ \sigma$.*

Proof. Let $y = \sigma(x) = (\eta_1, \ldots, \eta_s)$ for $x \in \mathbb{R}^s$. Since σ is linear, we have $\eta_i = u_i x$ for suitable points $u_i \in \mathbb{R}^s$. Let $p \in \Pi_k(\mathbb{R}^s)$. Then

$$p(x) = \sum_{|\alpha| \leq k} c_\alpha x^\alpha$$

Hence, by the Multinomial Theorem (Chapter 4, page 27) we have

$$p(\sigma(x)) = p(y) = \sum_{|\alpha| \leq k} c_\alpha y^\alpha = \sum_{|\alpha| \leq k} c_\alpha \prod_{i=1}^{s} \eta_i^{\alpha_i}$$

$$= \sum_{|\alpha| \leq k} c_\alpha \prod_{i=1}^{s} (u_i x)^{\alpha_i} = \sum_{|\alpha| \leq k} c_\alpha \prod_{i=1}^{s} \sum_{|\beta| = \alpha_i} \frac{\alpha_i!}{\beta!} u_i^\beta x^\beta$$

In this last expression, the innermost sum contains monomials of degree $|\beta| = \alpha_i$. Hence the sum is a polynomial of degree at most α_i. The product of these polynomials is therefore a polynomial of degree at most $\sum_{i=1}^{s} \alpha_i$, and this is at most k.   ∎

A map $\sigma : \mathbb{R}^s \longrightarrow \mathbb{R}^s$ is called a **translation** if it has the form $\sigma(x) = x + w$, where w is a prescribed point of $\mathbb{R}^s$. There are also **translation operators** which act on functions. These are often written T_w, and their effect on a function is

$$(T_w f)(x) = f(x + w)$$

When we say that a space V of functions is **invariant** under translations, we mean that

$$f \in V \quad \Longrightarrow \quad T_w f \in V \text{ for all } w.$$

Expressed otherwise: $T_w(V) \subset V$ for all w. Notice that $T_w f = f \circ \sigma$ and that T_w is a linear operator.

LEMMA 4. *The polynomial spaces $\Pi_k(\mathbb{R}^s)$ are invariant under the map $p \mapsto p \circ \sigma$, where σ is any translation.*

Proof. Let $p \in \Pi_k(\mathbb{R}^s)$ and let $w \in \mathbb{R}^s$. We must prove that $T_w p \in \Pi_k(\mathbb{R}^s)$. If $p(x) = \sum_{|\alpha| \le k} c_\alpha x^\alpha$ then, by the Binomial Theorem (Chapter 4, page 26),

$$p(x + w) = \sum_{|\alpha| \le k} c_\alpha (x + w)^\alpha = \sum_{|\alpha| \le k} c_\alpha \sum_{0 \le \beta \le \alpha} \binom{\alpha}{\beta} x^\beta w^{\alpha - \beta}$$

Since w is fixed, $w^{\alpha - \beta}$ is a scalar. As a function of x, $p(x + w)$ is thus a linear combination of terms x^β in which each β satisfies $\beta \le \alpha$. Since $|\beta| \le |\alpha| \le k$, this function is in $\Pi_k(\mathbb{R}^s)$. ∎

An **affine** map on a linear space is the sum of a linear map and a constant map. Thus an affine map has the form $\sigma(x) = Ax + b$, where A is a linear transformation on the space, and b is an element in the space.

THEOREM 1. *Let σ be an affine map of $\mathbb{R}^s$ to $\mathbb{R}^s$. The polynomial space $\Pi_k(\mathbb{R}^s)$ is invariant under the map $p \mapsto p \circ \sigma$.*

Proof. Let p be any element of $\Pi_k(\mathbb{R}^s)$. Let σ be given by the equation $\sigma(x) = Ax + b$. Then

$$p \circ \sigma = p(Ax + b) = q(Ax)$$

where $q(x) = p(x + b)$. By Lemma 4, $q \in \Pi_k(\mathbb{R}^s)$. By Lemma 3, $p \circ \sigma \in \Pi_k(\mathbb{R}^s)$. ∎

A set of $s + 1$ points in $\mathbb{R}^s$, say $\{x_0, x_1, \ldots, x_s\}$, is said to be **in general position** if the set

$$\{x_1 - x_0, x_2 - x_0, \ldots, x_s - x_0\}$$

is linearly independent. The following lemma shows that this is a property of the *set* and is not affected by the labeling of the points.

LEMMA 5. *The set* $\{x_0, x_1, \ldots, x_s\}$ *in* $\mathbb{R}^s$ *is in general position if and only if the conditions* $\sum_{i=0}^{s} \lambda_i x_i = 0$ *and* $\sum_{i=0}^{s} \lambda_i = 0$ *imply that* $\sum_{i=0}^{s} |\lambda_i| = 0$.

Proof. Assume that our set is in general position and that $\sum_{i=0}^{s} \lambda_i x_i = 0$, $\sum_{i=0}^{s} \lambda_i = 0$. Then $\sum_{i=0}^{s} \lambda_i (x_i - x_0) = 0$. Since $x_i - x_0 = 0$ when $i = 0$, we have $\sum_{i=1}^{s} \lambda_i (x_i - x_0) = 0$. Since the set $\{x_i - x_0 : 1 \le i \le s\}$ is linearly independent, $\sum_{i=1}^{s} |\lambda_i| = 0$. Since $\sum_{i=0}^{s} \lambda_i = 0$ we conclude that $\lambda_0 = 0$. Hence $\sum_{i=0}^{s} |\lambda_i| = 0$. The remainder of the proof is left as a problem. ∎

THEOREM 2. *If* $\{x_0, x_1, \ldots, x_s\}$ *is in general position in* $\mathbb{R}^s$, *and if* $v_0, v_1, \ldots, v_s$ *are any points of* $\mathbb{R}^s$, *then there is an affine map* σ *such that* $\sigma(x_i) = v_i$ *for* $0 \le i \le s$.

Proof. The set $\{x_i - x_0 : 1 \le i \le s\}$ is a basis for $\mathbb{R}^s$. Define A by requiring $A(x_i - x_0) = v_i - v_0$ for $1 \le i \le s$. Then this equation is true for $i = 0$. Hence $Ax_i + (v_0 - Ax_0) = v_i$, $0 \le i \le s$. ∎

COROLLARY. *In* $\mathbb{R}^2$, *let* T *and* S *be two triangles, of which the first is nondegenerate. Then there is an affine map* σ *such that* $\sigma(T) = S$.

Proof. Select σ so that the vertices of T are mapped to the vertices of S. This is possible by Theorem 2. Then use the fact that an affine map satisfies the equation

$$\sigma(\theta x + (1 - \theta)y) = \theta \sigma(x) + (1 - \theta)\sigma(y)$$

for all points x and y, and for all real θ. (See Problem 3.) ∎

THEOREM 3. *Suppose that interpolation by* $\Pi_k(\mathbb{R}^s)$ *is uniquely possible on a node set* $\mathcal{N}$, *where* $\#\mathcal{N} = \dim \Pi_k(\mathbb{R}^s) = \binom{s+k}{s}$. *If* σ *is an invertible affine map of* $\mathbb{R}^s$ *onto* $\mathbb{R}^s$ *then interpolation is also possible on* $\sigma(\mathcal{N})$.

Proof. Use Lemma 2 and Theorem 1. ∎

To illustrate how Theorem 3 can be used, we shall prove a result about $\Pi_3(\mathbb{R}^2)$. This space of polynomials has dimension 10. Let $\{x_1, x_2, \ldots, x_{10}\}$ be a set of 10 nodes in $\mathbb{R}^2$ such that

a. x_7, x_8, x_9, x_{10} are on a line L_1
b. x_4, x_5, x_6, x_7 are on a line L_2
c. $L_1 \ne L_2$
d. x_1, x_2, x_3 are not colinear and are not in $L_1 \cup L_2$.

The situation is shown in the figure.

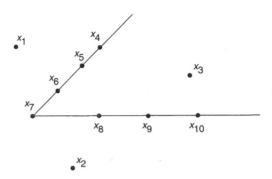

Figure 5.1   *A set of nodes for Theorem 4.*

THEOREM 4. *If the nodes satisfy the conditions (a)–(d), above, then interpolation of arbitrary data at the nodes is possible by a unique polynomial in* $\Pi_3(\mathbb{R}^2)$.

Proof. Use Problem 13 to obtain an affine map σ such that $\sigma(x_7), \ldots, \sigma(x_{10})$ are on the horizontal axis, whereas $\sigma(x_4), \ldots, \sigma(x_7)$ are on the vertical axis. By Theorem 3 it suffices to prove that if $p \in \Pi_3(\mathbb{R}^2)$ and $p(\sigma(x_i)) = 0$ for $1 \le i \le 10$ then $p = 0$. We write $p(x) = \sum_{i+j \le 3} c_{ij} \xi_1^i \xi_2^j$. Since p vanishes at 4 points on the horizontal axis and since

$$p(\xi_1, 0) = c_{00} + c_{10}\xi_1 + c_{20}\xi_1^2 + c_{30}\xi_1^3$$

we conclude that $c_{i0} = 0$ for $0 \le i \le 3$. Similarly we have $c_{0i} = 0$ for $0 \le i \le 3$. Thus p has the form

$$p(\xi_1, \xi_2) = c_{11}\xi_1\xi_2 + c_{12}\xi_1\xi_2^2 + c_{21}\xi_1^2\xi_2$$
$$= \xi_1\xi_2(c_{11} + c_{12}\xi_2 + c_{21}\xi_1)$$

Since p vanishes at $\sigma(x_i)$, $1 \le i \le 3$, and since these points are not colinear and not on the coordinate axes, we conclude that $c_{11} = c_{12} = c_{21} = 0$. Hence $p = 0$.   ∎

For discussions of sets in general position, consult [Pon], Section 1 or [Al], Appendix 1. The Multinomial Theorem is in [AS], page 823.

Problems

1. Fix a multi-index β in $\mathbb{Z}_+^s$. Denote by V the polynomial space spanned by

$$\{x \longmapsto x^\alpha : \alpha \le \beta\} \qquad (x \in \mathbb{R}^s)$$

If $\sigma : \mathbb{R}^s \to \mathbb{R}^s$ is a translation, is it true that $v \circ \sigma \in V$ whenever $v \in V$? Answer the same question when σ is an arbitrary linear map.

2. Complete the proof of Lemma 5.

3. Prove that a map σ is affine if and only if it satisfies the equation

$$\sigma(\theta x + (1 - \theta)y) = \theta\sigma(x) + (1 - \theta)\sigma(y) \qquad (\theta \in \mathbb{R})$$

   Prove that if σ is affine and K is convex, then $\sigma(K)$ is convex.

4. Let V be a linear space of functions defined on a set X. Let $\mathcal{N} \subset X$. Prove that $\dim(V|\mathcal{N}) \leq \#\mathcal{N}$ and $\dim(V|\mathcal{N}) \leq \dim(V)$. Remember to include the cases when one or more of these numbers is infinite. Give an example in which $\#X = \infty$ and $\dim(V|\mathcal{N}) = \#\mathcal{N}$ for all $\mathcal{N}$.

5. Which of these subspaces is invariant under translation?
   a. span $\{f_1, \ldots, f_n\}$ where $f_i(x) = (x - v_i)^{-1}$, $\quad x, v_i \in \mathbb{R}$, $1 \leq i \leq n$.
   b. span $\{f_1, \ldots, f_n\}$ where $f_i(x) = \exp(\lambda_i x)$, $\quad x, \lambda_i \in \mathbb{R}$, $1 \leq i \leq n$.
   c. $U = \text{span}\{f_0, \ldots, f_n\}$ where $f_i(x) = \cos ix$, $\quad x \in \mathbb{R}$, $0 \leq i \leq n$.
   d. $V = \text{span}\{f_1, \ldots, f_n\}$ where $f_i(x) = \sin ix$, $\quad 1 \leq i \leq n$.
   e. $U + V$, where U and V are given in parts (c) and (d).
   f. span of the four functions $x \to 1, \xi_1, \xi_2, \xi_1\xi_2$. Here $x = (\xi_1, \xi_2) \in \mathbb{R}^2$.

6. Prove that each translation operator T_a is an isometry on the spaces $L^2(\mathbb{R})$ and $C_b(\mathbb{R})$. The latter is the space of bounded continuous functions on $\mathbb{R}$, and is given the supremum norm.

7. Let $C_b(\mathbb{R})$ be as in the preceding problem, but define the norm

$$\|f\| = \sup |f(x)e^{-x^2}|$$

   Is T_a an isometry?

8. In $\mathbb{R}^s(s \geq 2)$ let T and S be two parallelograms. Is there an affine map $\sigma : \mathbb{R}^s \to \mathbb{R}^s$ such that $\sigma(T) = S$?

9. Is the affine map described in Theorem 2 unique? Find a theorem concerning this question.

10. Let $\sigma : \mathbb{R}^s \to \mathbb{R}^s$ be a translation. Prove that for any polynomial p, $\deg(p) = \deg(p \circ \sigma)$. Prove the same equality when σ is any nonsingular linear transformation.

11. An affine map, $x \mapsto Ax + b$, is invertible if and only if A is invertible.

12. Prove that if L_1 and L_2 are 2 nonparallel lines in $\mathbb{R}^2$ then there is an affine map $\sigma : \mathbb{R}^2 \to \mathbb{R}^2$ such that $\sigma(L_1)$ is the horizontal axis and $\sigma(L_2)$ is the vertical axis in $\mathbb{R}^2$.

13. Prove that if L_1 and L_2 are distinct lines in $\mathbb{R}^s$ having a point in common then there is an affine map $\sigma : \mathbb{R}^s \to \mathbb{R}^s$ such that all points on $\sigma(L_1)$ have the form $(\xi, 0, 0, \ldots, 0)$, and all points on $\sigma(L_2)$ have the form $(0, \xi, 0, \ldots, 0)$.

14. The bivariate functions $u(x) = u(\xi_1, \xi_2) = a + b\xi_1 + c\xi_2 + d\xi_1\xi_2$ are called **bilinear.** Prove that the 4-dimensional space of bilinear functions is invariant under these maps: $\sigma(x) = x + a$, $\tau(x) = Ax$, in which A is a diagonal 2×2 matrix. What happens if A is a more general 2×2 matrix? What additional hypotheses are needed if we want to use Lemma 2?

15. Consider the space described in Problem 1:

$$P_\beta = \text{span}\{x \longmapsto x^\alpha : \alpha \le \beta\}$$

Prove that if σ is a linear map of $\mathbb{R}^s$ to $\mathbb{R}^s$ whose matrix is diagonal, then $p \circ \sigma \in P_\beta$ whenever $p \in P_\beta$. (The matrix corresponds to σ when we use the usual basis for domain and range of σ.)

16. Prove that a functional ϕ is affine if and only if the map $x \mapsto \phi(x) - \phi(0)$ is linear.

References

[Al]   Alexsandrov, P. S. *Combinatorial Topology.* Graylock Press, Rochester, N.Y., 1956.

[AS]   Abramowitz, M., and I. A. Stegun. *Handbook of Mathematical Functions.* National Bureau of Standards, Washington, 1964. Reprint: Dover Publications, New York.

[Pon]   Pontryagin, L. S. *Foundations of Combinatorial Topology.* Graylock Press, Rochester, N.Y., 1952.

6
Projections

A **projection** is a bounded, linear, idempotent operator defined on, and taking values in, a single normed linear space. Recall that an operator A is **idempotent** if $A^2 = A$.

If $P : E \to E$ is a projection and if U is its range, we say that P is a projection of E onto U. A surjective map is indicated by the notation $\twoheadrightarrow$.

THEOREM 1. *If $P : E \twoheadrightarrow U$ is a projection, then U is a closed subspace of E. Furthermore, U is the set of fixed points of P :*

$$U = \{f \in E : Pf = f\}$$

Proof. The first assertion will follow from the second, since $P - I$ is continuous. If $u \in U$ then $u = Pf$ for some f, whence $Pu = P^2f = Pf = u$. If $Pf = f$, then obviously f is in the range of P, which is U. ∎

THEOREM 2. *If P is a projection, then so is $I - P$. Furthermore,*

$$\operatorname{range}(P) = \ker(I - P)$$

Proof. Observe that $(I - P)^2 = (I - P)(I - P) = I - P - P + P^2 = I - P$. Now use Theorem 1. ∎

Definition. A subspace U of a normed space E is said to be **complemented** if there is a projection $P : E \twoheadrightarrow U$.

If $P : E \twoheadrightarrow U$ is a projection, then the "Peirce decomposition" $I = P + (I - P)$ leads to an equation $X = U \oplus V$, where V is the range of $I - P$. This is a direct sum, because if $f \in U \cap V$ then $f = Pf = (I - P)f$ by Lemma 1, and $f = 0$.

A famous example of an uncomplemented subspace in a Banach space is the subspace c_0 of sequences tending to 0 in the space ℓ^∞ of all bounded sequences. The norm in ℓ^∞ is $\|x\| = \sup_n |x_n|$.

For any projection P, there are two sets of interest, namely, the range of P and the range of P^*. Recall that if E is a normed linear space, its **conjugate** space E^* consists of all continuous linear functionals on E. If $A : E \rightarrow E$ is any operator, then the **adjoint** of A is denoted by A^* and is defined by

$$A^*\phi = \phi \circ A \qquad (\phi \in E^*)$$

Thus $A^* : E^* \rightarrow E^*$. It is elementary to prove that if P is a projection, then so is P^*:

$$(P^*P^*)\phi = P^*(P^*\phi) = (P^*\phi) \circ P = \phi \circ P \circ P = \phi \circ P = P^*\phi$$

Let's see what all this means in a simple case. Let U be an n-dimensional subspace in E, and let $P : E \twoheadrightarrow U$ be a projection. Select any basis $\{u_1, u_2, \ldots, u_n\}$ for U. Obviously, for any $f \in E$ we must have

$$Pf = \sum_{i=1}^{n} \lambda_i u_i = \sum_{i=1}^{n} \lambda_i(f)u_i$$

for suitable real numbers $\lambda_i(f)$. An elementary argument shows that the functionals λ_i are linear:

$$\sum \lambda_i(\alpha f + \beta g)u_i = P(\alpha f + \beta g) = \alpha Pf + \beta Pg$$
$$= \alpha \sum \lambda_i(f)u_i + \beta \sum \lambda_i(g)u_i = \sum [\alpha\lambda_i(f) + \beta\lambda_i(g)]u_i$$

Since the u_i form a basis, we conclude that

$$\lambda_i(\alpha f + \beta g) = \alpha\lambda_i(f) + \beta\lambda_i(g)$$

Note that this argument *requires* the linear independence of the set $\{u_1, u_2, \ldots, u_n\}$.

A further consequence is that $\lambda_i(u_j) = \delta_{ij}$. To see this, just write the following equation and use again the unicity in representing a vector in terms of a basis:

$$\sum_{i=1}^{n} \delta_{ij}u_i = u_j = Pu_j = \sum_{i=1}^{n} \lambda_i(u_j)u_i$$

The continuity of each λ_i follows from the continuity of P. Indeed, if $f \rightarrow 0$, then $Pf \rightarrow 0$, $\sum \lambda_i(f)u_i \rightarrow 0$ and $\lambda_i(f) \rightarrow 0$ for $1 \leq i \leq n$. (Again, we need a basis in this argument.)

Now that the structure of P is known, we can find the structure of P^*. Let $\phi \in E^*$. Then $P^*\phi = \phi \circ P$, by definition. Hence for any $f \in E$,

$$(P^*\phi)(f) = \phi(Pf) = \phi\left(\sum \lambda_i(f)u_i\right) = \sum \lambda_i(f)\phi(u_i) = \left(\sum \phi(u_i)\lambda_i\right)(f)$$

Hence $P^*\phi = \sum \phi(u_i)\lambda_i$. The range of P^* is spanned by $\{\lambda_1, \ldots, \lambda_n\}$. This set is linearly independent because if $\sum c_i\lambda_i = 0$ then $\sum_i c_i\lambda_i(u_j) = 0$ for $j = 1, \ldots, n$, and so $c_j = 0$, $j = 1, \ldots, n$. Thus $\{\lambda_1, \lambda_2, \ldots, \lambda_n\}$ is a basis for the range of P^*.

Finally, we can interpret P as a "generalized interpolation" operator, because for any $f \in E$ we have

$$\lambda_j(Pf) = \lambda_j\left(\sum_i \lambda_i(f)u_i\right) = \sum_i \lambda_i(f)\lambda_j(u_i) = \lambda_j(f)$$

The tensor-product notation for P and P^* is

$$P = \sum \lambda_i \otimes u_i \quad \text{i.e.} \quad Pf = \sum \lambda_i(f)u_i$$

$$P^* = \sum u_i \otimes \lambda_i \quad \text{i.e.} \quad P^*\phi = \sum \phi(u_i)\lambda_i$$

The next chapter explains the tensor-product notation. The range of P has the basis $\{u_1, \dots, u_n\}$, and the range of P^* has $\{\lambda_1, \lambda_2, \dots, \lambda_n\}$ as a basis. The property $\lambda_i(u_j) = \delta_{ij}$ is called **biorthogonality.**

Projections need not have finite-dimensional ranges; the only requirement (besides linearity and continuity) is idempotency: $P^2 = P$. We can understand projections quite well without having a formula such as $P = \sum \lambda_i \otimes u_i$. For example, we have this result:

THEOREM 3. *A projection is uniquely determined by its range and the range of its adjoint.*

Proof. Let P and Q be projections such that their ranges are the same and the ranges of their adjoints are the same. Then $P(Qf) = Qf$, because $Qf \in \text{range}(Q) = \text{range}(P)$, and the application of P leaves invariant each element of its range. Similarly, $Q^*P^* = P^*$. It follows that $Q^* = (PQ)^* = Q^*P^* = P^*$. If f and ϕ are arbitrary, then

$$\phi(Pf) = (P^*\phi)(f) = (Q^*\phi)(f) = \phi(Qf)$$

Hence $Pf = Qf$ and $P = Q$. (This argument requires a corollary of the Hahn-Banach Theorem: If $\phi(f) = 0$ for all $\phi \in E^*$, then $f = 0$.)   ∎

The interpolation properties of a projection can also be established without the assumption of finite dimensionality. The following statement summarizes the situation.

THEOREM 4. *If P is a projection defined on a normed linear space E, then for any f in E and any ϕ in the range of P^* we have the generalized interpolation property: $\phi(Pf) = \phi(f)$.*

Proof. Since P^* is a projection (defined on E^*), it leaves invariant each element of its range. Thus,

$$\phi(Pf) = (\phi \circ P)f = (P^*\phi)(f) = \phi(f)$$ ∎

Example. Here we illustrate the concept of a "generalized interpolation operator." Consider the polynomial space $\Pi_{n-1}(\mathbb{R})$. We know that if $x_1, x_2, \dots, x_n$ are distinct points in $\mathbb{R}$, then the set of point-evaluation functionals $\{x_1^*, x_2^*, \dots, x_n^*\}$ is linearly independent on Π_{n-1}. One way to demonstrate this is to construct the corresponding ("cardinal") basis $\{v_1, v_2, \dots, v_n\}$ for Π_{n-1} such that $x_i^*(v_j) = \delta_{ij}$. Now, for a given $\varepsilon > 0$, define n functionals ϕ_i by the equation

$$\phi_i(f) = \frac{1}{2\varepsilon} \int_{x_i-\varepsilon}^{x_i+\varepsilon} f(x)\, dx \qquad (1 \le i \le n)$$

For a continuous function, f, $\lim_{\varepsilon \to 0} \phi_i(f) = f(x_i)$. For sufficiently small ε, $\{\phi_1, \phi_2, \ldots, \phi_n\}$ is linearly independent on Π_n. Indeed, for small ε, $\det(\phi_i(v_j)) \approx \det(x_i^*(v_j)) = 1$. It follows that Π_n has a basis $\{u_1, u_2, \ldots, u_n\}$ for which $\phi_i(u_j) = \delta_{ij}$. The corresponding projection is then defined by $Pf = \sum_{i=1}^{n} \phi_i(f)u_i$. Thus, for any continuous function f, Pf is a polynomial of degree at most $n - 1$ that "matches" f, in the sense that $\phi_i(Pf) = \phi_i(f)$ for $1 \le i \le n$. ∎

Let E be a normed linear space. A subset B in E^* is called a **norm-determining set** if, for each $f \in E$,

$$\|f\| = \sup\{|\phi(f)| : \phi \in B\}$$

Since $\|f\| \ge |\phi(f)|$ if $\phi \in B$ and $f \in E$, we have $\|\phi\| \le 1$ for each $\phi \in B$.

> **AUERBACH'S THEOREM.** *If E is an n-dimensional normed space and B is a closed norm-determining set in E^*, then there exist $f_1, \ldots, f_n$ in E and $\phi_1, \ldots, \phi_n \in B$ such that $\|f_i\| = 1$, $\|\phi_i\| = 1$, and $\phi_i(f_j) = \delta_{ij}$ for $1 \le i, j \le n$.*

Proof. Let $\{g_1, \ldots, g_n\}$ be any basis for E. By continuity and compactness, there exist $\phi_1, \ldots, \phi_n \in B$ such that $|\det(\phi_i(g_j))|$ is a maximum. We note that B contains a basis for E^*. Indeed, if this were not true, then B would lie in a hyperplane of the form $\{\phi \in E^* : \phi(h_0) = 0\}$ for some $h_0 \in E \backslash 0$. This condition would be incompatible with the requirement $\|h_0\| = \sup\{|\phi(h_0)| : \phi \in B\}$. Hence the number $\lambda = \det(\phi_i(g_j))$ is not zero. Let A be the matrix with elements $A_{ij} = \phi_i(g_j)$, and let C be the inverse of A. Set $f_j = \sum_{k=1}^{n} C_{kj} g_k$. Then

$$\phi_i(f_j) = \phi_i\left(\sum_{k=1}^{n} C_{kj} g_k\right) = \sum_{k=1}^{n} C_{kj} A_{ik} = \delta_{ij}$$

Consider the equation $\lambda = \sum_{k=1}^{n} A_{ik} \lambda C_{ki} = \sum_{k=1}^{n} \phi_i(g_k) \lambda C_{ki}$. This is the formula for computing the determinant of A by means of elements in the i-th row. The number λC_{ki} is the cofactor of the entry A_{ik}. By the choice of $\phi_1, \ldots, \phi_n$, the modulus of the determinant is not increased if we replace ϕ_i by any other element ψ in B. Hence

$$\left|\sum_{k=1}^{n} \psi(g_k) \lambda C_{ki}\right| \le |\lambda|$$

Here we have used the crucial fact that the cofactors associated with the entries of row i in the matrix do not actually depend on the elements in that row. It follows that $|\psi(\sum C_{ki} g_k)| = |\psi(f_i)| \le 1$. By taking a supremum for ψ in B, we obtain $\|f_i\| \le 1$. Since $\phi_i(f_i) = 1$, we conclude that $\|f_i\| = 1$ and $\|\phi_i\| = 1$. ∎

> **COROLLARY 1.** *Let U be an n-dimensional subspace in a normed space E. Then there is a projection of E onto U having norm at most n.*

Proof. By Auerbach's Theorem, there exist points $u_1, u_2, \ldots, u_n$ in U and functionals $\phi_1, \phi_2, \ldots, \phi_n$ in U^* such that $\|\phi_i\| = 1$, $\|u_i\| = 1$, and $\phi_i(u_j) = \delta_{ij}$. By the Hahn-

Banach Theorem, each ϕ_i can be extended to E without increasing its norm. Denote the extended functionals by ϕ_i also, and define $P = \sum_{i=1}^{n} \phi_i \otimes u_i$. Then

$$\|Pf\| = \left\|\sum \phi_i(f)u_i\right\| \leq \sum |\phi_i(f)|\,\|u_i\|$$
$$= \sum |\phi_i(f)| \leq \sum \|\phi_i\|\,\|f\| \leq n\|f\|$$

It is elementary to prove that $\{u_1, u_2, \ldots, u_n\}$ is a basis for U and that P is a projection onto U. ∎

COROLLARY 2. *Let X be a compact Hausdorff space, and let U be an n-dimensional subspace in $C(X)$. Then there exists an interpolating projection of norm at most n from $C(X)$ onto U.*

Proof. With each point x in X we associate a point-evaluation functional x^*, whose definition is $x^*(f) = f(x)$ for each f in $C(X)$. Since the norm in $C(X)$ is

$$\|f\| = \sup_x |f(x)| = \sup_x |x^*(f)|$$

we see that the point evaluations provide a norm-determining set. By Auerbach's Theorem, there exist elements $u_1, u_2, \ldots, u_n$ in U and points $x_1, x_2, \ldots, x_n$ in X such that $\|u_i\| = 1$ and $x_i^*(u_j) = \delta_{ij}$. Let $Pf = \sum_{i=1}^{n} f(x_i)u_i$. Then

$$\|Pf\| \leq \sum |f(x_i)|\,\|u_i\| \leq n\|f\|$$ ∎

A theorem of Kadec and Snobar asserts that an n-dimensional subspace has a projection of norm $\sqrt{n}$. Examples show that no upper bound of the form n^α exists with $\alpha < \frac{1}{2}$.

For surveys of the theory of projections, see [CP] and [C3]. For Auerbach's Lemma in its original form see [Rus]. The generalized form presented here is from [CP]. The result of Kadec and Snobar is in [KSn]. For a different proof of that theorem, see [Cha]. An entire book, by Brezinski, is devoted to biorthogonality, [Brez].

Problems

1. Let P be a projection. Prove that the ranges of P and $I - P^*$ obey
$$\mathcal{R}(P)^\perp = \mathcal{R}(I - P^*)$$
(Recall that $U^\perp = \{\phi \in E^* : \phi(u) = 0 \text{ for all } u \in U\}$.)

2. Prove that if P and Q are projections such that $\mathcal{R}(Q) \subset \mathcal{R}(P)$ and $\mathcal{R}(P^*) \subset \mathcal{R}(Q^*)$, then $P = Q$.

3. Consider the subspace Π_{n-1} in $C[-1, 1]$. Prove that there exist nodes $t_1 < \cdots < t_n$ in $[-1, 1]$ such that the corresponding Lagrange functions ℓ_i are all of supremum norm 1.

4. Use the spaces in Problem 3. Let $-1 \le t_0 < \cdots < t_n \le 1$. For each $f \in C[-1, 1]$, let $Pf = p$, where p is the element in Π_{n-1} for which the quantity $\max_i |f(t_i) - p(t_i)|$ is a minimum. Prove that P is a projection. (*Hint:* Use the Chebyshev theory.)

5. Prove that if $P : E \twoheadrightarrow U$ is a projection, and if $\|I - P\| = 1$ then for each $f \in E$, Pf is a best approximation of f in U.

6. If V is a subspace of a normed space E, the projection constant of V is

$$p(V, E) = \inf \{\|P\| : P \text{ is a projection of } E \text{ onto } V\}$$

Prove that if $V \ne 0$ then $1 \le p(V, E)$. If V is n-dimensional, then $p(V, E) \le \sqrt{n}$. Prove that $p(\Pi_n, C[-1, 1]) \le c \log n$.

7. Let $P : E \twoheadrightarrow V$ be a projection having the form $Pf = \sum_{i=1}^{n} \phi_i(f)v_i$, where $\{v_1, \ldots, v_n\}$ is a basis for V and $\phi_i \in E^*$. Define a pseudo-innerproduct in E by writing $\langle f, g \rangle = \sum_{i=1}^{n} \phi_i(f)\phi_i(g)$. Prove that P is an orthogonal projection and that $\{v_1, \ldots, v_n\}$ is an orthonormal set. Prove also that $f - Pf$ is orthogonal to all elements of E and that $\|f - Pf\| = 0$. Here, the pseudonorm is given by $\|f\|^2 = \langle f, f \rangle$.

8. Let $E = C[-1, 1]$. Fix knots $t_1 < \cdots < t_n$ with $t_1 = -1$ and $t_n = +1$. Let V be the space of first-degree splines having these knots. Prove that $\dim(V) = n$. Prove that $p(V, E) = 1$. (See the definition in Problem 6.)

9. Let V be a two-dimensional subspace in $C[-1, 1]$. Prove that if all constant functions are in V, then $p(V, E) = 1$. (See Problem 6 for the definition of $p(V, E)$.) Auerbach's Theorem will be helpful.

10. Consider the n linear functionals on $C[0, 1]$ defined as follows

$$\phi_i(f) = \int_{x_{i-1}}^{x_i} f(x)\, dx \qquad (1 \le i \le n)$$

where $x_i = i/n$ for $0 \le i \le n$. Prove the existence of a projection $P : C[0, 1] \twoheadrightarrow \Pi_{n-1}$ such that

$$\phi_i(f) = \phi_i(Pf) \qquad (1 \le i \le n, \quad f \in C[0, 1])$$

11. In the example discussed in the text, prove that $\lim_{\varepsilon \downarrow 0} \phi_i(f) = f(x_i)$ if f is continuous at x_i.

12. What is a minimal norm-determining set for the polynomial space Π_1 considered as a subspace of $C[a, b]$? (The norm is the sup-norm.)

13. Is a projection completely and uniquely determined by its range and its null space?

14. Find the projection $P : C[0, 1] \to \Pi_1$ such that $(Pf)(0) = f(0)$ and $\int_0^1 (Pf)(x)\, dx = \int_0^1 f(x)\, dx$ for all f in $C[0, 1]$.

References

[Ban]   Banach, S. *Théorie des Opérationes Linéaires,* Warsaw, 1932. Reprint, Haffner Publ. Co., New York, 1952.

[Blo1]   Bloom, T. "The Lebesgue constant for Lagrange interpolation in the simplex," *J. Approx. Theory* 54 (1988), 338–353.

[Bos2]   Bos, L. "Some remarks on the Fejér problem for Lagrange interpolation in several variables," *J. Approx. Theory* 60 (1990), 133–140.

[Brez]   Brezinski, C. *Biorthogonality and Its Applications to Numerical Analysis,* vol. 156 in Pure and Applied Mathematics, Dekker, Basel, 1992.

[Cha]   Chalmers, B. L. "A natural simple projection . . . ." *J. Approx. Theory* 32 (1981), 226–232.

[CP]   Cheney, E. W., and K. H. Price. "Minimal projections." In *Approximation Theory*. Ed. by A. Talbot. Academic Press, New York, 1970, pp. 261–289.

[C3]   Cheney, E. W. "Projection operators in approximation theory." In *Studies in Functional Analysis*. Ed. by R. G. Bartle. Math. Assoc. of America, Washington, D.C., 1980. pp. 50–80.

[Fek]   Fekete, M. "Über die Verteilung der Wurzeln bei gewissen algebraischen Gleichungen mit ganzahligen Koeffizienten." *Math. Zeit.* 17 (1923), 228–249.

[KSn]   Kadec, M. I. and M. G. Snowbar. "Some functionals over a compact Minkowski space." *Math. Notes* 10 (1971), 694–696.

[Rus]   Ruston, A. F. "Auerbach's theorem." *Proc. Camb. Phil. Soc.* 58 (1964), 476–480.

7
Tensor-Product Interpolation

Here we consider some very efficient methods of multivariate interpolation that are suitable on a Cartesian grid. At first we proceed in a general setting, but specialize later to polynomial and spline interpolation as examples.

Let X and Y be two domains on which various real-valued functions may be defined. Their **Cartesian product** is, by definition, the set

$$X \times Y = \{(x, y) : x \in X, y \in Y\}$$

Suppose next that two sets of nodes have been prescribed, one in X and the other in Y:

$$\{x_1, x_2, \ldots, x_n\} \subset X \qquad \{y_1, y_2, \ldots, y_m\} \subset Y$$

The **Cartesian grid** generated by these two sets is the set

$$\mathcal{N} = \{x_1, x_2, \ldots, x_n\} \times \{y_1, y_2, \ldots, y_m\}$$
$$= \{(x_i, y_j) : 1 \leq i \leq n, \, 1 \leq j \leq m\}$$

Obviously $\#\mathcal{N} = nm$. In order to carry out interpolation on $\mathcal{N}$, we suppose that we have interpolation operators P and Q as follows

$$P = \sum_{i=1}^{n} x_i^* \otimes u_i \qquad Q = \sum_{i=1}^{m} y_i^* \otimes v_i$$

Here we have introduced standard tensor-product notation, and employed the point-evaluation functionals, which are defined by $x^*(f) = f(x)$, etc. The equations above have this meaning:

$$Pf = \sum_{i=1}^{n} x_i^*(f) u_i \qquad Qg = \sum_{i=1}^{m} y_i^*(g) v_i$$

In order to have interpolation, we assume that $u_i(x_j) = \delta_{ij}$ and $v_i(y_j) = \delta_{ij}$. In even greater detail, our equations can be written as

$$(Pf)(x) = \sum_{i=1}^{n} f(x_i)u_i(x) \qquad (Qg)(y) = \sum_{i=1}^{m} g(y_i)v_i(y)$$

Familiar examples are the Lagrange interpolation operators in the classical (polynomial) case.

In order to be able to operate on functions of two variables (that is, functions on $X \times Y$) we extend the operators by defining $\overline{P}$ and $\overline{Q}$ as follows:

$$(\overline{P}F)(x, y) = \sum_{i=1}^{n} F(x_i, y)u_i(x) \qquad (\overline{Q}F)(x, y) = \sum_{i=1}^{m} F(x, y_i)v_i(y)$$

LEMMA 1. *The function $\overline{P}F$ interpolates F on the set $\{x_1, x_2, \ldots, x_n\} \times Y$, and the function $\overline{Q}F$ interpolates F on $X \times \{y_1, y_2, \ldots, y_m\}$.*

Proof. For $1 \le j \le n$ and for $y \in Y$,

$$(\overline{P}F)(x_j, y) = \sum_{i=1}^{n} F(x_i, y)u_i(x_j) = F(x_j, y) \qquad \blacksquare$$

Since X and Y are usually continua, these operators were given the whimsical name "transfinite" interpolation operators.

Example 1. Let $X = [0, 1] = Y$, and let node sets be chosen as before. We can use the classical Lagrange functions:

$$u_i(x) = \prod_{\substack{j=1 \\ j \ne i}}^{n} \frac{x - x_j}{x_i - x_j} \qquad v_i(y) = \prod_{\substack{j=1 \\ j \ne i}}^{m} \frac{y - y_j}{y_i - y_j}$$

The interpolants $\overline{P}F$ and $\overline{Q}F$ will be "hybrid" functions—partly polynomial and partly not. For example, $(\overline{P}F)(x, y)$ will be a polynomial in x but not (generally) a polynomial in y. We can say that it is a polynomial in x whose coefficients are functions of y. Similar remarks apply to $\overline{Q}F$. $\blacksquare$

Incidentally, the process of defining $\overline{P}$ and $\overline{Q}$ is a standard construction in tensor-product theory. The notation used there is $\overline{P} = P \otimes I$ and $\overline{Q} = I \otimes Q$.

To have an interpolation process for the node set $\mathcal{N}$, we simply use $\overline{P}\overline{Q}$.

LEMMA 2. *The function $\overline{P}\overline{Q}F$ interpolates F on $\mathcal{N}$.*

Proof.

$$(\overline{P}\overline{Q}F)(x_\nu, y_\mu) = \sum_i (\overline{Q}F)(x_i, y_\mu)u_i(x_\nu)$$

$$= (\overline{Q}F)(x_\nu, y_\mu) = \sum_i F(x_\nu, y_i)v_i(y_\mu)$$

$$= F(x_\nu, y_\mu) \qquad \blacksquare$$

A detailed formula for $\overline{P}\overline{Q}F$ is

$$(\overline{P}\overline{Q}F)(x, y) = \sum_{i=1}^{n} \sum_{j=1}^{m} F(x_i, y_j)u_i(x)v_j(y)$$

Notice that from this formula we deduce that $\overline{P}\overline{Q} = \overline{Q}\overline{P}$. In the polynomial case discussed in Example 1, $u_i \in \Pi_{n-1}(\mathbb{R})$ and $v_j \in \Pi_{m-1}(\mathbb{R})$. The functions ultimately used in the interpolation are

$$(x, y) \longmapsto u_i(x)v_j(y) \qquad (1 \leq i \leq n, \ 1 \leq j \leq m)$$

LEMMA 3. *If $\{u_i : 1 \leq i \leq n\}$ is any linearly independent set of functions on a set X and $\{v_j : 1 \leq j \leq m\}$ is any linearly independent set of functions on a set Y, then $\{u_i v_j : 1 \leq i \leq n, \ 1 \leq j \leq m\}$ is a linearly independent set of functions on $X \times Y$.*

Proof. Suppose $\sum_{i=1}^{n} \sum_{j=1}^{m} c_{ij} u_i v_j = 0$. Then for each x,

$$\sum_{j=1}^{m} \left(\sum_{i=1}^{n} c_{ij} u_i(x) \right) v_j = 0$$

Hence $\sum_{i=1}^{n} c_{ij} u_i(x) = 0$ for each j and for each x. Equivalently, $\sum_{i=1}^{n} c_{ij} u_i = 0$ for each j. Thus $c_{ij} = 0$ for each i and for each j. ∎

If U is a linear space whose elements are functions from X to $\mathbb{R}$, and if V is a linear space of functions from Y to $\mathbb{R}$ then $U \otimes V$ is defined to be the linear space generated by

$$\{uv : u \in U, \ v \in V\}$$

where uv is the function on $X \times Y$ defined by

$$(uv)(x, y) = u(x)v(y)$$

Hence,

$$U \otimes V = \left\{ \sum_{i=1}^{N} c_i u_i v_i : N \in \mathbb{N}, \ c_i \in \mathbb{R}, \ u_i \in U, \ v_i \in V \right\}$$

The tensor-product notation is also used in this way:

$$(u \otimes v)(x, y) = u(x)v(y)$$

We call $u \otimes v$ a **dyad.** Thus, $U \otimes V$ is generated by dyads.

LEMMA 4. *If $\{u_1, u_2, \ldots, u_n\}$ is a basis for U and $\{v_1, v_2, \ldots, v_m\}$ is a basis for V, then*

$$\{u_i v_j : 1 \leq i \leq n, \ 1 \leq j \leq m\}$$

is a basis for $U \otimes V$.

Proof. Lemma 3 asserts the linear independence of the given set. To show that it spans $U \otimes V$, notice that each generator uv is expressible as $(\sum a_i u_i)(\sum b_j v_j) = \sum c_{ij} u_i v_j$. ∎

Example 2. The tensor-product space $\Pi_n(\mathbb{R}) \otimes \Pi_m(\mathbb{R})$ has as one basis the set

$$\{(x, y) \longmapsto x^i y^j : 0 \le i \le n, 0 \le j \le m\}$$

Thus, it contains the monomial $x^n y^m$, for example. This is a term of degree $n + m$. We see immediately that

$$\Pi_n(\mathbb{R}) \otimes \Pi_m(\mathbb{R}) \subset \Pi_{n+m}(\mathbb{R}^2)$$

It is a *proper* inclusion, as can be seen by considering x^{n+m} or y^{n+m}. ∎

THEOREM. *Arbitrary data can be interpolated uniquely by the tensor-product space $\Pi_n(\mathbb{R}) \otimes \Pi_m(\mathbb{R})$ on any set of nodes having the form $\mathcal{N} = \{x_0, \dots, x_n\} \times \{y_0, \dots, y_m\}$.*

Proof. In Lemma 2, take $X = \{x_0, \dots, x_n\}$ and $Y = \{y_0, \dots, y_m\}$. ∎

Other bases for $\Pi_n(\mathbb{R}) \otimes \Pi_m(\mathbb{R})$ can be created by using Lemma 4. For example, let $\{x_0, x_1, \dots, x_n\}$ and $\{y_0, y_1, \dots, y_m\}$ be two sets of nodes. Let $u_0, u_1, \dots, u_n$ be the cardinal functions for interpolation by $\Pi_n(\mathbb{R})$ at nodes x_i. Then $u_i(x_j) = \delta_{ij}$. Let $v_0, v_1, \dots, v_m$ be defined similarly. Then $\{u_i \otimes v_j : 0 \le i \le n, 0 \le j \le m\}$ is a basis for $\Pi_n \otimes \Pi_m$, and $(u_i \otimes v_j)(x_p, y_q) = \delta_{pi} \delta_{qj} = \delta_{(i,j),(p,q)}$.

The construction used for producing two-variable interpolation can be used for any number of variables. For example, to interpolate a function of three variables, $(x, y, z) \longmapsto F(x, y, z)$, we can select nodes $\{x_1, x_2, \dots, x_n\}$ in X, $\{y_1, y_2, \dots, y_m\}$ in Y, and $\{z_1, z_2, \dots, z_k\}$ in Z. Next, we require functions (polynomials, perhaps) such that

$$u_i(x_j) = v_i(y_j) = w_i(z_j) = \delta_{ij}$$

Corresponding operators then are

$$P = \sum_{i=1}^{n} x_i^* \otimes u_i \qquad Q = \sum_{i=1}^{m} y_i^* \otimes v_i \qquad R = \sum_{i=1}^{k} z_i^* \otimes w_i$$

They can be extended to operate on functions of three variables. For example, $\overline{P}$ is defined by

$$(\overline{P}F)(x, y, z) = \sum_{i=1}^{n} F(x_i, y, z) u_i(x)$$

The operator $\overline{P}\,\overline{Q}\,\overline{R}$ will interpolate F at every point in the three-dimensional grid:

$$\{(x_i, y_j, z_\ell) : 1 \le i \le n, 1 \le j \le m, 1 \le \ell \le k\}$$

The formula for $\overline{P}\,\overline{Q}\,\overline{R}F$ is

$$(\overline{P}\,\overline{Q}\,\overline{R}F)(x, y, z) = \sum_{i=1}^{n} \sum_{j=1}^{m} \sum_{\ell=1}^{k} F(x_i, y_j, z_\ell) u_i(x) v_j(y) w_\ell(z)$$

Problems

1. Prove that $\overline{P}\,\overline{Q} = \overline{Q}\,\overline{P}$.

2. Let $X = Y = \mathbb{R}$. If u is a first-degree spline function on X and v is a first-degree spline function on Y, what is $u \otimes v$?

3. Prove that if $x_1, x_2, \ldots, x_n$ are distinct nodes in $(0,1)$ and if $y_1, y_2, \ldots, y_m$ are distinct nodes in $(0, 1)$, then arbitrary data on the grid $\{(x_i, y_j) : 1 \le i \le n, 1 \le j \le m\}$ can be interpolated by linear combinations of the nm functions $(x, y) \mapsto (xy - iy - jx + ij)^{-1}$, where $1 \le i \le n$ and $1 \le j \le m$.

4. Prove Lemma 2 by applying Lemma 1 twice, first to $\overline{P}$ and then to $\overline{Q}$.

5. Is Theorem 3 in Chapter 5 true for the polynomial spaces $\Pi_n \otimes \Pi_m$?

6. Prove that the set of functions
$$\{(x, y) \mapsto x^i y^j : i, j \in \mathbb{Z}_+\}$$
is linearly independent on any subset of $\mathbb{R}^2$ that has the form $A \times B$, where A and B are infinite subsets of $\mathbb{R}$. Generalize by considering finite sets of monomials and to Chebyshev systems.

References

[DU]   Diestel, J., and J. J. Uhl, Jr. *Vector Measures*. Math. Surveys. no. 15, *Amer. Math. Soc.* 1977.

[LS]   Lancaster, P., and K. Salkauskas. *Curve and Surface Fitting*. Academic Press, New York, 1986.

[Schat]   Schatten, R. *A Theory of Cross-Spaces*. Annals of Math. Studies, vol. 26, Princeton University Press, Princeton, 1950.

8

The Boolean Algebra of Projections

In Chapter 7 it was explained how bivariate approximation operators can be constructed by taking tensor products of univariate ones. For example, starting with operators P and Q, we formed $\overline{P} = P \otimes I$, $\overline{Q} = I \otimes Q$, and then $\overline{P}\,\overline{Q}$. Now we introduce another way of combining $\overline{P}$ and $\overline{Q}$.

For any two linear operators A and B, we define the **Boolean sum** by writing

$$A \oplus B = A + B - AB$$

For this to be meaningful, there should be a single linear space, E, such that $A : E \rightarrow E$ and $B : E \rightarrow E$. If A and B commute with each other (i.e., $AB = BA$) then $A \oplus B = B \oplus A$. Otherwise, these two Boolean sums will differ (and will have different approximation-theoretic properties).

THEOREM 1. *Let A and B be two projections defined on a normed linear space E. Let U and V be the ranges of A and B, respectively. If $ABA = BA$, then $A \oplus B$ is a projection onto $U + V$. The latter is therefore complemented and closed. Furthermore, BA is a projection of E onto $U \cap V$.*

Proof. Since $A \oplus B = A + B - AB$, it is clear that $A \oplus B$ maps E into $U + V$. From the equation

$$(A \oplus B)A = (A + B - AB)A = A^2 + BA - ABA = A + BA - BA = A$$

it is clear that, for each $u \in U$, $(A \oplus B)u = (A \oplus B)Au = Au = u$. Similarly,

$$(A \oplus B)B = AB + B^2 - AB^2 = AB + B - AB = B$$

and this shows that $(A \oplus B)v = v$ for each $v \in V$. Hence $(A \oplus B)w = w$ for all w in $U + V$, and $A \oplus B$ is a projection. To see that BA is a projection, just write $(BA)^2 = BABA = BBA = BA$. The range of BA is $U \cap V$, as we now verify. Let us use $\mathcal{R}$ to denote the range of an operator. If $f \in U \cap V$, then $BAf = Bf = f$, and $f \in \mathcal{R}(BA)$. Conversely, if $f \in \mathcal{R}(BA)$, then $f = BAf$, so $f \in \mathcal{R}(B) = V$. Also, $f = ABAf$, so $f \in \mathcal{R}(A) = U$. ∎

Theorem 1 provides a possible strategy for proving that the vector sum $U + V$ of two subspaces is closed. In general, such a sum will not be closed, even if both of U and V are closed. However, if there exist projections $A : E \twoheadrightarrow U$ and $B : E \twoheadrightarrow V$ such that $ABA = BA$ then $U + V$ is closed.

Example 1. Let P and Q be the interpolating projections introduced at the beginning of Chapter 7 (page 46). Let $\overline{P}$ and $\overline{Q}$ be the extensions defined subsequently. By Problem 1 in Chapter 7 (page 50), $\overline{P}\,\overline{Q} = \overline{Q}\,\overline{P}$. Then $(\overline{P} \oplus \overline{Q})F$ interpolates F on the union

$$(\{x_1, \ldots, x_n\} \times Y) \cup (X \times \{y_1, \ldots, y_m\})$$

To see that this is so, consider first any point (x_i, y) where $1 \le i \le n$ and $y \in Y$. Then, by Lemma 1 in Chapter 7 (page 47),

$$
\begin{aligned}
[(\overline{P} \oplus \overline{Q})F](x_i, y) &= (\overline{P}F)(x_i, y) + (\overline{Q}F)(x_i, y) - (\overline{P}\,\overline{Q}F)(x_i, y) \\
&= F(x_i, y) + (\overline{Q}F)(x_i, y) - (\overline{Q}F)(x_i, y) \\
&= F(x_i, y)
\end{aligned}
$$

Now consider any point (x, y_i) where $x \in X$ and $1 \le i \le m$. Then we have, again by Lemma 1 of Chapter 7, and by the formula for $\overline{P}\,\overline{Q}$ given just after Lemma 2 in Chapter 7,

$$
\begin{aligned}
[(\overline{P} \oplus \overline{Q})F](x, y_i) &= (\overline{P}F)(x, y_i) + \overline{Q}F(x, y_i) - (\overline{P}\,\overline{Q}F)(x, y_i) \\
&= (\overline{P}F)(x, y_i) + F(x, y_i) - (\overline{P}F)(x, y_i) \\
&= F(x, y_i)
\end{aligned}
$$

The formula for $\overline{P} \oplus \overline{Q}$, with notation as in Chapter 7, is

$$[(\overline{P} \oplus \overline{Q})F](x, y) = \sum_{i=1}^{n} F(x_i, y)u_i(x) + \sum_{j=1}^{m} F(x, y_j)v_j(y) - \sum_{i=1}^{n}\sum_{j=1}^{m} F(x_i, y_j)u_i(x)v_j(y)$$

It is easy to see why we cannot use $\overline{P} + \overline{Q}$, but must use instead $\overline{P} + \overline{Q} - \overline{P}\,\overline{Q}$: the operator $\overline{P} + \overline{Q}$ will give $2F(x_i, y_j)$ at the nodes. Hence by subtracting $\overline{P}\,\overline{Q}F$, we restore the correct values $F(x_i, y_j)$. ∎

The interpolation processes described here and in Chapter 7 are used in automotive design. A panel in a car body may be described by a finite number of cross-sections. These correspond to values of a function, $F(x_i, y)$, $1 \le i \le n$. The function F can be recovered by a formula $\sum_{i=1}^{n} F(x_i, y)u_i(x)$, where $u_i(x_j) = \delta_{ij}$. If cross-sections $F(x, y_j)$ are also available, F is recovered by the operator $\overline{P} \oplus \overline{Q}$ described in Example 1.

For projections, the commutative property, $PQ = QP$, or its two weaker versions, $PQP = QP$ and $PQP = PQ$, are decisive in determining the nature of the product (PQ) and the Boolean sum $(P \oplus Q)$. For example, if $PQP = QP$, then QP is a projection, for

$$(QP)^2 = (QP)(QP) = Q(PQP) = QQP = QP$$

However, PQ may fail to be a projection.

> **THEOREM 2.** *Let $A : E \to E$ and $B : E \to E$ be Banach space projections. If $ABA = AB$ then $(B \oplus A)^*$ is a projection of E^* onto $\mathcal{R}(A^*) + \mathcal{R}(B^*)$. Furthermore, $(AB)^*$ is a projection of E^* onto $\mathcal{R}(A^*) \cap \mathcal{R}(B^*)$.*

Proof. Our hypothesis $ABA = AB$ leads to $A^*B^*A^* = B^*A^*$. Hence Theorem 1 applies, and we conclude that $A^* \oplus B^*$ is a projection of E^* onto $\mathcal{R}(A^*) + \mathcal{R}(B^*)$. Also, B^*A^* is a projection of E^* onto $\mathcal{R}(A^*) \cap \mathcal{R}(B^*)$. Since $(B \oplus A)^* = A^* \oplus B^*$, our proof is complete. ∎

Example 2. An interesting example of Boolean sums arises in interpolating on the sides of a triangle. We take a standard triangle whose vertices are $(0, 0)$, $(0, 1)$, and $(1, 0)$. The first projection is defined by

$$(PF)(\xi_1, \xi_2) = F(\xi_1, 0)\frac{1 - \xi_1 - \xi_2}{1 - \xi_1} + F(\xi_1, 1 - \xi_1)\frac{\xi_2}{1 - \xi_1}$$

For each fixed value of ξ_1, this function interpolates F by a linear function of ξ_2 at the nodes 0 and $1 - \xi_1$. These nodes are on the horizontal leg and the hypotenuse of the triangle. See Figure 8.1.

Notice that PF interpolates F on these two sides of the triangle. In a similar way, we define Q:

$$(QF)(\xi_1, \xi_2) = F(0, \xi_2)\frac{1 - \xi_1 - \xi_2}{1 - \xi_2} + F(1 - \xi_2, \xi_2)\frac{\xi_1}{1 - \xi_2}$$

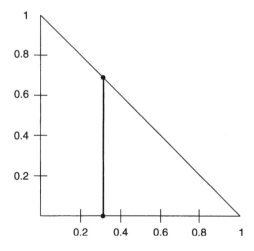

Figure 8.1 *Interpolation in the Triangle*

One can verify that $PQ \neq QP$, but $PQP = PQ$. Two steps involved in this are

$$(PQF)(\xi_1, \xi_2)$$

$$= F(0, 0)(1 - \xi_1 - \xi_2) + F(1, 0)\frac{\xi_1(1 - \xi_1 - \xi_2)}{1 - \xi_1} + F(\xi_1, 1 - \xi_1)\frac{\xi_2}{1 - \xi_1}$$

$$(QPF)(\xi_1, \xi_2)$$

$$= F(0, 0)(1 - \xi_1 - \xi_2) + F(0, 1)\frac{\xi_2(1 - \xi_1 - \xi_2)}{1 - \xi_2} + F(1 - \xi_2, \xi_2)\frac{\xi_1}{1 - \xi_2}$$

Now Theorem 2 is applicable, and we conclude that $Q \oplus P$ is a projection. It has the property that $(Q \oplus P)F$ interpolates F on all three sides of the triangle. This follows from Theorem 2, according to which $\mathcal{R}(Q^* \oplus P^*) = \mathcal{R}(Q^*) + \mathcal{R}(P^*)$. By our previous remarks concerning P and Q, we know that $\mathcal{R}(P^*)$ contains all the point-evaluation functionals corresponding to points on the horizontal leg and hypotenuse of the triangle. Similarly, $\mathcal{R}(Q^*)$ contains all the point functionals for the vertical leg and hypotenuse. ∎

The preceding discussion indicates the importance of commutativity among the projections we wish to combine. In Chapter 7, the projections $\overline{P}$ and $\overline{Q}$ were of a special structure that led to commutativity. Let us describe this important special case. For simplicity, we work in spaces of continuous functions, $C(X)$ and $C(Y)$, where X and Y are compact Hausdorff spaces. Let two operators be given (not necessarily projections):

$$A : C(X) \longrightarrow C(X) \qquad\qquad B : C(Y) \longrightarrow C(Y)$$

If $F \in C(X \times Y)$, we define its *sections* by

$$F_x(y) = F(x, y) = F^y(x)$$

(See Rudin [Ru0] for this notation.) Proceeding as in Chapter 7, we write

$$(\overline{A}F)(x, y) = (AF^y)(x) \qquad (\overline{B}F)(x, y) = (BF_x)(y)$$

These have been given the (bad) name of "parametric extensions" of A and B.

THEOREM 3. *The parametric extensions of two univariate operators commute.*

Proof. Fix an $x \in X$, and let x^* denote the corresponding point-evaluation functional. Thus

$$x^*(f) = f(x) \qquad (f \in C(X))$$

Then $x^* \circ A$ is a linear functional on $C(X)$. By the Riesz Representation Theorem [Ru0], there is a signed Borel measure μ such that

$$x^*(Af) = \int_X f(s)\, d\mu(s) \qquad (f \in C(X))$$

Similarly, we have (for a fixed y)

$$y^*(Bg) = \int_Y g(t)\, d\nu(t) \qquad (g \in C(Y))$$

Hence

$$(\overline{A}\overline{B}F)(x, y) = [A(\overline{B}F)^y](x) = (x^* \circ A)(\overline{B}F)^y = \int_X (\overline{B}F)^y(s) \, d\mu(s)$$

$$= \int_X (\overline{B}F)(s, y) \, d\mu(s) = \int_X (BF_s)(y) \, d\mu(s) = \int_X y^* BF_s \, d\mu(s)$$

$$= \int_X \int_Y F_s(t) \, dv(t) \, d\mu(s) = \int_X \int_Y F(s, t) \, dv(t) \, d\mu(s)$$

Similarly,

$$(\overline{B}\overline{A}F)(x, y) = \int_Y \int_X F(s, t) \, d\mu(s) \, dv(t)$$

By the Fubini theorem [Ru0] these two integrals are equal. The version of the Fubini theorem needed here involves *signed* measures, or, more generally, *complex* measures. The theorem is not explicitly given in [Ru0] but is an easy consequence of his Theorem 7.8. Its statement is as follows. Let μ and v be complex measures on measurable spaces $(X, \mathcal{A})$ and $(Y, \mathcal{B})$ respectively. If f is a complex-valued measurable function on $X \times Y$ such that

$$\int_Y \int_X |f(x, y)| \, d|\mu|(x) \, d|v|(y) < \infty$$

then

$$\int_Y \int_X f(x, y) \, d\mu(x) \, dv(y) = \int_X \int_Y f(x, y) \, dv(y) \, d\mu(x)$$

See also Dunford and Schwartz [DS], page 183, or the book of Bartle [Bart1].   ∎

Problems

1. Let $P : E \twoheadrightarrow U$ and $Q : E \twoheadrightarrow V$ be Banach space projections, and assume that $U \subset V$. What are $PQ, QP, P \oplus Q$ and $Q \oplus P$?

2. For two projections $P : E \to E, Q : E \to E$, prove that
   a. $\mathcal{R}(P) \cap \mathcal{R}(Q) = \mathcal{R}(PQ) \cap \mathcal{R}(QP)$.
   b. $\mathcal{R}(P) + \mathcal{R}(Q) = \mathcal{R}(P \oplus Q) + \mathcal{R}(Q \oplus P)$.

3. Assume the hypotheses of Problem 2, and let $QPQ = PQ$. Prove that $\mathcal{R}(PQ) = \mathcal{R}(P) \cap \mathcal{R}(Q)$.

4. For linear operators that are not necessarily projections prove or disprove that
   a. $\mathcal{R}(A) \subset \mathcal{R}(B \oplus A)$.
   b. $\mathcal{R}(A) \cap \mathcal{R}(B) \subset \mathcal{R}(AB) \cap \mathcal{R}(BA)$.

5. Adopt the hypotheses of Problem 2. Prove that in order to have $PQP = PQ$ it suffices to have $\phi \circ (Q - QP) = 0$ for all $\phi \in \mathcal{R}(P^*) \setminus \mathcal{R}(Q^*)$.

6. Let A and B be projections defined on a space E. Prove that in order for $A \oplus B$ to be a projection it is necessary and sufficient that $(I - A)BA(I - B) = 0$. Hence, conclude that either of the conditions $BA = ABA$ or $BA = BAB$ is sufficient.

7. In the example of interpolation on the sides of a triangle, determine whether the equation $QPQ = QP$ is valid.

8. Let X and Y be compact Hausdorff spaces, and let U and V be complemented subspaces, $U \subset C(X)$, $V \subset C(Y)$. Prove that $\{uv : u \in U, v \in V\}$ is closed in $C(X \times Y)$. Prove that the vector sum $\{F : F^y \in U, \text{ for all } y \in Y\} + \{F : F_x \in V, \text{ for all } x \in X\}$ is closed.

9. Establish these further conclusions to Theorem 1: $(A \oplus B)^*$ is a projection onto $\mathcal{R}(A^*) + \mathcal{R}(B^*)$, and $(BA)^*$ is a projection onto $\mathcal{R}(A^*) \cap \mathcal{R}(B^*)$.

10. Establish this further conclusion to Theorem 2: $B \oplus A$ is a projection.

11. Repeat Problem 4 when the operation $\mathcal{R}$ is replaced by the operation $\mathcal{F}$, where $\mathcal{F}(A)$ means the set of fixed points of A.

References

[BBG] Barnhill, R. E., G. Birkhoff, and W. J. Gordon. "Smooth interpolation in triangles." *J. Approx. Theory 8* (1973), 214–225.

[Bart1] Bartle, R. G. *Elements of Integration Theory.* Wiley, New York, 1966. Reprinted, enlarged, and retitled *Elements of Integration and Lebesgue Measure, 1995.*

[BasD] Baszenski, G., and F.-J. Delvos. "Boolean algebra and multivariable interpolation." In *Approximation and Function Spaces.* Banach Center Publications, vol. 22. PWN-Polish Scientific Publishers, Warsaw, 1989, 25–44.

[Bir] Birkhoff, G. "Interpolation to boundary data in triangles." *J. Math. Anal. Appl. 42* (1973), 199–208.

[Bir2] Birkhoff, G. "The algebra of multivariate interpolation." In *Constructive Approaches to Mathematical Models,* ed. by C. V. Coffman and G. J. Fix. Academic Press, New York, 1979, 345–363.

[DDS] Dahmen, W., R. DeVore, and K. Scherer. "Multi-dimensional spline approximation." *SIAM J. Numer. Anal. 17* (1980), 380–402.

[De3] Delvos, F.-J. "*d*-Variate Boolean interpolation." *J. Approx. Theory 34* (1982), 99–114.

[DeS] Delvos, F.-J., and W. Schempp. *Boolean Methods in Interpolation and Approximation.* Pitman Research Notes in Mathematics, vol. 230. Longmans (Wiley), New York, 1989.

[DS] Dunford, N., and J. T. Schwartz. *Linear Operators, Part I, General Theory.* Interscience, New York, 1958.

[GC] Gordon, W. J., and E. W. Cheney. "Bivariate and multivariate interpolation with non-commutative projectors." In *Linear Spaces and Approximation,* ed. by P. L. Butzer and B. Sz.-Nagy. ISNM, vol. 40, Birkhäuser, Basel, 1978, 381–387.

[GW] Gordon, W. F., and J. A. Wixom. "Pseudo–harmonic interpolation on convex domains." *SIAM J. Numer. Anal. 11* (1974), 909–933.

[Jia2] Jia, R. Q. "Approximation by multivariate splines: An application of Boolean methods." In *Numerical Methods of Approximation Theory,* ed. by D. Braess and L. L. Schumaker. International Series on Numerical Analysis, vol. 105, 1992, 117–134.

[Ru0] Rudin, W. *Real and Complex Analysis.* 2nd ed., McGraw-Hill, New York, 1974.

[Wa] Watkins, D. S. "Error bounds for polynomial blending function methods." *SIAM J. Numer. Anal. 14* (1977), 721–734.

9

The Newton Paradigm for Interpolation

Newton's algorithm for univariate polynomial interpolation has this essential character: Let g be a polynomial that interpolates a function f at nodes $x_1, x_2, \ldots, x_n$. Let $h(x) = (x - x_1)(x - x_2) \cdots (x - x_n)$. Then for suitable c, $g + ch$ will interpolate f at $x_1, x_2, \ldots, x_{n+1}$. Of course, we assume the nodes to be different from each other. The algorithm is applied repeatedly, starting with $n = 1$. The polynomial g in the above description can be of degree $n - 1$, but need not be.

Example. The Newton algorithm can be used to obtain the polynomial in Π_2 that assumes these values:

x	3	-2	0
$f(x)$	4	-6	16

The successive polynomials are $p_0(x) = 4$, $p_1(x) = 4 + 2(x - 3)$, and $p_2(x) = 4 + 2(x - 3) - 3(x - 3)(x + 2)$. For practical computation, the forms given are to be recommended over the "power" forms, which are $p_0(x) = 4$, $p_1(x) = 2x - 2$, $p_2(x) = -3x^2 + 5x + 16$. ∎

As a first step in generalizing the Newton algorithm, we notice that the nodes can be points in any "space," X. Again, g will be a function (not necessarily a polynomial) that interpolates a function f at nodes $x_1, \ldots, x_n$. Let x_{n+1} be a new node. We require a function h that takes the value 0 at $x_1, \ldots, x_n$, but has a nonzero value at x_{n+1}. Then the same conclusion can be drawn, namely, that for an appropriate value of c, $g + ch$ will interpolate f at $x_1, \ldots, x_{n+1}$.

At the next level of generalization, we replace $\{x_1, \ldots, x_n\}$ by any "nodal set," $\mathcal{N}$. It need not be finite. Assume that g interpolates f on $\mathcal{N}$, in symbols, $g|\mathcal{N} = f|\mathcal{N}$. Let y be a "new" node, that is, a point not in $\mathcal{N}$. We require an h such that $h|\mathcal{N} = 0$ and $h(y) \neq 0$. In fact, why not assume $h(y) = 1$? Then $g + f(y)h$ interpolates f on $\mathcal{N} \cup \{y\}$.

The next step of generalization is not so obvious. We intend to use $g + rh$ as an interpolant, but permit r to be a function more general than simply a constant. We use the notation $Z(h) = \{x : h(x) = 0\}$. Again $\mathcal{N}$ is a set of nodes. Let g interpolate f on $\mathcal{N} \cap Z(h)$. Let r interpolate $(f - g)/h$ on $\mathcal{N} \backslash Z(h)$. Then $g + rh$ interpolates f on $\mathcal{N}$, as is easily verified.

This last generalization of Newton's algorithm will be needed below in Theorem 2. First we need some more elementary results.

THEOREM 1. *The space $\Pi_k(\mathbb{R}^s)$ can interpolate arbitrary data on any set of $k + 1$ distinct nodes in $\mathbb{R}^s$.*

Proof. Let the nodes be x_i, for $0 \leq i \leq k$. By the lemma that follows, we can find a vector u in $\mathbb{R}^s$ such that the $k + 1$ inner products $t_i = ux_i$ are all different. If the datum at x_i is λ_i, select $q \in \Pi_k(\mathbb{R})$ so that $q(t_i) = \lambda_i$ $(0 \leq i \leq k)$. This is the classical one-variable interpolation problem. Put $p(x) = q(ux)$. Then $p \in \Pi_k(\mathbb{R}^s)$ and $p(x_i) = \lambda_i$ for $0 \leq i \leq n$. Here we need the Multinomial Theorem from Chapter 4, page 27, which shows that the function $x \mapsto (ux)^j$ belongs to $\Pi_j(\mathbb{R}^s)$. ∎

LEMMA. *If $x_1, x_2, \ldots, x_n$ are distinct points in $\mathbb{R}^s$, then for some u in $\mathbb{R}^s$ the n inner products ux_i are all different.*

Proof. We want to choose u so that

$$i \neq j \implies ux_i \neq ux_j$$

If we define hyperplanes by writing

$$H_{ij} = \{y \in \mathbb{R}^s : y(x_i - x_j) = 0\} \qquad (1 \leq i < j \leq n)$$

then we simply select $u \notin \bigcup H_{ij}$. ∎

THEOREM 2 (Gasca-Maeztu). *Let $\mathcal{N}$ be a set of $\frac{1}{2}(k + 1)(k + 2)$ nodes in $\mathbb{R}^s$, where $s \geq 2$. Suppose that there exist hyperplanes $H_0, H_1, \ldots, H_k$ in $\mathbb{R}^s$ such that*

  a. $\mathcal{N} \subset H_0 \bigcup H_1 \bigcup \cdots \bigcup H_k$

  b. $\#(\mathcal{N} \bigcap H_i) = i + 1 \qquad (0 \leq i \leq k)$

Then arbitrary data on $\mathcal{N}$ can be interpolated by $\Pi_k(\mathbb{R}^s)$. (Notice that only when $s = 2$ is the dimension of $\Pi_k(\mathbb{R}^s)$ equal to the number of nodes.)

Proof. The sets $\mathcal{N} \cap H_i$ are pairwise disjoint, for otherwise

$$\frac{1}{2}(k+1)(k+2) = \#\mathcal{N} < \sum_{i=0}^{k} \#(\mathcal{N} \cap H_i) = \sum_{i=0}^{k}(i+1) = \frac{1}{2}(k+1)(k+2)$$

Each hyperplane is of the form $H_i = \ell_i^{-1}(\{0\})$ for an appropriate $\ell_i \in \Pi_1(\mathbb{R}^s)$. Let f be the function whose values are to be interpolated. In step 1 of the construction we use Theorem 1 to obtain $p_k \in \Pi_k(\mathbb{R}^s)$ that interpolates f on $\mathcal{N} \cap H_k$. Proceeding inductively downward, suppose that we have found $p_i \in \Pi_k(\mathbb{R}^s)$ that interpolates f on $\mathcal{N} \cap (H_k \cup H_{k-1} \cup \cdots \cup H_i)$. Use Theorem 1 to find $r \in \Pi_{i-1}(\mathbb{R}^s)$ that interpolates $(f - p_i)/(\ell_k \ell_{k-1} \cdots \ell_i)$ on $\mathcal{N} \cap H_{i-1}$. Notice that $(\ell_k \ell_{k-1} \cdots \ell_i)(x) \neq 0$ on $\mathcal{N} \cap H_{i-1}$ by the disjoint nature of the sets $\mathcal{N} \cap H_j$. Notice also that $\#(\mathcal{N} \cap H_{i-1}) = i$, so that Theorem 1 does apply at this step. Now define $p_{i-1} = p_i + r(\ell_k \ell_{k-1} \cdots \ell_i)$. The degree of the added term is $(i-1) + (k-i+1) = k$. Hence $p_{i-1} \in \Pi_k(\mathbb{R}^s)$. Finally, p_{i-1} interpolates f on $\mathcal{N} \cap (H_k \cup H_{k-1} \cup \cdots \cup H_{i-1})$. Indeed, the added term vanishes on $H_k \cup H_{k-1} \cup \cdots \cup H_i$, so the interpolation property of p_i is not disturbed. For $x \in \mathcal{N} \cap H_{i-1}$ we have

$$
\begin{aligned}
p_{i-1}(x) &= p_i(x) + r(x)(\ell_k \ell_{k-1} \cdots \ell_i)(x) \\
&= p_i(x) + \frac{f(x) - p_i(x)}{(\ell_k \ell_{k-1} \cdots \ell_i)(x)}(\ell_k \ell_{k-1} \cdots \ell_i)(x) \\
&= f(x) \qquad\qquad\qquad\qquad\qquad \blacksquare
\end{aligned}
$$

Some node sets conforming to Theorem 2 are shown in Figure 9.1. Here $s = 2$ and $k = 2$.

The Newton paradigm, as described in the opening paragraphs of this chapter, can be applied with arbitrary linear functionals (not only point-evaluation functionals). The description in the general case is as follows. The setting should be a linear space, E. Let $\phi_1, \phi_2, \ldots$ be linear functionals defined on E. Let f be an element of E to be "interpolated." As before, this can be done inductively. In the inductive step, we assume that an element g is available in E such that $\phi_i(g) = \phi_i(f)$ for $1 \leq i \leq n$. Next, select h in E so that $\phi_i(h) = 0$ for $1 \leq i \leq n$ and $\phi_{n+1}(h) = 1$. The new interpolant will be of the form $g + ch$. In fact, $c = \phi_{n+1}(f)$. A concrete example of this process is suggested in Problem 8.

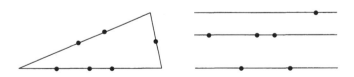

Figure 9.1   *Nodes for Theorem 2*

Problems

> *You react, Critias, as if I professed to know*
> *the answers to the questions I ask, and could*
> *explain it all to you if I wished. However, the*
> *opposite is true, for I also am inquiring into*
> *the truth of these matters.*
>
> Socrates, quoted in Plato's *Charmides*

1. For each pair (k, s), we ask, What is the largest $n = n(k, s)$ such that interpolation at any n nodes in $\mathbb{R}^s$ is possible by $\Pi_k(\mathbb{R}^s)$? For example, we know that $n(k, 1) = k + 1$ and $n(k, s) \geq k + 1$. Also $n(0, s) = 1$. Add to this knowledge.

2. For $x = (\xi_1, \xi_2) \in \mathbb{R}^2$, define $r(x) = \xi_1$. Let $n = \frac{1}{2}(k + 1)(k + 2)$, and let $\mathcal{N} = \{x_1, \dots, x_n\}$ be a node set in $\mathbb{R}^2$ such that
$$\#[\mathcal{N} \cap r^{-1}(r(x_i))] = i \qquad (1 \leq i \leq k + 1)$$
Prove that $\Pi_k(\mathbb{R}^2)$ can interpolate arbitrary data on the given nodes.

3. Let $L_0, L_1, \dots, L_k$ be vertical lines in $\mathbb{R}^2$. Put $n = \frac{1}{2}(k + 1)(k + 2)$. Let $\mathcal{N}$ be a set of n nodes in $\mathbb{R}^2$ such that $\#(L_i \cap \mathcal{N}) = i + 1$ for each i. Prove that arbitrary data on $\mathcal{N}$ can be interpolated by a function of the form $(\xi_1, \xi_2) \mapsto \sum_{i=0}^{k} p_i(\xi_1)q_i(\xi_2)$, in which (for each i) p_i and q_i are in $\Pi_i(\mathbb{R})$.

4. Let $x_0, x_1, \dots, x_k$ be nodes in $\mathbb{R}^s$ such that $\|x_0\| < \|x_1\| < \cdots < \|x_k\|$. Prove that arbitrary data on the nodes can be uniquely interpolated by a function $x \mapsto p(\|x\|)$, with $p \in \Pi_k(\mathbb{R})$. Generalize this result.

5. Let $x_0, x_1, \dots, x_n$ be distinct points in $\mathbb{R}^s$. Prove that there exists a point $z \in \mathbb{R}^s$ such that the numbers $\|z - x_i\|$ are all different. Prove that arbitrary data at the nodes x_i can be interpolated by $p(\|z - x\|)$, where $p \in \Pi_n(\mathbb{R})$.

6. Refer to the lemma given in this chapter. Is there an optimal choice for the vector u? Is there an algorithm for finding an optimal u?

7. Count the number of multiplications and divisions needed to carry out polynomial interpolation by the Newton algorithm. Use the algorithm described in the first paragraph of this chapter.

8. Refer to Problem 5. Make a simple modification in the interpolation process described there so that the interpolating function will be a polynomial.

9. (Newton's algorithm for a set of functionals.) Use the Newton paradigm to find $p \in \Pi_3(\mathbb{R})$ such that $p(0) = 3, p'(1) = 4, \int_0^1 p(x) \, dx = 5$, and $\int_0^1 x^2 p(x) \, dx = 6$.

10. Is the lemma in this chapter valid for a sequence $[x_1, x_2, \dots]$ in $\mathbb{R}^s$?

11. Generalize the result in Problem 5 so that a function of the form
$$x \longmapsto p\big(g(\|z - x\|)\big) \text{ can be used.}$$

12. In the lemma, we used the fact that $\mathbb{R}^s$ is not the union of a finite set of hyperplanes. Prove that $\mathbb{R}^s$ is not the union of a countable set of hyperplanes. Two proofs are wanted, one using measure theory and another using a Baire category argument.

13. Let $t_0 < t_1 < \cdots < t_n$. Define first-degree spline functions $u_1, u_2, \ldots, u_n$ as follows:

$$u_i(x) = \begin{cases} 0 & x < t_{i-1} \\ (x - t_{i-1})/(t_i - t_{i-1}) & t_{i-1} \le x \le t_i \\ 1 & x > t_i \end{cases}$$

Show that $\{u_1, u_2, \ldots, u_n\}$ is a Newtonian base for interpolation at nodes $t_1, t_2, \ldots, t_n$.

14. Generalize the result and procedures in Problem 3. (The lines L_i need not be vertical but should be parallel.)

15. Prove that in Problem 3 the resulting polynomials are unique if we add the condition that $p_i(\xi_1) = 1$ whenever (ξ_1, ξ_2) is on the line L_i.

16. In Problem 5, the Euclidean norm was to be used. Is the result true for the norm $\|x\|_\infty = \max_i |\xi_i|$?

17. In Problem 9, change the fourth requirement to read $\int_0^1 (1 - x)p'(x)\,dx = 6$. Then solve for the cubic polynomial. The purpose of this problem is to indicate that a polynomial of degree k cannot always interpolate the values of $k + 1$ linear functionals.

18. In Theorem 2, change statement (2) to read $\#(\mathcal{N} \cap H_i) = k + 1 - i$. Then prove the theorem by using induction in an increasing order, starting with a polynomial p_0 that interpolates f on $\mathcal{N} \cap H_0$.

References

[BR2]   de Boor, C., and A. Ron. "Computational aspects of polynomial interpolation in several variables." *Math. Comp. 58* (1992), 705–727.

[BR1]   de Boor, C., and A. Ron. "On multivariate polynomial interpolation." *Constr. Approx. 6* (1990), 287–302.

[Bo1]   de Boor, C. "Polynomial interpolation in several variables." In *Proceedings of a Conference Honoring S. D. Conte,* ed. by R. DeMillo and J. R. Rice. Plenum Press, New York.

[GM]   Gasca, M., and J. I. Maeztu. "On Lagrange and Hermite interpolation in R^k." *Numer. Math. 39* (1982), 1–14.

[GaMu] Gasca, M., and G. Mühlbach, "Multivariate polynomial approximation under projectivities." Part I, *Numer. Algorithms 1* (1991), 375–400. Part II, ibid. 2 (1992), 255–278.

[Ga]   Gasca, M. "Multivariate polynomial interpolation." In *Computation of Curves and Surfaces,* ed. by W. Dahmen, M. Gasca, and C. Micchelli. Kluwer Academic, 1990, 215–235.

[LS]   Lancaster, P., and K. Salkauskas. *Curve and Surface Fitting.* Academic Press, New York, 1986.

[M2]   Micchelli, C. A. "Algebraic Aspects of Interpolation." In *Approximation Theory,* ed. by C. de Boor, *Amer. Math. Soc. Proc. Symp. Appl. Math.* vol. 36. Amer. Math. Soc., Providence, RI, 1986, 81–102.

[We]   Werner, H., "Remarks on Newton type multivariate interpolation for subsets of grids." *Computing 25* (1980), 181–191.

10

The Lagrange Paradigm for Interpolation

As in the preceding chapter, we seek to generalize classical methods of interpolation to a wider setting, in particular to multivariate problems.

Recall the Lagrange procedure for interpolation at a set of nodes $x_1, x_2, \ldots, x_n$ in $\mathbb{R}$. We define some basic polynomials $v_i \in \Pi_{n-1}(\mathbb{R})$ by putting

$$(1) \qquad v_i(x) = \prod_{\substack{j=1 \\ j \neq i}}^{n} \frac{x - x_j}{x_i - x_j} \qquad (1 \leq i \leq n)$$

These functions have the "cardinal property," $v_i(x_j) = \delta_{ij}$. Hence, a polynomial interpolant to a function f is given by an element $Pf \in \Pi_{n-1}(\mathbb{R})$ as follows:

$$(2) \qquad Pf = \sum_{i=1}^{n} f(x_i) v_i$$

It is clear that the principle of Lagrange's method can be extracted and applied in a more general setting. We allow X to be any set. (No algebraic structure is needed on X.) A function $\phi : X \times X \to \mathbb{R}$ is introduced, subject to the sole requirement that

$$(3) \qquad \phi(x, y) = 0 \iff x = y$$

If nodes $x_1, \ldots, x_n$ are prescribed in X, then we define

$$(4) \qquad v_i(x) = \prod_{\substack{j=1 \\ j \neq 1}}^{n} \frac{\phi(x, x_j)}{\phi(x_i, x_j)} \qquad (1 \leq i \leq n)$$

It follows that $v_i(x_j) = \delta_{ij}$, and an interpolating projection is defined by the equation

$$(5) \qquad Pf = \sum_{i=1}^{n} f(x_i)v_i$$

The range of this projection is an n-dimensional space having $\{v_1, \ldots, v_n\}$ as a basis. Notice that the denominators in Equation (4) are scalars that are present only to secure the condition $v_i(x_i) = 1$.

Example 1. In the preceding discussion, we recover the classical case by letting $X = \mathbb{R}$ and $\phi(x, y) = x - y$. ∎

Example 2. If (X, d) is a metric space, we can use $\phi(x, y) = d(x, y)$. ∎

Example 3. If $X = \mathbb{R}^s$ and if $\phi(x, y) = \|x - y\|^2$, with the Euclidean norm, then we find that $v_i \in \Pi_{2n-2}(\mathbb{R}^s)$. To verify this, first, let $x = (\xi_1, \xi_2, \ldots, \xi_s)$. Then $\|x\|^2 = \sum_{i=1}^{s} \xi_i^2 = q(x)$, where $q \in \Pi_2(\mathbb{R}^s)$. Hence

$$(6) \qquad v_i(x) = c_i \prod_{\substack{j=1 \\ j \neq i}}^{n} q(x - x_j)$$

where c_i is a constant. ∎

In the classical Lagrange process, we form v_i as a product of $n - 1$ affine factors, and therefore v_i is a polynomial of degree $n - 1$. In higher-dimensional cases, the same strategy can be used, because a product of k affine functions on $\mathbb{R}^s$ is still a polynomial of degree k on $\mathbb{R}^s$. Here, an "affine" function is simply a member of $\Pi_1(\mathbb{R}^s)$. It has the form

$$(7) \qquad x \longmapsto \alpha_0 + \alpha_1 \xi_1 + \alpha_2 \xi_2 + \cdots + \alpha_s \xi_s \qquad (x = (\xi_1, \ldots, \xi_s))$$

These remarks prepare the way for the next result.

THEOREM 3 (Chung and Yao). *Given k and s in $\mathbb{N}$, set $n = \dim \Pi_k(\mathbb{R}^s)$. Let nodes $x_1, x_2, \ldots, x_n$ be given in $\mathbb{R}^s$. If there exist hyperplanes H_{iv}, (where $1 \le i \le n, 1 \le v \le k$) such that*

$$(8) \qquad x_j \in \bigcup_{v=1}^{k} H_{iv} \iff j \neq i \qquad (1 \le i, j \le n)$$

then arbitrary data at the nodes can be interpolated by $\Pi_k(\mathbb{R}^s)$.

Proof. For each hyperplane H_{iv} there is an affine functional ℓ_{iv} such that $H_{iv} = \ell_{iv}^{-1}(\{0\})$. We set

$$(9) \qquad q_i(x) = \prod_{v=1}^{k} \ell_{iv}(x) \qquad (1 \le i \le n, x \in \mathbb{R}^s)$$

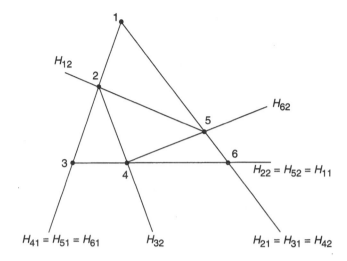

Figure 10.1   *Illustrating Theorem 3*

Obviously, $q_i \in \Pi_k(\mathbb{R}^s)$. Our hypothesis (8) implies that x_i does not belong to any of the hyperplanes $H_{i1}, \ldots, H_{ik}$. Hence $\ell_{iv}(x_i) \neq 0$ for $1 \leq v \leq k$, and $q_i(x_i) \neq 0$. Similar reasoning shows that $q_i(x_j) = 0$ if $i \neq j$. Putting $v_i(x) = q_i(x)/q_i(x_i)$, we obtain the "cardinal" property: $v_i(x_j) = \delta_{ij}$. Consequently, an interpolant to f is given by the formula $\sum_{i=1}^{n} f(x_i)v_i$. ∎

An illustration of Theorem 3, with $s = 2$, $k = 2$, $n = 6$ is shown in Figure 10.1. To verify that it is correct, one can use the hypothesis in Equation (8). This asserts that, for each i, $H_{i1} \cup H_{i2}$ should contain all nodes except x_i. These six assertions can be quickly verified.

Recall the notion of a set of points being in "general position" in $\mathbb{R}^s$, as discussed in Chapter 5.

LEMMA. *If $\{y_0, y_1, \ldots, y_s\}$ is a set of $s + 1$ points in general position in $\mathbb{R}^s$, then each point x in $\mathbb{R}^s$ has a unique representation $x = \sum_{i=0}^{s} \lambda_i y_i$ in which $\sum_{i=0}^{s} \lambda_i = 1$.*

Proof. By the definition of the term "general position," the set $\{y_i - y_0 : 1 \leq i \leq s\}$ is linearly independent, and is therefore a basis for $\mathbb{R}^s$. Hence, for any x there are unique coefficients λ_i for which $x - y_0 = \sum_{i=1}^{s} \lambda_i(y_i - y_0)$. Define $\lambda_0 = 1 - \sum_{i=1}^{s} \lambda_i$. Then $\sum_{i=0}^{s} \lambda_i = 1$ and

$$x = y_0 + \sum_{i=1}^{s} \lambda_i(y_i - y_0) = \sum_{i=1}^{s} \lambda_i y_i + y_0\left(1 - \sum_{i=1}^{s} \lambda_i\right) = \sum_{i=0}^{s} \lambda_i y_i$$

For any representation $x = \sum_{i=0}^{s} \mu_i y_i$ with $\sum_{i=0}^{s} \mu_i = 1$ we have

$$x - y_0 = \sum_{i=0}^{s} \mu_i(y_i - y_0) = \sum_{i=1}^{s} \mu_i(y_i - y_0)$$

It follows that $\mu_i = \lambda_i$ for $1 \leq i \leq s$, and then $\mu_0 = \lambda_0$ follows as well. ∎

THEOREM 4. *Let $\{y_0, y_1, \ldots, y_s\}$ be a set of $s + 1$ points in general position in $\mathbb{R}^s$, and let $k \in \mathbb{N}$. Interpolation by $\Pi_k(\mathbb{R}^s)$ is possible on the node set $\{\sum_{i=0}^{s} \lambda_i y_i : \sum_{i=0}^{s} \lambda_i = 1, k\lambda_i \in \mathbb{Z}_+\}$.*

Proof. By the preceding lemma, each point x in $\mathbb{R}^s$ has a representation $x = \sum_{i=0}^{s} \lambda_i(x) y_i$ in which $\sum_{i=0}^{s} \lambda_i(x) = 1$. Elementary arguments show that the functionals λ_i are affine. With any node v, we associate a set of integer pairs:

$$J(v) = \{(i, j) : 0 \le i \le s, \, 0 \le j < k\lambda_i(v)\}$$

In this set, we notice that for each i there are $k\lambda_i(v)$ values of j. Hence

$$\# J(v) = \sum_{i=0}^{s} k\lambda_i(v) = k$$

With the node v we associate a family of k hyperplanes $\{H_{ij} : (i, j) \in J(v)\}$, where

$$H_{ij} = \{x \in \mathbb{R}^s : k\lambda_i(x) = j\}$$

It is now to be shown that the hypotheses of Theorem 3 are fulfilled, namely, for any two nodes, u and v,

(10) $$v \ne u \iff u \in \bigcup \{H_{ij} : (i, j) \in J(v)\}$$

First we prove that if $u = v$ then $u \notin \bigcup \{H_{ij} : (i, j) \in J(v)\}$. This means, in simpler terms, that each node v satisfies

$$v \notin \bigcup \{H_{ij} : (i, j) \in J(v)\}$$

This follows at once by observing that if $(i, j) \in J(v)$, then $k\lambda_i(v) > j$ and $v \notin H_{ij}$, by the definition H_{ij}.

Now let u and v be two different nodes. We cannot have $\lambda_i(u) \ge \lambda_i(v)$ for all i because at least one of these inequalities would have to be strict (since $u \ne v$), and then we would have $\sum \lambda_i(u) > \sum \lambda_i(v) = 1$, a contradiction. Thus, there must be an index i_0 for which $k\lambda_{i_0}(u) < k\lambda_{i_0}(v)$. Letting $j_0 = k\lambda_{i_0}(u)$, we have $j_0 < k\lambda_{i_0}(v)$, and so $(i_0, j_0) \in J(v)$. Also, $u \in H_{i_0 j_0}$. This proves that $u \in \bigcup \{H_{ij} : (i, j) \in J(v)\}$. We have therefore completed the proof of the equivalence expressed in (10). Now apply Theorem 3 to draw the desired conclusion. ∎

COROLLARY. *Fix s and k in $\mathbb{N}$. The Vandermonde matrix whose generic term is α^β, where $\alpha, \beta \in \mathbb{Z}_+^s$, $|\alpha| \le k$, and $|\beta| \le k$, is nonsingular.*

Proof. Interpret the α in the set $\{\alpha \in \mathbb{Z}_+^s : |\alpha| \le k\}$ as nodes. They arise from the construction in Theorem 4 if we let $y_i = ke_i$ $(1 \le i \le s)$ and $y_0 = 0$. Here e_i is the vector whose j-th coordinate is δ_{ij}. ∎

Remarks. In Theorem 4 and its corollary, the number of nodes is equal to the dimension of $\Pi_k(\mathbb{R}^s)$, i.e., $\binom{k + s}{s}$. The set of nodes α considered in the Corollary has the appearance shown in Figure 10.2 when $k = 4$ and $s = 2$.

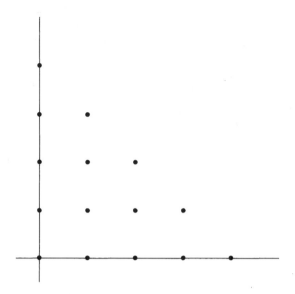

Figure 10.2   *Node Set* $\{\alpha \in \mathbb{Z}_+^2 : |\alpha| \leq 4\}$

In the next theorem, we have a set of nodes that arises by intersecting hyperplanes. Fixing s and k, we consider hyperplanes $H_1, H_2, \ldots, H_{s+k}$ in $\mathbb{R}^s$ having these two properties:

1. If $J \subset \{1, 2, \ldots, s + k\}$ and $\#J = s$, then $\bigcap_{i \in J} H_i$ is a single point, which we may call x_J.

2. The points x_J obtained as in (1) are all distinct.

THEOREM 5.   *The set* $\{x_J : J \subset \{1, 2, \ldots, s + k\}, \#J = s\}$ *is a node set on which interpolation by* $\Pi_k(\mathbb{R}^s)$ *is possible.*

Proof.   Let us enumerate all the sets J as $J_1, J_2, \ldots, J_n$. Thus $n = \binom{s+k}{s}$. For each i we write x_i instead of x_{J_i}. Thus, x_i is the unique point in $\bigcap_{v \in J_i} H_v$. Define $K_i = \{1, 2, \ldots, s+k\} \setminus J_i$. Then $\#K_i = k$. With each node x_i we associate the k hyperplanes H_v for $v \in K_i$. Now we must show that the hypotheses of Theorem 3 are fulfilled. Thus we must prove

$$i \neq j \iff x_j \in \bigcup_{v \in K_i} H_v$$

First, let $i = j$. We want to prove that

$$x_i \notin \bigcup_{v \in K_i} H_v$$

If this is false, then $x_i \in H_\mu$ for some $\mu \in K_i$. We know that $x_i \in \bigcap_{v \in J_i} H_v$. So now $x_i \in H_\mu \cap (\bigcap_{v \in J_i} H_v)$. We can replace one v in J_i by μ and get a different J_j so that $x_i = x_j$, contrary to hypothesis (2) above.

Now suppose $x_j \notin \bigcup_{v \in K_i} H_v$. Since $x_j \in \bigcap_{v \in J_j} H_v$, the sets K_i and J_j must be disjoint. By their definition, this implies $i = j$.   ∎

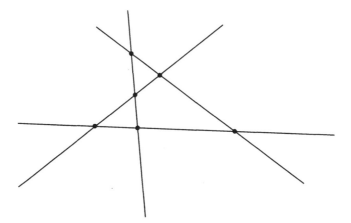

Figure 10.3   *Illustrating Theorem 5*

Example 4. Let $s = 2$ and $k = 2$. We require four hyperplanes (i.e., lines in $\mathbb{R}^2$). There will be six pairs of lines whose intersections will be a set of six nodes suitable for interpolation by $\Pi_2(\mathbb{R}^2)$. See Figure 10.3 for a typical configuration.   ∎

The interpolation process defined by Equations (3), (4), and (5) has a variation known as **Shepard's method.** We assume that ϕ obeys Equation (3) and is nonnegative. In other terms:

    1. $\phi(x, y) > 0$ if $x \neq y$
    2. $\phi(x, x) = 0$

Define

$$u_i(x) = \prod_{\substack{j=1 \\ j \neq 1}}^{n} \phi(x, x_j) \qquad (1 \leq i \leq n)$$

(11)
$$u(x) = \sum_{i=1}^{n} u_i(x)$$

(12)
$$v_i = u_i / u \qquad (1 \leq i \leq n)$$

We see that u_i vanishes at each node except x_i, and is positive elsewhere. Therefore $u(x) > 0$ for all x, and $u(x_i) = u_i(x_i)$. Also $\sum_{i=1}^{n} v_i = 1$. Hence $0 \leq v_i \leq 1$. The function v_i has a global maximum at x_i because $v_i(x_i) = 1$. It has a global minimum at the other nodes because $v_i(x_j) = 0$ for $j \neq i$. One unfortunate consequence of these remarks is that if each v_i is differentiable at every node, then $(Dv_i)(x_j) = 0$. Here D is the Frechet derivative operator. (It is the ordinary derivative if we are working in $\mathbb{R}$; it is the gradient if we are working in $\mathbb{R}^s$.)

The interpolating operator is again

(13)
$$Pf = \sum_{i=1}^{n} f(x_i) v_i$$

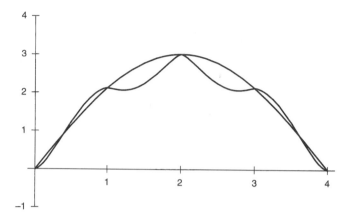

Figure 10.4   *Interpolation by Shepard's Method*

Because $\sum_{i=1}^{n} v_i = 1$, P has the favorable property of reproducing constant functions. Furthermore, we have

(14) $\alpha \leq f \leq \beta \implies \alpha \leq Pf \leq \beta$ $(\alpha, \beta \in \mathbb{R})$

This also shows that $Pf = f$ when f is a constant function. Another consequence is that for $f \in C(X)$, $\|Pf\|_\infty \leq \|f\|_\infty$. Thus $\|P\| = 1$. But if ϕ is differentiable, the preceding remarks indicate that the graph of Pf will have a flat spot at each node. Figure 10.4 shows the interpolant to $f(x) = 3\sin(\pi x/4)$ when $\phi(x, y) = |x - y|^{3/2}$ and five equally spaced nodes are employed.

Problems

1. Prove that if $x_0, \ldots, x_n \in \mathbb{R}^s$ and if $f(x_i) \geq 0$ for $0 \leq i \leq n$, then there is a polynomial $p \in \Pi_{2n}(\mathbb{R}^s)$ that interpolates f at the points x_i and is nonnegative everywhere in $\mathbb{R}^s$.

2. Consider an abstract operator of the form given in Equation (13) where $v_i(x_j) = \delta_{ij}$. Assume that P has the property in Equation (14). Prove that if the v_i are differentiable, then Pf has derivative 0 at each node.

3. In Theorem 4, suppose that $y_0 = 0$. Prove that the node set is

$$\left\{ \sum_{i=1}^{s} \lambda_i y_i : k\lambda_i \in \mathbb{Z}_+ \quad \text{and} \quad \sum_{i=1}^{s} \lambda_i \leq 1 \right\}$$

4. Prove carefully that the product of an affine function with a polynomial of degree k is a polynomial of degree at most $k + 1$. (These are functions on $\mathbb{R}^s$.)

5. Prove that if $\{y_0, y_1, \ldots, y_s\}$ is in general position in $\mathbb{R}^s$, then in the equations

$$x = \sum_{i=0}^{s} \lambda_i y_i \qquad \sum_{i=0}^{s} \lambda_i = 1$$

the λ_i are affine functionals of x. Prove that $\lambda_i(y_j) = \delta_{ij}$.

6. Give formulas for a sequence of lines $L_1, L_2, \ldots$ in $\mathbb{R}^2$ such that for any pair of integers $1 \leq i < j$, $L_i \cap L_j$ is a single point x_{ij} in $\mathbb{R}^2$, and the points so obtained are all different from each other. If possible, describe an uncountable family of such lines.

7. Let $t_1 < t_2 < \cdots < t_n$. For $i = 1, 2, \ldots, n$, let v_i be the continuous piecewise linear function such that $v_i(t_j) = \delta_{ij}$ $(1 \leq i, j \leq n)$. The domain of v_i is $[t_1, t_n]$. Show that this is a "Lagrangian" base for first-degree spline interpolation at the nodes t_i.

8. Discuss Shepard's interpolation on $\mathbb{R}$ when the function $\phi(x, y) = |x - y|$ is employed.

References

[BDL] Barnhill, R. E., R. P. Dube, and F. F. Little. "Properties of Shepard's surfaces." *Rocky Mt. J. Math. 13* (1983), 365–382.

[BR2] de Boor, C., and A. Ron. "Computational aspects of polynomial interpolation in several variables." *Math. Comp. 58* (1992), 705–727.

[BR1] de Boor, C., and A. Ron. "On multivariate polynomial interpolation." *Constr. Approx. 6* (1990), 287–302.

[Bo1] de Boor, C. "Polynomial interpolation in several variables." In *Proceedings of a Conference Honoring S. D. Conte,* ed. by R. DeMillo and J. R. Rice. Plenum Press, New York.

[Boo2] de Boor, C. "Multivariate approximation." In *State of the Art in Numerical Analysis*, ed. by A. Iserles and M. J. D. Powell. Oxford University Press, 1987, 87–109.

[Bos5] Bos, L. "On certain configurations of points in R^n which are unisolvent for polynomial interpolation." *J. Approx. Theory 64* (1991), 271–280.

[ChuLa] Chui, C. K., and M. J. Lai. "Vandermonde determinant and Lagrange interpolation in R^s. In *Nonlinear and Convex Analysis*, ed. by B. L. Lin and L. Simons. Marcel Dekker, New York, 1988, pp. 23–36.

[CY] Chung, K. C., and T. H. Yao. "On lattices admitting unique Lagrange interpolation." *SIAM J. Numer. Anal. 14* (1977), 735–741.

[DMV] Della Vecchis, B., G. Mastroianni, and P. Vertesi. "Direct and converse theorems for Shepard rational approximation." *Numer. Func. Anal. Optim. 17* (1996), 537–561.

[Fa] Farwig, R. "Rate of convergence of Shepard's global interpolation formula." *Math. Comp. 46* (1986), 577–590.

[Fr] Franke, R. "Scattered data interpolation: tests of some methods." *Math. Comp. 38* (1982), 381–200.

[GM] Gasca, M., and J. I. Maeztu. "On Lagrange and Hermite interpolation in R^k." *Numer. Math. 39* (1982), 1–14.

[GaMu] Gasca, M., and G. Mühlbach. "Multivariate polynomial approximation under projectivities." Part I, *Numer. Algo. 1* (1991), 375–400. Part II, ibid. 2 (1992), 255–278.

[Ga] Gasca, M. "Multivariate polynomial interpolation." In *Computation of Curves and Surfaces,* ed. by W. Dahmen, M. Gasca, and C. Micchelli. Kluwer Academic, 1990, 215–235.

[GW] Gordon, W. F., and J. A. Wixom. "Pseudo-harmonic interpolation on convex domains." *SIAM J. Numer. Anal. 11* (1974), 909–933.

[Jet]   Jetter, K. "Some contributions to bivariate interpolation and cubature." In *Approximation Theory IV*, ed. by C. Chui, L. Schumaker, and J. Ward. Academic Press, New York, 1983, 533–538.

[LS]   Lancaster, P., and K. Salkauskas. "Curve and Surface Fitting." Academic Press, New York, 1986.

[M2]   Micchelli, C. A. "Algebraic aspects of interpolation." In *Approximation Theory*, ed. by C. de Boor. Amer. Math. Soc. Proc. Symp. Appl. Math., vol. 36. Providence, RI, 1986.

[Sh]   Shepard, D. "A two-dimensional interpolation function for irregularly spaced data." *Proc. 23rd Nat. Conf. ACM* (1968), 517–523.

[We]   Werner, H. "Remarks on Newton type multivariate interpolation for subsets of grids." *Computing 25* (1980), 181–191.

[ZhZh]   Zhang, Ren Jiang, and Songping Zhou. "Three conjectures on Shepard interpolatory operators." *J. Approx. Theory 93* (1998), 399–414.

11

Interpolation by Translates of a Single Function

As in many of the preceding chapters, we consider here the general linear problem of interpolation. A set X is specified, and one wishes to interpolate a function $f : X \to \mathbb{R}$. Certain nodes $x_1, x_2, \ldots, x_n$ are prescribed. If a function g has the property $g(x_i) = f(x_i)$ for $1 \le i \le n$, then we say that g **interpolates** f at the nodes. The function g is customarily restricted to an n-dimensional linear space of well-behaved and easily computed functions. The question raised now is whether g can be a linear combination of n **translates** of a single function ϕ:

$$g(x) = \sum_{i=1}^{n} a_i \phi(x - v_i)$$

Of course, this equation requires some algebraic structure on X: If x and y belong to X then $x - y$ must be defined and must belong to X. An (additive) group or a linear space would be a suitable setting.

Example 1. Let $X = \mathbb{R}$ and $\phi(x) = e^x$. This function will not serve, for the translates of ϕ generate only a one-dimensional subspace:

$$\phi(x - v) = e^{x-v} = e^{-v} e^x = e^{-v} \phi(x)$$

Example 2. Let $X = \mathbb{R}$ and $\phi(x) = x^{-1}$. This function does serve well. The next theorem establishes this assertion.

> **THEOREM 1.** *Any n translates of the function $x \mapsto x^{-1}$ span an n-dimensional Haar space on any interval that contains none of the translation points.*

Proof. Let $v_1, v_2, \ldots, v_n$ be n distinct real numbers, and let $[a, b]$ be an interval containing none of the points v_i. Our basic functions are then $u_j(x) = (x - v_j)^{-1}$. To test the

Haar property, let $x_1, x_2, \ldots, x_n$ be n distinct points in $[a, b]$. We must prove that the determinant $\det\left(u_j(x_i)\right)$ is not zero. By Cauchy's Lemma [C1], page 195,

$$\det\left((x_i - v_j)^{-1}\right) = \prod_{1 \le j < i \le n} (x_i - x_j)(v_j - v_i) \Big/ \prod_{i=1}^{n} \prod_{j=1}^{n} (x_i - v_j)$$

This is certainly not zero.   ∎

Example 3. Let $X = \mathbb{R}$ and $\phi(x) = x^n$. This function serves our purpose too. See the next theorem.

> **THEOREM 2.** *Any $n + 1$ translates of the function $x \mapsto x^n$ form a basis for $\Pi_n(\mathbb{R})$.*

Proof. Consider a typical element of $\Pi_n(\mathbb{R})$, say $\sum_{k=0}^{n} b_k x^{n-k}$. Let $v_0, v_1, \ldots, v_n$ be distinct points in $\mathbb{R}$. We seek coefficients $c_0, c_1, \ldots, c_n$ so that

$$\sum_{j=0}^{n} c_j (x + v_j)^n = \sum_{k=0}^{n} b_k x^{n-k}$$

Expanding by the Binomial Theorem (Chapter 4, page 26), we have

$$\sum_{j=0}^{n} c_j \sum_{k=0}^{n} \binom{n}{k} x^{n-k} v_j^k = \sum_{k=0}^{n} b_k x^{n-k}$$

Since the set of monomials $x \mapsto 1, x, x^2, \ldots$ is linearly independent, we can equate the coefficients of x^{n-k} and arrive at

$$\sum_{j=0}^{n} \binom{n}{k} v_j^k c_j = b_k \qquad (0 \le k \le n)$$

or

$$\sum_{j=0}^{n} v_j^k c_j = b_k \Big/ \binom{n}{k} \qquad (0 \le k \le n)$$

This is a system of $n + 1$ linear equations in the $n + 1$ unknowns $c_0, c_1, \ldots, c_n$. The coefficients v_j^k form a Vandermonde matrix, which is nonsingular. (See the Example in Chapter 1, page 4.) Hence the unknown c_i are uniquely determined.   ∎

Example 4. Let $X = \mathbb{R}^s$ and $\phi(x) = x^\alpha$, where α is a fixed multi-index. What space is generated by the translates of ϕ? By the Binomial Theorem (Chapter 4, page 26),

$$(x + v)^\alpha = \sum_{0 \le \beta \le \alpha} \binom{\alpha}{\beta} x^\beta v^{\alpha - \beta}$$

This shows that all translates of ϕ lie in the space generated by $\{x \mapsto x^\beta : 0 \le \beta \le \alpha\}$. This set of monomials is linearly independent (see Theorem 1 in Chapter 4, page 28). It generates the polynomial space

$$P_\alpha(\mathbb{R}^s) = \left\{ x \longmapsto \sum_{0 \le \beta \le \alpha} c_\beta x^\beta : c_\beta \in \mathbb{R} \right\}$$ ∎

THEOREM 3. *The translates of the function $x \mapsto x^\alpha$, ($x \in \mathbb{R}^s$), generate $P_\alpha(\mathbb{R}^s)$ (as defined earlier).*

Proof. Let $\sum_{0 \le \beta \le \alpha} e_\beta x^{\alpha - \beta}$ be any element of $P_\alpha(\mathbb{R}^s)$. We seek $c_\gamma \in \mathbb{R}$ and $v_\gamma \in \mathbb{R}^s$ so that

$$\sum_{0 \le \gamma \le \alpha} c_\gamma (x + v_\gamma)^\alpha = \sum_{0 \le \beta \le \alpha} e_\beta x^{\alpha - \beta}$$

Applying the Binomial Theorem, we obtain

$$\sum_{0 \le \gamma \le \alpha} c_\gamma \sum_{0 \le \beta \le \alpha} \binom{\alpha}{\beta} x^{\alpha - \beta} v_\gamma^\beta = \sum_{0 \le \beta \le \alpha} e_\beta x^{\alpha - \beta}$$

Since the functions $x^{\alpha - \beta}$ form a linearly independent set, we can equate coefficients of $x^{\alpha - \beta}$ to obtain

$$\sum_{0 \le \gamma \le \alpha} c_\gamma \binom{\alpha}{\beta} v_\gamma^\beta = e_\beta \qquad (0 \le \beta \le \alpha)$$

Or

$$\sum_{0 \le \gamma \le \alpha} v_\gamma^\beta c_\gamma = e_\beta \Big/ \binom{\alpha}{\beta} \qquad (0 \le \beta \le \alpha)$$

This is a square system of linear equations for the unknowns c_γ. The coefficient matrix is again a Vandermonde matrix, having entries v_γ^β. Since the monomials $v \mapsto v^\beta$ form a linearly independent set, there exist particular points v_γ in $\mathbb{R}^s$ for which the matrix (v_γ^β) is nonsingular. See Problem 8, page 8. For this choice of translations we obtain a unique set of coefficients c_γ. ∎

Example 5. To illustrate what is happening in Theorem 3, let us take $s = 2$ and $\alpha = (3, 1)$. Write $x = (\xi_1, \xi_2)$. The space P_α is of dimension 8, and one basis is

$$x \longmapsto 1, \quad \xi_1, \quad \xi_2, \quad \xi_1 \xi_2, \quad \xi_1^2, \quad \xi_1^2 \xi_2, \quad \xi_1^3, \quad \xi_1^3 \xi_2$$

Let us use this set of eight translation vectors v_j:

$$(0, 0), (0, 1), (0, 2), \ldots, (0, 7)$$

The functions $(x + v_j)^\alpha$ are then of the form

$$(\xi_1, \xi_2)^{(3,1)}, \quad (\xi_1, \xi_2 + 1)^{(3,1)}, \ldots, \quad (\xi_1, \xi_2 + 7)^{(3,1)}$$

In other terms, these are

$$\xi_1^3 \xi_2, \quad \xi_1^3 (\xi_2 + 1), \quad \xi_1^3 (\xi_2 + 2), \ldots, \quad \xi_1^3 (\xi_2 + 7)$$

These functions span a two-dimensional space. If we use as $\{v_j\}_1^8$ a Cartesian product $\{a_0, a_1, a_2, a_3\} \times \{b_0, b_1\}$, we will obtain a base for P_α of the form $x \mapsto (\xi_1 + a_i)^3 (\xi_2 + b_j)$. ∎

The space $P_\alpha(\mathbb{R}^s)$ is actually a tensor product of s simple factors:

$$P_\alpha(\mathbb{R}^s) = \Pi_{\alpha_1}(\mathbb{R}) \otimes \Pi_{\alpha_2}(\mathbb{R}) \otimes \cdots \otimes \Pi_{\alpha_s}(\mathbb{R})$$

The vector space on the right side of this equation is generated by functions of the form

$$x \longmapsto p_1(\xi_1)p_2(\xi_2) \cdots p_s(\xi_s)$$

where, for each i, p_i is a univariate polynomial of degree at most α_i. This is in harmony with the definitions in Chapter 7, where the case $s = 2$ was considered. By the general case of Lemma 4 in Chapter 7 (page 48), the space in question is generated by monomials

$$x \longmapsto \xi_1^{\beta_1}\xi_2^{\beta_2} \cdots \xi_s^{\beta_s} \equiv x^\beta \qquad (0 \leq \beta \leq \alpha)$$

Example 6. Let $X = \mathbb{R}$ and $\phi(x) = |x|$. This example leads us to first-degree spline functions. Indeed, a function of the form $g(x) = \sum_{j=1}^n a_j|x - v_j|$ is obviously continuous and piecewise linear. Its *knots* (points where the linear pieces are joined) are the points v_j. The following theorem shows that the functions $x \mapsto |x - v_j|$ form a linearly independent set, and moreover that they can be used for interpolation at the knots. The proof is constructive. See [LS], pages 68–71. ∎

THEOREM 4. *Let $x_1 < x_2 < \cdots < x_n$ be in $\mathbb{R}$. Let ordinates λ_i be given. Set*

$$m_j = (\lambda_j - \lambda_{j-1})/(x_j - x_{j-1}) \qquad (2 \leq j \leq n)$$
$$m_{n+1} = -m_1 = (\lambda_n + \lambda_1)/(x_n - x_1)$$
$$a_j = (m_{j+1} - m_j)/2 \qquad (1 \leq j \leq n)$$

Then the function $\sum_{j=1}^n a_j|x - x_j|$ interpolates data $\lambda_1, \ldots, \lambda_n$ given at $x_1, \ldots, x_n$.

Proof. Let the coefficients a_j be determined by the algorithm, and define a function g by $g(x) = \sum_{j=1}^n a_j|x - x_j|$. We must prove that $g(x_i) = \lambda_i$, for $1 \leq i \leq n$. We write

$$2g(x) = \sum_{j=1}^n (m_{j+1} - m_j)|x - x_j| = \sum_{j=2}^{n+1} m_j|x - x_{j-1}| - \sum_{j=1}^n m_j|x - x_j|$$

$$= \sum_{j=2}^n m_j\{|x - x_{j-1}| - |x - x_j|\} + m_{n+1}|x - x_n| - m_1|x - x_1|$$

If $x = x_1$ in this equation, the result is

$$2g(x_1) = \sum_{j=2}^n m_j\{(x_{j-1} - x_1) - (x_j - x_1)\} + m_{n+1}(x_n - x_1)$$

$$= \sum_{j=2}^n m_j(x_{j-1} - x_j) + m_{n+1}(x_n - x_1)$$

$$= \sum_{j=2}^n (\lambda_{j-1} - \lambda_j) + (\lambda_n + \lambda_1) = 2\lambda_1$$

Similarly, one finds that $2g(x_n) = 2\lambda_n$. If $1 < i < n$,

$$
\begin{aligned}
2g(x_i) &= \sum_{j=2}^{i} m_j\{(x_i - x_{j-1}) - (x_i - x_j)\} + \sum_{j=i+1}^{n} m_j\{(x_{j-1} - x_i) - (x_j - x_i)\} \\
&\quad + m_{n+1}(x_n - x_i) - m_1(x_i - x_1) \\
&= \sum_{j=2}^{i} m_j(x_j - x_{j-1}) - \sum_{j=i+1}^{n} m_j(x_j - x_{j-1}) + m_{n+1}(x_n - x_i) \\
&\quad + m_{n+1}(x_i - x_1) \\
&= \sum_{j=2}^{i} (\lambda_j - \lambda_{j-1}) - \sum_{j=i+1}^{n} (\lambda_j - \lambda_{j-1}) + m_{n+1}(x_n - x_1) \\
&= (\lambda_i - \lambda_1) - (\lambda_n - \lambda_i) + (\lambda_n + \lambda_1) = 2\lambda_i \qquad \blacksquare
\end{aligned}
$$

Remark. Stronger versions of Theorem 4 can be proved, in which the nodes of interpolation are permitted to be different from the knots. The Schoenberg-Whitney Theorem in the theory of splines governs this situation. See, for example, [Bo4], page 200; [Sch2], page 165; or [KinC], page 407.

Problems

1. What can be said about the spaces generated by translates of $x \mapsto \cos x$? Answer the same question for the function $x \mapsto \cosh x$.

2. Let p be any polynomial of degree n on $\mathbb{R}$. Is it true that any $n + 1$ translates of p generate $\Pi_n(\mathbb{R})$?

3. Let α be a multi-index in $\mathbb{Z}_+^s$. Let $n = \#\{\beta : 0 \le \beta \le \alpha\}$. Is it true that, for any n points x_i in $\mathbb{R}^s$, the matrix (x_i^β) is nonsingular? Here $1 \le i \le n, 0 \le \beta \le \alpha, \beta \in \mathbb{Z}_+$.

4. Prove that if $v_1, \dots, v_n$ are distinct nonzero numbers in the interval $(-1, 1)$ then the functions $x \mapsto (1 + v_j x)^{-1}$ generate an n-dimensional Haar space in $C[-1, 1]$.

5. (Continuation) Define $q_k(x) = \Pi_{j=1, j \ne k}^{n}(x - v_j)/\Pi_{j=1}^{n}(1 + v_j x)$. Show that these functions provide a base for the Haar space in Problem 4. Prove that $q_k(x)/q_k(v_k)$ form a cardinal base for interpolation at $v_1, v_2, \dots, v_n$.

6. For each $y \in \mathbb{R}^2$, define f_y by the equation $f_y(x) = \|x + y\|^2$, where $x \in \mathbb{R}^2$. Find the dimension of the linear span of $\{f_y : y \in \mathbb{R}^2\}$.

7. For $y \in \mathbb{R}^s$, define $f_y(x) = \|x + y\|^2$, where $x \in \mathbb{R}^s$. Find the dimension of the linear space spanned by $\{f_y : y \in \mathbb{R}^s, \|y\| = 1\}$.

8. Define f_y as in Problem 7. Prove that the span of $\{f_y : y \in \mathbb{R}^s\}$ is $\Pi_1(\mathbb{R}^s) \oplus U$, where U is the one-dimensional space generated by the single function $u(x) = \|x\|^2$.

9. Fix a multi-index $\alpha \in \mathbb{Z}_+^s$. Fix node sets $\mathcal{N}_i \subset \mathbb{R}$, for $1 \le i \le s$, and assume that

$$
\#\mathcal{N}_i = \alpha_i + 1 \qquad (1 \le i \le s)
$$

Let $\mathcal{N} = \mathcal{N}_1 \times \mathcal{N}_2 \times \cdots \times \mathcal{N}_s$. Then $\mathcal{N} \subset \mathbb{R}^s$. Prove that interpolation by $P_\alpha(\mathbb{R}^s)$ is uniquely possible on $\mathcal{N}$.

10. Does the space $\Pi_k(\mathbb{R}^s)$ have a basis consisting of functions $x \mapsto (x + v_i)^\alpha$, where α is a fixed multi-index?

11. Explore this procedure for polynomial interpolation on $\mathbb{R}$: Let nodes $x_0, x_1, \ldots, x_n$ be given. We seek a polynomial p of degree at most n such that $p(x_i) = b_i, 0 \le i \le n$. We seek p in the form

$$p(x) = \sum_{j=0}^{n} c_j(x - x_j)^n$$

Prove that the interpolation matrix is nonsingular.

12. Let $\{u_1, u_2, \ldots, u_n\}$ be a basis for a vector space U, and let $u_0 \in U$. Find the exact conditions under which $\{u_i - u_0 : 1 \le i \le n\}$ is a basis for U.

Research Problems

1. Find all functions $\phi : \mathbb{R} \to \mathbb{R}$ having the property that for all n and for any set of n points $v_1, v_2, \ldots, v_n$ the set of functions

$$\{x \longmapsto \phi(x - v_i) : 1 \le i \le n\}$$

is a Chebyshev system on some interval. (One example is $\phi(x) = 1/x$.)

2. Alter Problem 1 by asking only that the set of functions be linearly independent. (One example is $\phi(x) = |x|$.)

3. For a fixed n, find all functions $\phi : \mathbb{R} \to \mathbb{R}$ such that for any n points x_i the $n \times n$ matrix $(\phi(x_i - x_j))$ is nonsingular.

References

[Bo4]   de Boor, C. *A Practical Guide to Splines.* Springer-Verlag, New York, 1978.

[C1]   Cheney, E. W. *Introduction to Approximation Theory.* McGraw-Hill, New York, 1966. 2nd ed., Chelsea, New York, 1982. Amer. Math. Soc. 1998. Japanese edition, Kyoritsu Shuppan, Tokyo, 1978. Chinese edition, Beijing, 1981.

[KinC]   Kincaid, D., and W. Cheney. *Numerical Analysis.* 2nd ed. Brooks/Cole, Pacific Grove, CA, 1996.

[LS]   Lancaster, P., and K. Salkauskas. *Curve and Surface Fitting.* Academic Press, New York, 1986.

[LC1]   Light, W. A., and E. W. Cheney. "Interpolation by periodic radial basis functions." *J. Math. Analysis and Appl. 168* (1992), 111–130.

[Ph1]   Phillips, G. M. "Algorithms for piecewise straight line approximations." *Comput. J. 11* (1968), 211–212.

[Sch2]   Schumaker, L. L. *Spline Functions: Basic Theory.* Wiley, New York, 1981.

[XC1]   Xu, Yuan, and E. W. Cheney. "Interpolation by periodic radial functions." *Computers Math. Applic. 24* (1992), 201–215.

12

Positive Definite Functions

In this chapter we study another class of functions whose translates can be useful in multivariate interpolation. Recall from Chapter 11 that with a function $f : \mathbb{R}^s \to \mathbb{R}$ in hand, we can ask whether arbitrary data at specified nodes $x_1, x_2, \ldots, x_n$ in $\mathbb{R}^s$ can be interpolated by a function of the form $\sum_{j=1}^{n} a_j f(x - x_j)$. The answer is yes if and only if the matrix $A_{ij} = f(x_i - x_j)$ is nonsingular. Our goal is to characterize the functions f for which this matrix A is invariably positive definite. Along the way we prove some famous theorems and review a bit of matrix theory. The reader may wish to consult Chapter 15, where the Fourier transform is discussed.

Definition. Let X be a linear space (over the real or complex field). A function $f : X \to \mathbb{C}$ is said to be **positive definite** on X if for any finite set of points $x_1, x_2, \ldots, x_n$ in X the $n \times n$ matrix $A_{ij} = f(x_i - x_j)$ is nonnegative definite, i.e.,

$$u^*Au = \sum_{i=1}^{n} \sum_{j=1}^{n} \bar{u}_i u_j A_{ij} \geq 0$$

for all $u \in \mathbb{C}^n$. If $u^*Au > 0$ whenever the points x_i are distinct and $u \neq 0$, then we say that f is **strictly positive definite.** In these equations, u is a column vector and u^* is a row vector, the conjugate transpose of u.

The nomenclature in the definition has been standard for many years, and we have no choice but to endure it. Naturally, we are more interested in the strictly positive definite functions, for these lead to the nonsingular interpolation problems. We begin with a long list of examples.

Example 1. Let X be any real inner-product space, and let $y \in X$. The function $x \mapsto e^{iyx}$ is positive definite on X. To verify this, let $x_1, x_2, \ldots, x_n$ be n points in X. Let $u \in \mathbb{C}^n$. Then

$$\sum_{k=1}^{n} \sum_{j=1}^{n} \bar{u}_k u_j e^{iy(x_k - x_j)} = \sum_{k=1}^{n} \bar{u}_k e^{iyx_k} \sum_{j=1}^{n} u_j e^{-iyx_j} = \left| \sum_{j=1}^{n} u_j e^{-iyx_j} \right|^2$$

This last expression is obviously nonnegative. ∎

Example 2. Let X be any real inner-product space. Let $x_1, x_2, \ldots, x_n$ be any n different points in X. We shall see in Chapter 14 that the function $t \mapsto (1 + t)^{-1}$ is completely monotone. By Schoenberg's Theorem (Chapter 15, page 101), we conclude that the matrix $A_{ij} = (1 + \|x_i - x_j\|^2)^{-1}$ is positive definite. Thus the function $f(x) = (1 + \|x\|^2)^{-1}$ is strictly positive definite on any inner-product space. In fact, it is radial as well. ∎

Example 3. A sum of two positive definite functions on X is also positive definite. ∎

Example 4. A nonnegative linear combination of positive definite functions on X is also positive definite. The set of all positive definite functions on X is therefore a cone. ∎

Example 5. The cosine function is positive definite on $\mathbb{R}$ because it is expressible as $\cos x = \frac{1}{2}(e^{ix} + e^{-ix})$, and Examples 2 and 3 can be applied. Alternatively, one can use a direct calculation involving the trigonometric identity $\cos(x - y) = \cos x \cos y + \sin x \sin y$. ∎

Example 6. If $f \in L^1(\mathbb{R}^s)$ and $f \geq 0$, then $\hat{f}$ (the Fourier transform of f) is positive definite on $\mathbb{R}^s$. This is a limiting case of Examples 2 and 4. A direct verification is as follows:

$$u^* Au = \sum_{k=1}^{n} \sum_{j=1}^{n} \bar{u}_k u_j \hat{f}(x_k - x_j) = \sum_{k=1}^{n} \sum_{j=1}^{n} \bar{u}_k u_j \int_{\mathbb{R}^s} f(y) e^{-2\pi i (x_k - x_j) y} \, dy$$

$$= \int_{\mathbb{R}^s} f(y) \sum_{k=1}^{n} \bar{u}_k e^{-2\pi i x_k y} \sum_{j=1}^{n} u_j e^{2\pi i x_j y} \, dy = \int_{\mathbb{R}^s} f(y) |g(y)|^2 \, dy \geq 0$$

where we have written

$$g(y) = \sum_{j=1}^{n} u_j e^{2\pi i x_j y} \qquad (y \in \mathbb{R}^s)$$ ∎

Example 7. The function $F(x) = (1 + ix)^{-1}$ is positive definite on $\mathbb{R}$. This is proved by using Example 6, with f taken to be $e^{-2\pi x}$ on $[0, \infty)$ but 0 on $(-\infty, 0)$. Here are the steps:

$$\hat{f}(x) = \int_0^\infty e^{-2\pi y} e^{-2\pi i x y} \, dy = \frac{e^{-2\pi(1+ix)y}}{-2\pi(1+ix)} \bigg|_{y=0}^{y=\infty} = \frac{1}{2\pi(1+ix)}$$ ∎

Example 8. If f is positive definite on $\mathbb{R}$, then we can define a positive definite function f^* on any real linear space X as follows. Select a linear functional ϕ on X, and define $f^* = f \circ \phi$. To verify positive definiteness, write

$$\sum \sum u_k \bar{u}_j f(\phi(x_k - x_j)) = \sum \sum u_k \bar{u}_j f(\phi(x_k) - \phi(x_j)) \geq 0$$ ∎

Before delving into some general theory of positive definite functions, we shall prove a few results in matrix theory.

LEMMA 1. *If A is an $n \times n$ complex matrix such that $u^*Au = 0$ for all $u \in \mathbb{C}^n$, then $A = 0$.*

Proof. Let A fulfill the hypotheses. For any u and v we have

$$0 = (u + v)^*A(u + v) = u^*Au + u^*Av + v^*Au + v^*Av$$

and consequently

(1) $$0 = u^*Av + v^*Au$$

In this last equation, replace u by iu to obtain $0 = -iu^*Av + iv^*Au$ or equivalently

(2) $$0 = -u^*Av + v^*Au$$

On adding Equations (1) and (2) we conclude that $v^*Au = 0$. Let $v = Au$ to get $(Au)^*(Au) = 0$. Thus $\|Au\|^2 = 0$ and $Au = 0$. Since u was arbitrary, $A = 0$. ∎

If A is a complex matrix, A^* will denote its conjugate transpose. Formally, $(A^*)_{ij} = \overline{A}_{ji}$. If $A = A^*$, we say that A is **Hermitian.** A Hermitian matrix is necessarily square.

THEOREM 1. *For a complex matrix A, these two properties are equivalent:*

   a. *A is Hermitian;*
   b. *u^*Au is real for all $u \in \mathbb{C}^n$.*

Proof. Assume (a). Then $\overline{(u^*Au)} = (u^*Au)^* = u^*A^*u = u^*Au$. Hence u^*Au is real, since it equals its own conjugate.

Now assume (b). Write $A = B + iC$, where $B = \frac{1}{2}(A + A^*)$ and $C = \frac{1}{2i}(A - A^*)$. We quickly verify that B and C are Hermitian:

$$B^* = \frac{1}{2}(A^* + A^{**}) = \frac{1}{2}(A^* + A) = B$$

$$C^* = \frac{-1}{2i}(A^* - A^{**}) = \frac{1}{2i}(A - A^*) = C$$

By the first part of this proof, u^*Bu and u^*Cu are real. Since $u^*Au = u^*Bu + iu^*Cu$, we conclude that $u^*Cu = 0$. By the preceding lemma, $C = 0$. Hence $A = B$, and A is Hermitian. ∎

COROLLARY. *A nonnegative definite matrix is Hermitian.*

It is noteworthy that from the condition $u^TAu = 0$ for all $u \in \mathbb{R}^n$ we cannot conclude that A is 0, nor Hermitian, nor symmetric. For example,

$$(u_1, u_2)\begin{bmatrix} 0 & 1 \\ -1 & 0 \end{bmatrix}\begin{pmatrix} u_1 \\ u_2 \end{pmatrix} = (u_1, u_2)\begin{pmatrix} u_2 \\ -u_1 \end{pmatrix} = u_1u_2 - u_2u_1 = 0$$

LEMMA 2. *If a matrix is positive definite, then its eigenvalues are positive and its determinant is positive. The statement remains true if we replace "positive" by "nonnegative."*

Proof. Let A be a positive definite matrix, and let λ be one of its eigenvalues. Then there exists a corresponding eigenvector $v \in \mathbb{C}^n$. Since $v \neq 0$,

$$0 < v^*Av = v^*\lambda v = \lambda\|v\|^2$$

Hence $\lambda > 0$. Since all eigenvalues of A are positive, their product is positive. But the product of the eigenvalues equals the determinant of A. Indeed, the latter is the constant term in the characteristic polynomial, i.e., $p(0)$, where

$$p(t) = \det(A - tI) = (-1)^n(t - \lambda_1)(t - \lambda_2) \cdots (t - \lambda_n)$$

The second assertion of the lemma is proved by making minor adjustments in the proof just given. ∎

LEMMA 3. *If f is a positive definite function on a linear space, then*

    a. $f(0) \geq 0$
    b. $f(-x) = \overline{f(x)}$
    c. $|f(x)| \leq f(0)$

Proof. In the definition of a positive definite function, let $n = 1$. The 1×1 matrix $f(0)$ is nonnegative definite, and $f(0) \geq 0$. Next, let $n = 2$, $x_2 = 0$, and $x_1 = x$. The resulting matrix is

$$A = \begin{bmatrix} f(x_1 - x_1) & f(x_1 - x_2) \\ f(x_2 - x_1) & f(x_2 - x_2) \end{bmatrix} = \begin{bmatrix} f(0) & f(x) \\ f(-x) & f(0) \end{bmatrix}$$

By the preceding corollary, A must be Hermitian, and therefore $\overline{f(x)} = f(-x)$. This establishes (b). To prove (c), use Lemma 2 to conclude that the matrix A has nonnegative determinant. Thus, with the help of part (b), we have

$$\det(A) = f(0)^2 - |f(x)|^2 \geq 0$$

Since $f(0) \geq 0$ by part (a), we obtain $f(0) \geq |f(x)|$. ∎

A high point in matrix theory is the spectral theorem for Hermitian matrices. It asserts that a Hermitian matrix A has a factorization $A = UDU^*$, in which U is a unitary matrix and D is a diagonal matrix. (A unitary matrix is, by definition, a square matrix such that $U^*U = I$.) Thus, $U^*AU = D$, and A is "unitarily similar" to a diagonal matrix. Two matrices that are similar to each other have the same characteristic polynomial and the same eigenvalues. Therefore, the eigenvalues of A are the same as the eigenvalues of D; these in turn are the diagonal elements of D.

LEMMA 4. *An $n \times n$ complex matrix A is nonnegative definite if and only if it can be expressed as $A = B^*B$, for some matrix B (which is not necessarily square).*

Proof. If A is nonnegative definite then it is Hermitian, by the preceding Corollary. By the spectral theorem, $A = U^*DU$ for some unitary matrix U and some diagonal matrix D. The eigenvalues of A are the diagonal elements of D. They are nonnegative, by Lemma 2.

Hence $D^{1/2}$ is real. The equation $A = U^* D^{1/2} D^{1/2} U = (D^{1/2}U)^* (D^{1/2}U)$ gives the desired factorization. On the other hand, if $A = B^*B$ then $u^*Au = u^*B^*Bu = (Bu)^*(Bu) = \|Bu\|^2 \geq 0$. (Compare this lemma to Lemma 2 in Chapter 16, page 114.) ∎

Definition. The **Schur product** (or **Hadamard product**) of two $n \times n$ matrices A and B is the matrix C having elements $C_{ij} = A_{ij} B_{ij}$. (This is the kindergarten way of multiplying matrices.)

SCHUR'S LEMMA. *The Schur product of two nonnegative definite matrices is nonnegative definite.*

Proof. Let A and B be nonnegative definite, and let C be their Schur product. By Lemma 4, A and B have factorizations $A = E^*E$, $B = F^*F$. The proof of Lemma 4 shows that we may take E and F to be $n \times n$ matrices. Now we have

$$C_{ij} = A_{ij} B_{ij} = (E^*E)_{ij} (F^*F)_{ij} = \sum_{\nu=1}^{n} \overline{E}_{\nu i} E_{\nu j} \sum_{\mu=1}^{n} \overline{F}_{\mu i} F_{\mu j}$$

This equation indicates that C has a factorization $C = G^*G$. We let G be the $n^2 \times n$ matrix whose elements are described thus: $G_{(\nu,\mu),j} = E_{\nu j} F_{\mu j}$, where all indices run over the set $\{1, 2, \ldots, n\}$. Lemma 4 then shows that C is nonnegative definite. For those who recoil from the concept of a matrix whose elements are indexed with *triples* of integers, an alternative is to write

$$u^*Cu = \sum_i \sum_j \overline{u}_i u_j \sum_\nu \sum_\mu \overline{E}_{\nu i} \overline{F}_{\mu i} E_{\nu j} F_{\mu j}$$

$$= \sum_\nu \sum_\mu \left(\sum_i \overline{u}_i \overline{E}_{\nu i} \overline{F}_{\mu i} \right) \left(\sum_j u_j E_{\nu j} F_{\mu j} \right)$$

$$= \sum_\nu \sum_\mu \left| \sum_j u_j E_{\nu j} F_{\mu j} \right|^2 \geq 0 \qquad ∎$$

COROLLARY. *The family of all positive definite functions defined on a prescribed linear space is algebraically closed under multiplication.*

Proof. Let f and g be positive definite functions on a linear space X. Let $x_1, x_2, \ldots, x_n \in X$. Let $A_{ij} = f(x_i - x_j) g(x_i - x_j)$. We wish to show that A is nonnegative definite. It is obviously the Schur product of two nonnegative definite matrices, and Schur's Lemma completes the proof. ∎

COROLLARY. *If f is positive definite on a linear space, and if ϕ is an analytic function having nonnegative Taylor coefficients in the disk $\{z : |z| \leq f(0)\}$, then $\phi \circ f$ is positive definite.*

Proof. Let $\phi(z) = \sum_{k=0}^{\infty} a_k z^k$, with $a_k \geq 0$. Since $|f(x)| \leq f(0)$, by Lemma 3, we can write $\phi(f(x)) = \sum a_k [f(x)]^k$. By Schur's Theorem, each power f^k is positive definite, and so is $\phi \circ f$. ∎

LEMMA 5. *If f is a positive definite function on a normed linear space X, and if f is continuous at 0, then f is uniformly continuous on X. In fact,*

$$|f(x) - f(y)|^2 \le 2f(0)\mathcal{R}[f(0) - f(y - x)]$$

Proof. Taking $n = 3$ in the definition of a positive definite function, we have

$$[\bar{u}_1\ \bar{u}_2\ \bar{u}_3] \begin{bmatrix} f(x_1 - x_1) & f(x_1 - x_2) & f(x_1 - x_3) \\ \overline{f(x_1 - x_2)} & f(x_2 - x_2) & f(x_2 - x_3) \\ \overline{f(x_1 - x_3)} & \overline{f(x_2 - x_3)} & f(x_3 - x_3) \end{bmatrix} \begin{bmatrix} u_1 \\ u_2 \\ u_3 \end{bmatrix} \ge 0$$

Let x and y be any two points in X. Let $x_1 = x$, $x_2 = 0$, and $x_3 = x - y$ in the preceding inequality, getting

$$[\bar{u}_1\ \bar{u}_2\ \bar{u}_3] \begin{bmatrix} f(0) & f(x) & f(y) \\ \overline{f(x)} & f(0) & f(y - x) \\ \overline{f(y)} & \overline{f(y - x)} & f(0) \end{bmatrix} \begin{bmatrix} u_1 \\ u_2 \\ u_3 \end{bmatrix} \ge 0$$

This leads to

$$(|u_1|^2 + |u_2|^2 + |u_3|^2)f(0) + 2\mathcal{R}\{\bar{u}_1 u_2 f(x) + \bar{u}_1 u_3 f(y) + \bar{u}_2 u_3 f(y - x)\} \ge 0$$

Select $\theta \in \mathbb{R}$ so that $e^{i\theta}[f(x) - f(y)] = |f(x) - f(y)|$. Then put $u_1 = t$, $u_2 = e^{i\theta}$, and $u_3 = -e^{i\theta}$, where t is a real parameter. The previous inequality now becomes

$$(t^2 + 2)f(0) + 2\mathcal{R}\{te^{i\theta}f(x) - te^{i\theta}f(y) - f(y - x)\} \ge 0$$

After some algebraic manipulation, this becomes

$$t^2 f(0) + 2t|f(x) - f(y)| + 2\mathcal{R}\{f(0) - f(y - x)\} \ge 0$$

This quadratic function of t, being nonnegative for all t, must have a nonpositive discriminant. Hence $4|f(x) - f(y)|^2 - 4f(0)2\mathcal{R}\{f(0) - f(y - x)\} \le 0$. This is equivalent to the asserted inequality. ∎

The Fourier transform of a measure μ on $\mathbb{R}^s$ is defined by the equation

$$\hat{\mu}(x) = \int_{\mathbb{R}^s} e^{-2\pi ixy}\, d\mu(y) \qquad (x \in \mathbb{R}^s)$$

The measure μ in this equation can be a signed measure. To ensure the existence of the integral, we should assume that $|\mu|(\mathbb{R}^s) < \infty$. If μ is a positive Borel measure on $\mathbb{R}^s$, then the computation in Example 6 can be extended to see that $\hat{\mu}$ is a positive definite continuous function on $\mathbb{R}^s$. Bochner established the converse [Bo3]:

BOCHNER'S THEOREM. *In order that a function $f : \mathbb{R}^s \to \mathbb{C}$ be positive definite and continuous, it is necessary and sufficient that it be the Fourier transform of a nonnegative finite-valued Borel measure on $\mathbb{R}^s$.*

The one-dimensional version of this theorem was given in 1932 by Bochner [Boc2]. He established the higher-dimensional version in [Boc3]. See [St] for the history and for generalizations.

For positive definite functions and the Bochner theorem, consult [Do2], page 184; [E1], page 715; [Ru1], page 17; [Loo], pages 97, 126, 141; [Ru2], page 290; [SW], page 32; [St]; [Ka], page 137; [RS], page 330; [HJ1], page 400; [Boc3]; [Hu], page 184; [N], page 391; [E2], page 108; [Helg]; [Bo2]; [BCR], page 66; [Phe1], page 118; [Cho], vol. 2, page 114; [Yos], page 346; and [Gol], page 59.

Problems

1. Prove or disprove: If f is a positive definite function defined on a linear space, then so is the function g defined by $g(x) = f(x) - f(0)$.

2. Prove that for an $n \times n$ complex matrix A the equation $A^T = -A$ is equivalent to the condition $u^T A u = 0$ for all $u \in \mathbb{R}^n$.

3. Prove that if A is a positive definite matrix, then $A_{ii} A_{jj} > |A_{ij}|^2$ when $i \neq j$.

4. Prove that if A is a positive definite matrix, then so is the matrix B defined by $B_{ij} = A_{ij}(A_{ii} A_{jj})^{-1/2}$.

5. Prove that if f is a positive definite function on a linear space, then so are $\mathcal{R}f, \bar{f}$, and $|f|^2$. (The notation $\mathcal{R}f$ denotes the real part of f.)

6. Prove that if f is a positive definite function on a linear space X and if $g : X \to X$ is linear, then $f \circ g$ is positive definite.

7. Prove that if A is a nonnegative definite $n \times n$ matrix and if B is any $n \times m$ matrix, then $B^T A B$ is nonnegative definite. Find the analogous result for positive definite matrices.

8. What constant functions are positive definite? What linear functionals are positive definite?

9. Let u and v be elements of $\mathbb{C}^n$, both interpreted as $n \times 1$ matrices. Thus uv^* is an $n \times n$ matrix. Find the necessary and sufficient conditions for uv^* to be nonnegative definite.

10. Prove that the family of all positive definite functions on a prescribed linear space is closed in the topology of pointwise convergence.

11. Let X be any set, and let f be any function from X into $\mathbb{C}$. Let $x_1, x_2, \ldots, x_n$ be any points of X. Prove that the matrix A defined by $A_{ij} = f(x_i)\overline{f(x_j)}$ is nonnegative definite. (Cf. Problem 9.)

12. Let $\phi \in L^1(\mathbb{R}^s)$, and define f by the equation

$$ f(x) = \int_{\mathbb{R}^s} \phi(x + y)\overline{\phi(y)} \, dy \qquad (x \in \mathbb{R}^s) $$

Prove that f is positive definite.

13. Prove this partial converse of Lemma 2. If the eigenvalues of a Hermitian matrix are positive, then it is positive definite. If the eigenvalues are nonnegative, then it is nonnegative definite.

14. Give an example of a matrix that has positive eigenvalues but is not nonnegative definite.

15. Prove that if A is an $n \times n$ Hermitian matrix, then there is an orthonormal base $\{u_1, u_2, \ldots, u_n\}$ for $\mathbb{C}^n$ and a set of n real numbers $\{c_1, c_2, \ldots, c_n\}$ such that $A = \sum_{j=1}^{n} c_j u_j u_j^*$.

16. Prove that these three functions are positive definite on $\mathbb{R}$: $f(x) = x^{-1} \sin ax$, $g(x) = x^{-2} \cos x \sin^2(x/2)$, $h(x) = e^{-a|x|}$.

17. Prove that these three functions are positive definite on $\mathbb{R}$: $g(x) = (a^2 + x^2)^{-1}$, $f(x) = \cos(ax)e^{-b|x|}$, $h(x) = e^{-x^2/a}$.

18. Prove that these three functions are positive definite on $\mathbb{R}$: $g(x) = \operatorname{sech}(ax)$, $f(x) = x^{-1}(e^{-a|x|}e^{-b|x|})$, $h(x) = \operatorname{Arctan}(a/|x|)$.

19. Without using Bochner's Theorem, prove that if μ is a finite-valued nonnegative Borel measure on $\mathbb{R}^s$, then $|\hat{\mu}(x)| \leq \hat{\mu}(0)$.

20. Let A be an $n \times n$ real, symmetric matrix such that $u^T A u \geq 0$ for all $u \in \mathbb{R}^n$. Prove that A is nonnegative definite, i.e., that $u^* A u \geq 0$ for all $u \in \mathbb{C}^n$.

21. Prove that if A is an $n \times n$ nonnegative definite matrix of rank r, then there is an orthogonal set of nonzero vectors $v_1, v_2, \ldots, v_r$ such that $A = \sum_{k=1}^{r} v_k v_k^*$. *Hint:* By the spectral theorem, there is an orthonormal set $\{u_1, u_2, \ldots, u_r\}$ such that $Ax = \sum \lambda_k \langle x, u_k \rangle u_k$.

22. Use Problem 21 to prove that the rank of the Schur product of two nonnegative definite matrices does not exceed the product of their ranks.

23. For the set of all $m \times n$ matrices, adopt the Schur product as multiplication. What is the Schur identity? What is the Schur inverse? Which matrices are invertible? If a positive definite matrix has a Schur inverse, is the inverse positive definite?

24. A function of two variables is said to be **separable** if it has the form $(x, y) \mapsto \sum_{i=1}^{N} u_i(x) v_i(y)$. What separable functions are positive definite?

25. Define a measure μ by $\mu(A) = 1$ if $\xi \in A$ and $\mu(A) = 0$ otherwise. Here ξ is a prescribed point in $\mathbb{R}^s$. What is $\hat{\mu}$ in familiar terms?

26. Let $a^* A u$ be real for all u in $\mathbb{C}^n$. Prove that the real and imaginary parts of $u^* A v + v^* A u$ are equal.

References

[BCR]   Berg, C., J. P. R. Christensen, and P. Ressel. *Harmonic Analysis on Semigroups*, Springer-Verlag, Berlin, 1984.

[Bere]   Berezanskii, Yu. M. "A generalization of a multidimensional theorem of Bochner." *Sov. Math. Doklady 2* (1961), 143–147.

[Boc1]   Bochner, S. "A theorem on Fourier-Stieltjes integrals." *Bull. Amer. Math. Soc. 40* (1934), 271–276.

[Boc2]   Bochner, S. *Vorlesungen über Fouriersche Integrale*, Leipzig, 1932.

[Boc3]   Bochner, S. "Monotone Funktionen, Stieltjes Integrale und Harmonische Analyse." *Math. Ann. 108* (1933), 378–410.

[Bo4]   Bochner, S. "Hilbert distances and positive definite functions." *Annals of Math. 42* (1941), 647–656.

[Cho]   Choquet, Gustave, *Lectures on Analysis* (in 3 vols). W. A. Benjamin, New York, 1969.

[Cr]   Crum, M. M. "On positive definite functions." *Proc. London Math. Soc. 6* (1956), 548–560.

[Do2]   Donoghue, W. F. *Distributions and Fourier Transforms.* Academic Press, New York, 1969. Pure and Applied Mathematics, vol. 32.

[E1]   Edwards, R. E. *Functional Analysis.* Holt, Rinehart and Winston, New York, 1965.

[E2]   Edwards, R. E. *Integration and Harmonic Analysis on Groups.* Cambridge University Press, 1972.

[FaH]   Falb, P. L., and U. Haussmann. "Bochner's theorem in infinite dimensions." *Pacific J. Math 43* (1972), 601–618.

[FH]   FitzGerald, C. H., and R. A. Horn. "On fractional Hadamard powers of positive definite matrices." *J. Math. Anal. Appl. 61* (1977), 633–642.

[Gasp2] Gasper, G. "Positivity and special functions." In *Theory and Applications of Special Functions.* ed. by R. A. Askey. Academic Press, New York, 1975, 375–433.

[Gold]   Goldberg, R. R. *Fourier Transforms.* Cambridge University Press, 1970.

[Gr]   Gross, L. *Harmonic Analysis on Hilbert Space.* Amer. Math. Soc. Memoirs No. 46. Amer. Math. Soc., Providence, RI, 1963.

[Helg]   Helgason, S. *Groups and Geometric Analysis.* Academic Press, New York, 1984.

[HJ1]   Horn, R. A., and C. R. Johnson. *Matrix Analysis.* Cambridge University Press, Cambridge, 1985.

[Hu]   Husain, T. *Introduction to Topological Groups.* W. B. Saunders, Philadelphia, 1966.

[Ka]   Katznelson, Y. *Harmonic Analysis.* Wiley, New York, 1968. Reprint, Dover, New York, 1976.

[Kol]   Koldobskii, A. L. "Schoenberg's problem on positive definite functions." *St. Petersburg Math. J. 3* (1992) 563–570.

[Kue]   Kuelbs, J. "Positive definite symmetric functions on linear spaces." *J. Math. Anal. Appl. 42* (1973), 413–426.

[Loo]   Loomis, L. *Abstract Harmonic Analysis.* Van Nostrand, New York, 1953.

[Mis]   Misiewicz, J. K. "Positive definite norm dependent functions on ℓ^∞." *Stat. and Probab. Letters 8* (1989), 255–260.

[N]   Naimark, M. A. *Normed Rings.* Wolters-Noordhoff, Groningen, 1964.

[Phe1]   Phelps, R. R. *Lectures on Choquet's Theorem.* Van Nostrand, New York, 1966. Rev. ed., *Ergebnisse der Math.,* Springer-Verlag, Berlin, 1984.

[RS]   Reed, M., and B. Simon. *Methods of Modern Mathematical Physics,* Vol. I, *Functional Analysis.* 2nd ed., Academic Press, New York, 1980.

[Ric]   Richards, D. St. P. "Positive definite symmetric functions on finite dimensional spaces, I." *J. Multivariate Analysis 19* (1986), 280–298. Part II, *Stat. Prob. Letters 3* (1985), 325–329.

[Ru1]   Rudin, W. *Fourier Analysis on Groups.* Interscience, New York, 1963.

[Ru2]   Rudin, W. *Functional Analysis.* McGraw-Hill, New York, 1973.

[Sas]   Sasvári, Z. *Positive Definite and Definitizable Functions.* Akademie Verlag, Berlin, 1994.

[S4]   Schoenberg, I. J. "Metric spaces and positive definite functions." *Trans. Amer. Math. Soc. 44* (1938), 522–536.

[St]   Stewart, J. "Positive definite functions and generalizations, an historical survey." *Rocky Mtn. J. Math. 6* (1976), 409–434.

[SW]   Stein, E. M., and G. Weiss. *Introduction to Fourier Analysis on Euclidean Spaces.* Princeton University Press, Princeton, NJ, 1971.

[Yos]   Yosida, K. *Functional Analysis.* 2nd. ed., Springer-Verlag, New York, 1968.

13

Strictly Positive Definite Functions

In the preceding chapter, we presented some of the classical theory of positive definite functions defined on linear spaces. From the standpoint of interpolation theory, the positive definiteness of a function f is not strong enough; it allows us to conclude only that a matrix $f(x_i - x_j)$ is nonnegative definite. This matrix may therefore be singular. We must require our functions to be *strictly positive definite,* so that the corresponding matrices will be positive definite. Thus, some strengthening of the theorems in Chapter 12 is required. Our first goal is a theorem such as Bochner's, for strict positive definiteness. Some lemmas are needed first. Recall that a point x is an "accumulation point" of a set S if every neighborhood of x contains a point of S other than x. (Notice that x itself may or may not be a member of S.)

> **LEMMA 1.** *If* $\lambda_1, \lambda_2, \ldots, \lambda_n$ *are* n *distinct complex numbers, then the* n *functions*
>
> $$g_k(z) = e^{\lambda_k z} \qquad (1 \le k \le n \, , \, z \in \mathbb{C})$$
>
> *form a linearly independent set (over the complex field) on any domain in* $\mathbb{C}$ *that has a point of accumulation.*

Proof. We proceed by induction. For $n = 1$ the result is obvious, since $g_1(z)$ is not zero anywhere in $\mathbb{C}$. Suppose now that the lemma has been proved for $n = 1, 2, \ldots, m - 1$, and let $f(z) = \sum_{k=1}^{m} c_k g_k(z)$. Suppose that $f(z) = 0$ on a set S in $\mathbb{C}$ that has a point of accumulation. Since f is an entire function, we conclude that $f(z) = 0$ for all $z \in \mathbb{C}$. (See, for example, [T], page 88, or [Ru0], page 225.) Consider the function

$$F(z) = \frac{d}{dz} \, [e^{-\lambda_m z} f(z)] = \frac{d}{dz} \sum_{k=1}^{m} c_k e^{(\lambda_k - \lambda_m)z} = \sum_{k=1}^{m-1} c_k (\lambda_k - \lambda_m) e^{(\lambda_k - \lambda_m)z}$$

Since $f = 0$, we have $F = 0$. By the induction hypothesis, it follows that the coefficients $c_k(\lambda_k - \lambda_m)$ are all zero, for $1 \le k \le m - 1$. Hence $c_k = 0$ for $1 \le k \le m - 1$, and the function f reduces to $f(z) = c_m e^{\lambda_m z}$. Since $f = 0$, $c_m = 0$. Hence all c_k are zero and the set $\{g_1, g_2, \ldots, g_m\}$ is linearly independent on S.   ∎

LEMMA 2. *Let $g(z) = \sum_{k=1}^n c_k e^{\lambda_k z}$, where $z \in \mathbb{C}$. Assume that the complex "frequencies" $\lambda_1, \lambda_2, \ldots, \lambda_n$ are all different, and the complex coefficients $c_1, c_2, \ldots, c_n$ are not all zero. Then the set of zeros of g has no point of accumulation.*

Proof. By Lemma 1, the functions $z \mapsto e^{\lambda_k z}$ form a linearly independent set. It follows that $g \neq 0$. Since g is an entire function (i.e., it is analytic at each point of $\mathbb{C}$), its set of zeros can have no accumulation point. See [T], page 88.   ∎

If X is a topological space, the σ-algebra of Borel sets is defined to be the smallest σ-algebra containing all open sets. A Borel measure is then a measure whose domain is the σ-algebra of Borel sets. In this section we often write "nonnegative Borel measure" because in other contexts we may consider signed measures or complex-valued measures.

In the next theorems, we use the **carrier** of a (nonnegative) Borel measure μ on a topological space X. This is defined to be the set

$$C = X \backslash \bigcup \{\mathcal{O} : \mathcal{O} \text{ is open and } \mu(\mathcal{O}) = 0\}$$

This concept is discussed in [R], page 308 and [HewS], page 122. It is obvious that C is closed, as it is the complement of an open set. It is proved in [HewS] that $\mu(X \backslash C) = 0$, provided that X is locally compact. If C is countable, then the measure is said to be **purely atomic** or **purely discontinuous** [HewS], page 334.

LEMMA 3. *Let X be a topological space, and let μ be a (nonnegative) Borel measure on X. If f is a continuous, nonnegative function on X such that $\int f \, d\mu = 0$, then $f(x) = 0$ on the carrier of μ.*

Proof. Let F denote the carrier of μ, and let $E = \{x : f(x) \neq 0\}$. We want to show that $E \cap F$ is empty. Suppose on the contrary that there exists a point x_0 in $E \cap F$. Select an open neighborhood $\mathcal{O}$ of x_0 and a positive ε such that $f(x) > \varepsilon$ on $\mathcal{O}$. If $\mu(\mathcal{O}) = 0$, then by the definition of the carrier of μ, $\mathcal{O} \subset X \backslash F$. This is contrary to the fact that $x_0 \in \mathcal{O} \cap F$. Thus, we have $\mu(\mathcal{O}) > 0$, and an immediate contradiction:

$$0 = \int_X f \, d\mu \ge \int_{\mathcal{O}} f \, d\mu \ge \varepsilon \int_{\mathcal{O}} d\mu = \varepsilon \mu(\mathcal{O}) > 0 \qquad ∎$$

LEMMA 4. *Let X be a topological space whose topology has a countable base. Let μ be a Borel measure on X. If C is the carrier of μ, then $\mu(X \backslash C) = 0$.*

Proof. Let $\{\mathcal{O}_\lambda : \lambda \in \Lambda\}$ be the collection of all open sets in X such that $\mu(\mathcal{O}) = 0$. By the definition of the carrier of μ, $X \backslash C = \bigcup \mathcal{O}_\lambda$. Let $\mathcal{B}$ be a countable base for the topology of X. Then each $\mathcal{O}_\lambda$ can be expressed as $\mathcal{O}_\lambda = \bigcup_{i=1}^\infty B_{\lambda i}$, where $B_{\lambda i} \in \mathcal{B}$. Since $\mu(\mathcal{O}_\lambda) = 0$, we have $\mu(B_{\lambda i}) = 0$ for all λ and all i. The set $\{B_{\lambda i} : \lambda \in \Lambda, i \in \mathbb{N}\}$ is countable (being a subset of $\mathcal{B}$). Since $X \backslash C = \bigcup \{B_{\lambda i} : \lambda \in \Lambda, i \in \mathbb{N}\}$, we have $\mu(X \backslash C)$ by the countable additivity of the measure μ. ∎

THEOREM 1. *Let μ be a (nonnegative) Borel measure on $\mathbb{R}$ such that $0 < \mu(\mathbb{R}) < \infty$. If the carrier of μ has at least one point of accumulation, then $\hat{\mu}$ is strictly positive definite on $\mathbb{R}$.*

Proof. Assume the hypotheses on μ. Note that $\left| e^{i\lambda y} \right| = 1$ when λ and y are real. Since $\mu(\mathbb{R}) < \infty$, $e^{i\lambda y}$ is μ-integrable. Hence $\hat{\mu}$ is well-defined. To verify that $\hat{\mu}$ is strictly positive definite, let $x_1, x_2, \ldots, x_n$ be distinct points in $\mathbb{R}$ and let $c_1, c_2, \ldots, c_n$ be complex numbers, not all zero. Then

$$\sum_{k=1}^{n} \sum_{j=1}^{n} \bar{c}_k c_j \hat{\mu}(x_k - x_j) = \sum_{k=1}^{n} \sum_{j=1}^{n} \bar{c}_k c_j \int e^{-2\pi i (x_k - x_j) y} \, d\mu(y)$$

$$= \int \sum_{k=1}^{n} \bar{c}_k e^{-2\pi i x_k y} \sum_{j=1}^{n} c_j e^{2\pi i x_j y} \, d\mu(y)$$

$$= \int \left| \sum_{j=1}^{n} c_j e^{2\pi i x_j y} \right|^2 d\mu(y)$$

$$= \int |g(y)|^2 \, d\mu(y)$$

where we have set $g(y) = \sum c_j \exp(2\pi i x_j y)$. Suppose that this final integral is 0. By Lemma 3, $g(y) = 0$ on the carrier of μ. By Lemma 2, the zero set of g has no points of accumulation, and thus the carrier of μ must have the same property, contrary to our hypotheses. ∎

THEOREM 2. *Let f be a nonnegative Borel function on $\mathbb{R}$. If $0 < \int_{\mathbb{R}} f < \infty$, then $\hat{f}$ is strictly positive definite.*

Proof. We reduce this to Theorem 1 by using the measure μ defined (for any Borel set A) by the equation

$$\mu(A) = \int_A f(x) \, dx$$

Let C be the carrier of μ. Since $\mathbb{R}$ has a countable base for its topology, $\mu(\mathbb{R} \backslash C) = 0$ by Lemma 4. Hence $\mu(C) = \mu(\mathbb{R})$. The set C must have a point of accumulation, for if not, C would be a countable set and have Lebesgue measure 0, whence this contradiction:

$$0 = \int_C f(x) \, dx = \mu(C) = \mu(\mathbb{R}) = \int_{\mathbb{R}} f(x) \, dx > 0$$

Now, by Theorem 1, $\hat{\mu}$ is strictly positive definite on $\mathbb{R}$. Furthermore, $\hat{\mu} = \hat{f}$, as we now prove. From the definition of μ, we see that if g is the characteristic function of a Borel set B, then

$$\int_{\mathbb{R}} g(x) \, d\mu(x) = \mu(B) = \int_B f(x) \, dx = \int_{\mathbb{R}} g(x) f(x) \, dx$$

This last equation is true, therefore, for all simple Borel functions and their limits. In particular, if $g(x)$ is $e^{-2\pi i\lambda x}$ we find that $\hat{\mu} = \hat{f}$.   ∎

We can find many strictly positive definite functions on $\mathbb{R}$ by using Theorem 2 and a table of Fourier transforms. The following pairs are taken from [O]. In that volume the Fourier transform is defined by

$$\hat{f}(x) = \int_{-\infty}^{\infty} f(y)e^{ixy}\,dy$$

$f(x) = \begin{cases} 1/2 & \lvert x\rvert \le 1 \\ 0 & \lvert x\rvert > 1 \end{cases}$	$\hat{f}(x) = x^{-1}\sin x$
$f(x) = (1 + x^2)^{-1}/\pi$	$\hat{f}(x) = e^{-\lvert x\rvert}$
$f(x) = e^{-\lvert x\rvert}/2$	$\hat{f}(x) = (1 + x^2)^{-1}$
$f(x) = \pi^{-1/2}e^{-x^2}$	$\hat{f}(x) = e^{-x^2/4}$
$f(x) = (1/2\pi)(1 + x^{-2})$	$\hat{f}(x) = \lvert x\rvert^{-1}(1 - e^{-\lvert x\rvert})$
$f(x) = \operatorname{sech}(\pi x)$	$\hat{f}(x) = \operatorname{sech}(y/2)$
$f(x) = (2x)^{-1}(x^{-1} - \operatorname{csch} x)$	$\hat{f}(x) = \log(1 + e^{-\pi/\lvert x\rvert})$

LEMMA 5. *Let $\{y_i\}$ be a finite set of nonzero points in $\mathbb{R}^s$, $s \ge 2$. Let Q be an orthogonal projection of $\mathbb{R}^s$ onto a subspace of dimension $s - 1$. Then there is an orthogonal matrix U such that $QUy_i \ne 0$ for all i.*

Proof. For a suitable w of norm 1, $Q = I - ww^T$. This map is the orthogonal projection of $\mathbb{R}^s$ onto $w^\perp$. To verify this, observe first that for any x, $Qx \in w^\perp$ because $w^TQx = w^Tx - w^Tww^Tx = w^Tx - w^Tx = 0$. Secondly, one verifies that $x - Qx \perp w^\perp$. Indeed, if $y \in w^\perp$ then $y^T(x - Qx) = y^T(ww^Tx) = 0$.

Next, select a vector v of norm 1 that is not parallel to any y_i nor equal to w. Set $u = \alpha(v - w)$ with $\alpha = \lVert v - w\rVert^{-1}$. Define the orthogonal matrix $U = I - 2uu^T$. One verifies the orthogonality by writing

$$UU^T = (I - 2uu^T)(I - 2uu^T)^T = I - 4uu^T + 4uu^Tuu^T = I$$

It transpires that $Uv = w$. To verify this, we write

$$\begin{aligned}
Uv - w &= v - 2uu^Tv - w \\
&= v - w - 2\alpha^2(v - w)(v - w)^Tv \\
&= (v - w)[1 - 2\alpha^2(v^Tv - w^Tv)] \\
&= (v - w)[1 - \alpha^2(2 - 2w^Tv)] \\
&= (v - w)[1 - \alpha^2(v^Tv + w^Tw - w^Tv - v^Tw)] \\
&= (v - w)[1 - \alpha^2(v - w)^T(v - w)] = 0
\end{aligned}$$

Now we observe that for any vector x, the condition $Qx = 0$ is equivalent to $x = ww^Tx$. Hence if $QUy_i = 0$ then $Uy_i = ww^TUy_i$ and

$$y_i = U^T w w^T U y_i = U^T w y_i^T U^T w = v y_i^T v$$

This shows that y_i is parallel to v, contrary to the definition of v. Hence $QUy_i \neq 0$ for all i. ∎

LEMMA 6. *Let $x_1, x_2, \ldots, x_n$ be n distinct points in $\mathbb{R}^s$. Let $c_1, c_2, \ldots, c_n$ be complex numbers, not all zero. Then the function $g : \mathbb{R}^s \to \mathbb{C}$ defined by*

$$g(x) = \sum_{k=1}^{n} c_k e^{ix_k x} \qquad (x \in \mathbb{R}^s)$$

is different from zero almost everywhere.

Proof. For $s = 1$, the result is contained in Lemma 2. For an inductive proof, suppose the result is true for dimensions $1, 2, \ldots, s - 1$. Let g be given as above. Let Q be the standard projection of $\mathbb{R}^s$ onto $\mathbb{R}^{s-1}$. (It simply deletes the last coordinate of a point.) By Lemma 5 applied to the set of differences $x_i - x_j$, there is an orthogonal matrix U such that QUx_i are all distinct points. Define $h(x) = g(U^T x)$. We shall show that $h(x) \neq 0$ almost everywhere. Write each point Ux_k as a pair (v_k, λ_k) where $v_k \in \mathbb{R}^{s-1}$ and $\lambda_k \in \mathbb{R}$. The points v_k are all distinct. Write $x = (w, \xi)$ with $w \in \mathbb{R}^{s-1}$ and $\xi \in \mathbb{R}$. Thus we have

$$h(x) = g(U^T x) = \sum c_k \exp\{ix_k^T U^T x\} = \sum c_k \exp\{i(Ux_k)^T x\}$$
$$= \sum c_k \exp\{iv_k w + i\lambda_k \xi\} = \sum c_k e^{i\lambda_k \xi} e^{iv_k w}$$

We interpret h as a function on $\mathbb{R}^{s-1} \times \mathbb{R}$, and write $h(x) = h(w, \xi)$. Since the vectors $v_1, v_2, \ldots, v_n$ are distinct elements of $\mathbb{R}^{s-1}$ and since $\sum |c_k e^{i\lambda_k \xi}| = \sum |c_k| > 0$, the induction hypothesis implies that for each ξ, $h(w, \xi) \neq 0$ almost everywhere in w. The following lemma shows that $h(w, \xi) \neq 0$ almost everywhere in (w, ξ). That is, $h(x) \neq 0$ almost everywhere on $\mathbb{R}^s$. ∎

LEMMA 7. *Let $(X, \mathcal{A}, \mu)$ and $(Y, \mathcal{B}, \nu)$ be two σ-finite measure spaces. We denote by $(X \times Y, \mathcal{A} \otimes \mathcal{B}, \mu \times \nu)$ the product measure space. Let f be a measurable function on $X \times Y$. Assume that for almost all x in X, $f_x(y) \neq 0$ almost everywhere on Y. Then $f(x, y) \neq 0$ almost everywhere on $X \times Y$.*

Proof. Let $E = \{(x, y) : f(x, y) = 0\}$. We wish to prove that E has measure 0. By the theory of product measures [HewS], page 384; [Ru0], page 149; and [R], page 257,

(1) $$(\mu \times \nu)(E) = \int \nu(E_x) \, d\mu(x)$$

Note that

$$E_x = \{y \in Y : (x, y) \in E\} = \{y : f_x(y) = 0\} = Z(f_x)$$

Since $f_x(y) \neq 0$ almost everywhere on Y for almost all x, we have $\nu(E_x) = \nu(Z(f_x)) = 0$ for almost all x. Hence the integrand in Equation (1) is 0 almost everywhere on X, and the integral is 0. ∎

THEOREM 3. *Let μ be a nonnegative, finite-valued Borel measure on $\mathbb{R}^s$ whose carrier is not a set of Lebesgue measure 0. Then $\hat{\mu}$ is strictly positive definite on $\mathbb{R}^s$.*

Proof. We proceed as in the proof of Theorem 1, *mutatis mutandis.* Thus

$$\sum\sum \bar{c}_k c_j \hat{\mu}(x_k - x_j) = \int \left|\sum c_j e^{2\pi i x_j y}\right|^2 d\mu(y) = \int_{\mathbb{R}^s} |g(y)|^2 \, d\mu(y)$$

This last expression is clearly nonnegative. To prove that it is positive, suppose that it is 0. Again we conclude that the carrier of μ is contained in the zero set of g. But the latter has Lebesgue measure 0, by Lemma 4. Hence the carrier of μ is of Lebesgue measure 0, contrary to our hypothesis. This contradiction completes the proof.   ■

Problems

1. Let f be a strictly positive definite function on $\mathbb{R}$. Let X be a linear space. Let $\phi_1, \phi_2, \dots, \phi_m$ be elements of X^*. Prove that $\sum_{i=1}^{m} f \circ \phi_i$ is positive definite on X but *not* strictly positive definite.

2. Is the inverse Fourier transform of a positive function (in L^1) strictly positive definite?

3. Let μ be a nonnegative Borel measure on $\mathbb{R}^s$, and let C be its carrier. Prove that for any μ-integrable function f on $\mathbb{R}^s$

$$\int_{\mathbb{R}^s} f \, d\mu = \int_C f \, d\mu$$

4. Let μ be a nonnegative Borel measure on $\mathbb{R}^s$. Prove that the carrier of μ is the smallest closed set whose complement has μ-measure 0. Explain why this assertion may fail for topological spaces other than $\mathbb{R}^s$.

5. Let f be continuous and positive definite on $\mathbb{R}^s$. If f is real-valued, then the Borel measure corresponding to it has a special property. Find the property and prove it.

6. For a function $f : \mathbb{R}^s \to \mathbb{C}$, two types of positive definiteness can be defined, called *PDr* and *PDc*. In both definitions, we consider $A_{kj} = f(x_k - x_j)$. In *PDr* we require $u^T A u \geq 0$ for all real vectors u, and in *PDc* we require $u^* A u \geq 0$ for all complex vectors u. Obviously $PDc \subset PDr$. Give an example of an f in $PDr \setminus PDc$.

7. Adopt the terminology of the preceding problem. Prove that if f belongs to *PDr* on $\mathbb{R}^s$ then for every x, $\frac{1}{2}[f(x) + f(-x)]$ is real and bounded in modulus by $f(0)$.

8. Let X be a topological space, and μ a Borel measure on X such that $\mu(\mathcal{O}) > 0$ for every nonvoid open set $\mathcal{O}$. What is the carrier of μ?

9. A subset S in a topological space X is said to be "discrete" if, for each $x \in S$, the singleton $\{x\}$ is open in the relative topology on S. Prove that this property is equivalent to the assertion that all points of accumulation of S lie in $X \setminus S$.

10. Prove that if a subset S in $\mathbb{R}$ has no point of accumulation, then S is countable.

11. Prove that the function $f(x) = (x^2 + 1)^{-1}$ is strictly positive definite on $\mathbb{R}^s$.

12. Let X and μ be as in Problem 11. Prove that if C is the carrier of μ, then

$$\int_X f(x)\,d\mu(x) = \int_C f(x)\,d\mu(x)$$

for all Borel functions f.

13. Let f be a nonnegative Borel function on $\mathbb{R}^s$. Define $\mu(A)$ for any Borel set A by $\mu(A) = \int_A f(x)\,dx$. Show that for any Borel function g,

$$\int_{\mathbb{R}^s} g(x)f(x)\,dx = \int_{\mathbb{R}^s} g(x)\,d\mu(x)$$

Hint: Prove it first when g is the characteristic function of a Borel set.

14. (Continuation) Let f and μ be as in Problem 13. Let g be a nonnegative continuous function on $\mathbb{R}^s$. Prove that if $\int_{\mathbb{R}^s} g(x)f(x)\,dx = 0$, then g vanishes on the carrier of μ.

15. (Continuation) Give a direct proof of Theorem 2 without invoking Theorem 1. (Lemma 2 will be useful.)

References

[Chan2] Chang, Kuei-Fang. "Strictly positive definite functions." *J. Approx. Theory 87* (1996), 148–158. Addendum, *J. Approx. Theory 88* (1997), 384.

[HewS] Hewitt, E., and K. Stromberg. *Real and Abstract Analysis.* Springer-Verlag, New York, 1965.

[O] Oberhettinger, F. *Fourier Transforms of Distributions and Their Inverses.* Academic Press, New York, 1973.

[R] Royden, H. L. *Real Analysis.* 2nd ed., Macmillan, New York, 1968.

[Ru0] Rudin, W. *Real and Complex Analysis.* 2nd ed., McGraw-Hill, New York, 1974.

[T] Titchmarsh, E. C. *The Theory of Functions.* 2nd ed., Oxford University Press, London, 1939.

14
Completely Monotone Functions

Some of the theory of completely monotone functions is necessary for our later work on radial basis functions in Chapter 15. We begin with a definition.

Definition. A function f is said to be **completely monotone** on $[0, \infty)$ if

    1. $f \in C[0, \infty)$
    2. $f \in C^\infty(0, \infty)$
    3. $(-1)^k f^{(k)}(t) \geq 0$ for $t > 0$ and $k = 0, 1, 2, \dots$

Such functions exist in great abundance. Here are some examples that can be quickly verified directly from the definition:

    1. $f(t) = a \qquad (a \geq 0)$
    2. $f(t) = (t + a)^b \qquad (a > 0 \geq b)$
    3. $f(t) = e^{-at} \qquad (a \geq 0)$

A famous theorem of Bernstein and Widder gives a complete characterization of this function class. This theorem states, in effect, that a function is completely monotone if and only if it is the Laplace transform of a nonnegative bounded Borel measure.

    The theorem has an equivalent formulation in terms of the Riemann-Stieltjes integral. If γ is a nondecreasing function, we define

$$\int_a^b f(t)\, d\gamma(t) = \lim \sum_{i=1}^n f(\xi_i)[\gamma(t_i) - \gamma(t_{i-1})]$$

The limit here is similar to the one used in the Riemann integral. Thus the limit is equal to a number L if for each $\varepsilon > 0$ there corresponds a $\delta > 0$ such that for any partition $a = t_0 < t_1 < \cdots < t_n = b$ satisfying $\max_i |t_i - t_{i-1}| < \delta$ and for any points ξ_i satisfying $t_{i-1} \leq \xi_i \leq t_i$ we have

$$\left| L - \sum_{i=1}^n f(\xi_i)[\gamma(t_i) - \gamma(t_{i-1})] \right| < \varepsilon$$

Of course, as in the Riemann integral, the defining limit may not exist in some cases. (It does exist if f is continuous.) Notice that when $\gamma(x) = x$, we recover the familiar Riemann integral. Introductions to the Riemann-Stieltjes integral can be found in Widder's book [W4] and in Hewitt-Stromberg [HewS].

> **THEOREM 1.** (Bernstein–Widder) *A function* $f : [0, \infty) \to [0, \infty)$ *is completely monotone if and only if there is a nondecreasing bounded function* γ *such that* $f(t) = \int_0^\infty e^{-st} \, d\gamma(s)$.

Proof. The easier half of the proof is to show that if γ is as stated then the integral defines a completely monotone function. The derivatives of f are obtained by differentiating under the integral. (The validity of this procedure is addressed in Theorem 5.) We obtain

$$f^{(k)}(t) = \int_0^\infty (-s)^k e^{-st} \, d\gamma(s)$$

The sign of $f^{(k)}(t)$ is clearly $(-1)^k$. To test the continuity of f at 0, note first that

$$f(0) = \int_0^\infty d\gamma(s) = \gamma(\infty) - \gamma(0)$$

On the other hand, by the Monotone Convergence Theorem,

$$\lim_{t \downarrow 0} f(t) = \lim_{t \downarrow 0} \int_0^\infty e^{-st} \, d\gamma(s) = \int_0^\infty \lim_{t \downarrow 0} e^{-st} \, d\gamma(s) = \int_0^\infty d\gamma(s) = \gamma(\infty) - \gamma(0)$$

For the other half of the proof, suppose that f is completely monotone on $[0, \infty)$. As explained later, the sequence $f(n/m)$, $(n = 0, 1, 2, \dots)$ is completely monotone for any $m \in \mathbb{N}$. By the Hausdorff Moment Theorem (Theorem 2 below), there is a nondecreasing bounded function β_m such that

$$f\left(\frac{n}{m}\right) = \int_0^1 s^n \, d\beta_m(s) \qquad (n = 0, 1, 2, \dots)$$

Also, from Theorem 2 we can assume that $\beta_m(0) = 0$ and that for every s, $\beta_m(s) = \frac{1}{2}[\beta_m(s+0) + \beta_m(s-0)]$. Replacing n by nm in an equation above, we have

$$f(n) = \int_0^1 s^{nm} \, d\beta_m(s) = \int_0^1 s^n \, d\beta_m(s^{1/m})$$

By the uniqueness part of Theorem 2, $\beta_m(s^{1/m}) = \beta_1(s)$. Putting $\gamma(s) = -\beta_1(e^{-s})$ and $s = e^{-\sigma}$, we have

$$f\left(\frac{n}{m}\right) = \int_0^1 s^{n/m} \, d\beta_1(s) = \int_{0+}^1 s^{n/m} \, d\beta_1(s)$$

$$= \int_\infty^0 e^{-\sigma n/m} \, d\beta_1(e^{-\sigma}) = \int_0^\infty e^{-ns/m} \, d\gamma(s)$$

By continuity, this leads to

$$f(t) = \int_0^\infty e^{-st} \, d\gamma(s) \qquad\qquad \blacksquare$$

The measure-theoretic version of the Bernstein-Widder Theorem states that f is completely monotone on $[0, \infty)$ if and only if it is the Laplace transform of a nonnegative, finite-valued, regular, Borel measure on $[0, \infty)$. The term **regular** when applied to a nonnegative measure ν means that

$$\nu(A) = \sup \nu(K) = \inf \nu(O)$$

where K ranges over the compact sets contained in A, and O ranges over the open sets containing A.

Some of the results needed in the preceding proof are given here. For a sequence $\mu = [\mu_0, \mu_1, \mu_2, \ldots]$, the forward difference operation is defined by

$$\Delta\mu = [\mu_1 - \mu_0, \mu_2 - \mu_1, \ldots]$$

This defines Δ as a linear operator, and its powers are defined in the usual way. If $(-1)^k \Delta^k \mu \geq 0$ for $k = 0, 1, 2, \cdots$, then μ is said to be a **completely monotone sequence.** Examples are $\mu_n = (n + 1)^{-1}$ and $\mu_n = \lambda^n$, if $0 < \lambda \leq 1$. The famous Moment Theorem of Hausdorff is as follows:

> **THEOREM 2.** (Hausdorff) *In order that a sequence* $[\mu_0, \mu_1, \ldots]$ *be completely monotone, it is necessary and sufficient that it be the moment sequence of a nondecreasing bounded function β on* $[0, 1]$; *that is,*
>
> $$\mu_n = \int_0^1 t^n \, d\beta(t)$$
>
> *If we insist that $\beta(0) = 0$ and $\beta(t) = \frac{1}{2}[\beta(t + 0) + \beta(t - 0)]$ for every t, then β is uniquely determined by the sequence μ.*

The forward difference operator is also defined for functions:

$$(\Delta f)(x) = f(x + 1) - f(x)$$

Its powers obey the equation

$$(\Delta^k f)(x) = \sum_{j=0}^{k} \binom{k}{j}(-1)^{k-j} f(x + j) = \sum_{j=0}^{k} \binom{k}{j}(-1)^j f(x + k - j)$$

See [AS], page 882 or [Stef], page 10. Since these operations are special cases of divided differences, we can write (using standard notation for divided differences)

$$(\Delta f)(x) = f[x, x + 1]$$
$$(\Delta^2 f)(x) = f[x, x + 1, x + 2], \text{ etc.}$$

If $f^{(k)}$ exists and is continuous, then for a suitable ξ between x and $x + k$,

$$(\Delta^k f)(x) = f[x, x + 1, \ldots, x + k] = \frac{1}{k!} f^{(k)}(\xi)$$

See, for example, [KinC], page 357. These remarks allow us to conclude that if f is completely monotone, then (for each m) the sequence $f(n/m)$ is completely monotone. Indeed, we can let $F(x) = f(x/m)$ and $\mu_n = f(n/m)$ so that

$$(\Delta^k \mu)_n = (\Delta^k F)(n) = \frac{1}{k!} F^{(k)}(\xi) = \frac{1}{k! m^k} f^{(k)}(\xi/m)$$

The sign of this last term is $(-1)^k$, by the complete monotonicity of f.

Examples. By taking various functions γ, we can generate interesting completely monotone functions:

    1. If $\gamma(s) = 1 - e^{-s}$ then $f(t) = (t + 1)^{-1}$.

    2. If $\gamma(s) = s$ for $0 \le s \le 1$ and $\gamma(s) = 1$ for $s > 1$, then $f(t) = (1 - e^{-t})/t$

One can obtain further examples by using the following theorems.

THEOREM 3. *The family of all completely monotone functions on* $[0, \infty)$ *is algebraically closed under the formation of linear combinations with positive coefficients.*

THEOREM 4. *If f and g are completely monotone on* $[0, \infty)$, *then so is fg.*

Proof. By Leibniz's rule,

$$D^k(fg) = \sum_{j=0}^{k} \binom{k}{j} (D^j f)(D^{k-j}g)$$

The sign of $(D^j f)(D^{k-j}g)$ is $(-1)^j(-1)^{k-j} = (-1)^k$.   ■

Now we address the question of whether "differentiation under the integral" is a valid procedure. Thus, we seek suitable hypotheses to make the following equation correct:

$$(1) \qquad \frac{d}{dt} \int_X f(x, t)\, d\mu(x) = \int_X \frac{\partial f}{\partial t}(x, t)\, d\mu(x)$$

The setting is as follows. A measure space $(X, \mathcal{A}, \mu)$ is prescribed. Thus X is a set, $\mathcal{A}$ is a σ-algebra of subsets of X, and $\mu : \mathcal{A} \to [0, \infty]$ is a measure. An open interval (a, b) is also prescribed. The function f is defined on $X \times (a, b)$ and takes values in $\mathbb{R}$. Select a point t_0 in (a, b) where $(\partial f/\partial t)(x, t_0)$ exists for all x. What further assumptions are needed in order that Equation (1) shall be true for $t = t_0$? Let us assume that

(2)   For each t in (a, b), the function $x \mapsto f(x, t)$ belongs to $L^1(X, \mathcal{A}, \mu)$.

(3)   There exists a function $g \in L^1(X, \mathcal{A}, \mu)$ such that

$$\left| \frac{f(x, t) - f(x, t_0)}{t - t_0} \right| \le g(x) \qquad (x \in X, a < t < b, t \ne t_0)$$

THEOREM 5. *Under the hypotheses given above, Equation* (1) *is true for the point $t = t_0$.*

Proof. By Hypothesis (2) we are allowed to define

$$\phi(t) = \int_X f(x, t)\, d\mu(x)$$

Now the derivative $\phi'(t_0)$ exists if and only if for each sequence t_n converging to t_0 we have

$$\phi'(t_0) = \lim_{n \to \infty} \frac{\phi(t_n) - \phi(t_0)}{t_n - t_0} = \lim_{n \to \infty} \int_X \frac{f(x, t_n) - f(x, t_0)}{t_n - t_0}\, d\mu(x)$$

By Hypothesis (3), the integrands in the preceding equation are bounded in magnitude by the single L^1-function g. The Lebesgue Dominated Convergence Theorem allows an interchange of limit and integral. Hence

$$\phi'(t_0) = \int_X \lim_{n \to \infty} \frac{f(x, t_n) - f(x, t_0)}{t_n - t_0} \, d\mu(x) = \int_X \frac{\partial f}{\partial t}(x, t_0) \, d\mu(x) \qquad \blacksquare$$

This proof is given by Bartle [Bart1]. Related theorems can be found in advanced calculus books, such as [W4], page 352, and in McShane's book [McS]. A useful corollary of Theorem 5 is as follows.

THEOREM 6. *Let $(X, \mathcal{A}, \mu)$ be a measure space such that $\mu(X) < \infty$. Let $f : X \times (a, b) \to \mathbb{R}$. Assume that for each n, $(\partial^n f / \partial t^n)(x, t)$ exists, is measurable, and is bounded on $X \times (a, b)$. Then*

$$(4) \qquad \frac{d^n}{dt^n} \int_X f(x, t) \, d\mu(x) = \int_X \frac{\partial^n f}{\partial t^n}(x, t) \, d\mu(x) \qquad (n = 1, 2, \ldots)$$

Proof. Since $\mu(X) < \infty$, any bounded measurable function on X is integrable. To see that Hypothesis (3) of the preceding theorem is true, use the mean value theorem:

$$\left| \frac{f(x, t) - f(x, t_0)}{t - t_0} \right| = \left| \frac{\partial f}{\partial t}(x, \xi) \right| \leq M$$

where M is a bound for $|\partial f / \partial t|$ on $X \times (a, b)$. By the preceding theorem, Equation (4) is valid for $n = 1$. The same argument can be repeated to give an inductive proof for all n.
$\blacksquare$

Further references on the Bernstein-Widder Theorem are [Be]; [Cho], vol. 2, page 239; [Phe1], page 11; [Phe2], page 155; [SG]; [W1]; [W3], page 157; and [W5], page 162. For the moment problem, see [W5], [ST], [Lan], and [Ak1]. For the Riemann-Stieltjes integral, consult [W4].

Problems

1. Prove that for any nondecreasing bounded function γ, the function $f(t) = \int_1^\infty (t + s)^{-1} \, d\gamma(s)$ is completely monotone on $[0, \infty)$. Generalize this result.

2. Prove that if $0 < a < b$ then the function

$$f(t) = \log \frac{t + b}{t + a}$$

is completely monotone on $[0, \infty)$.

3. Determine whether the function $f(t) = \pi/2 - \tan^{-1} t$ is completely monotone on $[0, \infty)$.

4. Prove that if g is a nonnegative member of $L^1[0, \infty)$, then its Laplace transform is completely monotone on $[0, \infty)$.

5. Prove that a polynomial of degree one or more cannot be completely monotone on $[0, \infty)$.

6. A function f is defined to be completely monotone on $(0, \infty)$ if $f \in C^\infty(0, \infty)$ and $(-1)^k f^{(k)}(t) \geq 0$ for all k and t. Verify that the function $f(t) = t^{-1}$ is completely monotone on $(0, \infty)$ but not on $[0, \infty)$.

7. (Continuation) Prove that if γ is nondecreasing on $[0, \infty)$ and if the integral $\int_0^\infty e^{-st}\, d\gamma(s)$ exists for all t, then the resulting function of t is completely monotone on $(0, \infty)$.

8. What must be assumed of the function g if the function $t \mapsto g(-t)$ is to be completely monotone on $[0, \infty)$?

9. What must be assumed of the function g if $f \circ g$ is to be completely monotone on $[0, \infty)$ whenever f is completely monotone on $[0, \infty)$?

10. If f is completely monotone on $[0, \infty)$, does it follow that the function $t \mapsto f(\sqrt{t})$ is completely monotone on $[0, \infty)$?

11. Let f be completely monotone on $[0, \infty)$. Prove that $\lim_{t \to \infty} t^k f^{(k)}(t) = 0$ for $k = 1, 2, 3, \ldots$. (Stronger results are known. For example, the function $t \mapsto t^k f^{(k)}(t)$ is integrable. See a paper by R. E. Williamson in *Duke J. Math. 23* (1956), 189–207, or the book [SG]. What is the value of the limit when $k = 0$?

12. By considering the definition of the Riemann-Stieltjes integral, show that if γ has a jump discontinuity of magnitude c at t_0, then the integral $\int_a^b f(t)\, d\gamma(t)$ will contain a term $cf(t_0)$.

13. (Continuation) Show that at a point of discontinuity of γ, say t_0, the integral $\int f(t)\, d\gamma(t)$ is not changed by redefining γ at t_0 by the equation $\gamma(t_0) = \frac{1}{2}[\gamma(t_0+) + \gamma(t_0-)]$.

14. Prove the formula for $\Delta^k f$ given in the text.

15. A function f is said to be **absolutely monotone** on an interval (a, b) if $f^{(k)}(t) \geq 0$ for $k = 0, 1, 2, \ldots$ and for all t in (a, b). Prove that a function g is completely monotone on $(0, \infty)$ if and only if the function $t \mapsto g(-t)$ is absolutely monotone on $(-\infty, 0)$.

16. Prove Theorem 3.

17. Carry out the inductive proof that is needed to establish Theorem 6.

18. Prove that the family CM of all completely monotone functions on $[0, \infty)$ is closed under translation by a positive number. Thus if $f \in CM$ and $c > 0$, then $t \mapsto f(t + c)$ is also in CM.

19. Prove that a function f is completely monotone on $(0, \infty)$ if and only if there exists a Borel measure μ on $(0, 1]$ such that $f(t) = \int_0^1 x^t\, d\mu(x)$. (The change of variable $x = e^{-y}$ is useful.)

References

[AS]   Abramowitz, M., and I. A. Stegun. *Handbook of Mathematical Functions.* National Bureau of Standards, Washington, 1964. Reprint, Dover, New York.

[Ak1]   Akhiezer, N. I. *The Classical Moment Problem.* New York, 1955.

[Bart1]   Bartle, R. G. *Elements of Integration Theory.* Wiley, New York, 1966. Reprinted, enlarged, and retitled: *Elements of Integration and Lebesgue Measure,* 1995.

[Be]   Bernstein, S. N. "Sur les fonctions absolument monotones." *Acta Math. 52* (1929), 1–66.

[Cho]   Choquet, G. *Lectures on Analysis* (3 vols). W. A. Benjamin, New York, 1969.

[HewS]   Hewitt, E., and K. Stromberg. *Real and Abstract Analysis.* Springer-Verlag, New York, 1965.

[KinC]   Kincaid, D., and W. Cheney. *Numerical Analysis.* 2nd ed., Brooks/Cole, Pacific Grove, CA, 1996.

[Lan]   Landau, H. J. (ed.). *Moments in Mathematics.* Amer. Math. Soc., Providence, RI, 1987.

[McS]   McShane, E. J. *Integration.* Princeton University Press, Princeton, NJ, 1944.

[Phe1]   Phelps, R. R., "Lectures on Choquet's Theorem." Van Nostrand, New York, 1966. Rev. ed., *Ergebnisse der Math.* Springer-Verlag, Berlin, 1984.

[Phe2]   Phelps, R. R. "Integral representation for elements of convex sets." In *Studies in Functional Analysis,* ed. by R. G. Bartle. Math. Assoc. of America, 1980, 115–157.

[SG]   Shilov, G. E., and B. L. Gurevich. *Integral, Measure and Derivative: A Unified Approach.* Prentice-Hall, Englewood Cliffs, NJ, 1966.

[ST]   Shohat, J. A., and J. D. Tamarkin. *The Problem of Moments.* Amer. Math. Soc., Providence, RI, 1943.

[Stef]   Steffensen, J. F. *Interpolation.* Chelsea, New York, 1950.

[W1]   Widder, D. V. "Necessary and sufficient conditions for the representation of a function as a Laplace integral." *Trans. Amer. Math. Soc. 33* (1932), 851–892.

[W3]   Widder, D. V. *An Introduction to Transform Theory.* Pure and Applied Mathematics series, vol. 42. ed. by P. A. Smith and S. Eilenberg. Academic Press, New York, 1971.

[W4]   Widder, D. V. *Advanced Calculus.* 2nd ed., Prentice-Hall, Englewood Cliffs, NJ, 1961. Reprint, Dover, New York.

[W5]   Widder, D. V. *The Laplace Transform.* Princeton University Press, Princeton, NJ, 1946.

15

The Schoenberg Interpolation Theorem

In this chapter we continue our study of positive definite functions, but concentrate now on those having radial symmetry. A real-valued function F on an inner-product space is said to be **radial** if $F(x) = F(y)$ whenever $\|x\| = \|y\|$. If this property is present, the value of $F(x)$ depends only on $\|x\|$, and consequently there exists another function $f : [0, \infty) \to \mathbb{R}$ such that

$$F(x) = f(\|x\|^2)$$

Our principal objective in this chapter is to establish the following theorem of I. J. Schoenberg (1938), published in [S1].

> **THEOREM 1.** *If f is completely monotone but not constant on $[0, \infty)$, then the function $x \mapsto f(\|x\|^2)$ is a radial, strictly positive definite function on any inner-product space. Thus, for any n distinct points $x_1, x_2, \ldots, x_n$ in such a space the matrix $A_{ij} = f(\|x_i - x_j\|^2)$ is positive definite (and therefore nonsingular).*

This theorem provides a rich source of functions that are suitable for interpolation of data in the Euclidean spaces $\mathbb{R}^1, \mathbb{R}^2, \ldots$. Before we can give the proof, some preliminary results will be needed. Many of these are of independent interest.

Recall that an inner-product space is a normed linear space X in which an inner product has been defined and obeys these axioms

(1) $$\langle x, y \rangle \in \mathbb{C}$$

(2) $$\langle x, y \rangle = \overline{\langle y, x \rangle}$$

(3) $$\langle x, x \rangle = \|x\|^2$$

101

(4) $$\langle x + y, z \rangle = \langle x, z \rangle + \langle y, z \rangle$$

(5) $$\langle \lambda x, y \rangle = \lambda \langle x, y \rangle$$

From the axioms it follows that $\langle x, \lambda y \rangle = \bar{\lambda}\langle x, y \rangle$. The space $\mathbb{R}^s$ is an inner-product space in which $\langle x, y \rangle = \sum_{i=1}^{s} \xi_i \eta_i$. Here, $x = (\xi_1, \xi_2, \ldots, \xi_s)$ and $y = (\eta_1, \eta_2, \ldots, \eta_s)$.

In these Euclidean spaces, we usually write xy in place of the more cumbersome notation $\langle x, y \rangle$.

LEMMA 1. $\int_{-\infty}^{\infty} e^{-x^2}\, dx = \sqrt{\pi}$.

Proof. (This is accomplished by a notorious trick.) Denote the integral by I. Then, with polar coordinates, we have

$$I^2 = \int_{-\infty}^{\infty} e^{-x^2}\, dx \int_{-\infty}^{\infty} e^{-y^2}\, dy = \int\int_{\mathbb{R}^2} e^{-(x^2+y^2)}\, dx\, dy = \int_0^{2\pi} \int_0^{\infty} e^{-r^2} r\, dr\, d\theta$$

$$= 2\pi \int_0^{\infty} \left(-\frac{1}{2}\right) e^{-r^2} (-2r\, dr) = -\pi \left[e^{-r^2}\right]_{r=0}^{r=\infty} = \pi$$

Hence $I = \sqrt{\pi}$. (One can avoid polar coordinates. See Problem 12.)   ∎

For univariate functions we adopt this definition of the Fourier transform:

$$\hat{f}(x) = \int_{-\infty}^{\infty} f(y) e^{-2\pi i xy}\, dy \qquad (x, y \in \mathbb{R})$$

This is the definition in the standard reference [SW]. Another standard reference, [Ru2], uses, however,

$$\hat{f}(x) = \frac{1}{\sqrt{2\pi}} \int_{-\infty}^{\infty} f(y) e^{-ixy}\, dy$$

Each definition has its advantages, disadvantages and adherents. Since these two (and other) definitions are often encountered, the reader must be alert to possible misunderstanding.

The next result states that the function $x \mapsto e^{-\pi x^2}$ is a fixed point of the Fourier transform. Functions of the form $x \mapsto e^{-\alpha x^2}$ are called **Gaussian** functions.

LEMMA 2. $\int_{-\infty}^{\infty} e^{-\pi x^2} e^{-2\pi i xy}\, dx = e^{-\pi y^2}$ (for all $y \in \mathbb{R}$)

Proof. This important formula is proved in [SW], page 6; [Ru2], page 170; [Gold], page 43; [RS], page 321; and [DMc], page 53. The proofs in [Ru2] and [DMc] are similar to the one presented here.

Define

$$F(x) = \int_{-\infty}^{\infty} e^{-y^2} e^{-2ixy}\, dy \qquad (x \in \mathbb{R})$$

Differentiation under the integral sign gives us

$$F'(x) = i \int_{-\infty}^{\infty} e^{-y^2} e^{-2ixy} (-2y) \, dy$$

Integration by parts can be used here if we first note that

$$\int e^{-y^2} (-2y \, dy) = e^{-y^2}$$

Hence

$$F'(x) = i \left\{ e^{-y^2} e^{-2ixy} \Big|_{-\infty}^{\infty} - \int_{-\infty}^{\infty} e^{-y^2} e^{-2ixy} (-2ix) \, dy \right\}$$

$$= -2x \int_{-\infty}^{\infty} e^{-y^2} e^{-2ixy} \, dy = -2x F(x)$$

From the definition of F, and with Lemma 1, we obtain

$$F(0) = \int_{-\infty}^{\infty} e^{-y^2} \, dy = \sqrt{\pi}$$

We easily solve the initial-value problem $F'(x) = -2xF(x)$, with $F(0) = \sqrt{\pi}$, obtaining the unique solution

$$F(x) = \sqrt{\pi} \, e^{-x^2}$$

Thus, we have proved

$$\int_{-\infty}^{\infty} e^{-y^2} e^{-2ixy} \, dy = \sqrt{\pi} \, e^{-x^2}$$

In this equation, substitute $x = \sqrt{\pi} \, u$ and $y = \sqrt{\pi} \, v$. The result is

$$\int_{-\infty}^{\infty} e^{-\pi v^2} e^{-2\pi i u v} \sqrt{\pi} \, dv = \sqrt{\pi} \, e^{-\pi u^2}$$

This is equivalent to the equation we sought to prove.   ∎

The multivariate Fourier transform is defined by the same formula used for univariate functions, but now x and y are vectors (points) in $\mathbb{R}^s$, and the notation xy signifies the inner product of x and y. Thus if $x = (\xi_1, \xi_2, \ldots, \xi_s)$ and $y = (\eta_1, \eta_2, \ldots, \eta_s)$, then $xy = \sum_{i=1}^{s} \xi_i \eta_i$. Consequently $x^2 = xx = \langle x, x \rangle = \|x\|^2$.

The next result asserts that the multivariate function $x \mapsto e^{-\pi \|x\|^2}$ is a fixed point of the Fourier transform. Since $x^2 = \|x\|^2$, this new result is notationally the same as Lemma 2.

LEMMA 3. *For $y \in \mathbb{R}^s$,*

$$\int_{\mathbb{R}^s} e^{-\pi \|x\|^2} e^{-2\pi i xy} \, dx = e^{-\pi \|y\|^2}$$

Proof. If we write $x = (\xi_1, \xi_2, \ldots, \xi_s)$ and $y = (\eta_1, \eta_2, \ldots, \eta_s)$, then the preceding integral becomes

$$\int_{\mathbb{R}^s} \exp\left\{-\pi \sum_{j=1}^{s} \xi_j^2\right\} \exp\left\{-2\pi i \sum_{j=1}^{s} \xi_j \eta_j\right\} d\xi_1\, d\xi_2 \cdots d\xi_s = \prod_{j=1}^{s} \int_{-\infty}^{\infty} e^{-\pi \xi_j^2} e^{-2\pi i \xi_j \eta_j}\, d\xi_j$$

By Lemma 2, this last expression becomes

$$\prod_{j=1}^{s} e^{-\pi \eta_j^2} = \exp\left\{-\pi \sum_{j=1}^{s} \eta_j^2\right\} = e^{-\pi \|y\|^2}$$

We have used a general principle given in Problem 9.   ∎

LEMMA 4. *For x and y in $\mathbb{R}^s$, and $\alpha > 0$,*

$$\int_{\mathbb{R}^s} \left(\frac{\pi}{\alpha}\right)^{s/2} e^{-(\pi^2/\alpha)\|x\|^2} e^{-2\pi i x y}\, dx = e^{-\alpha\|y\|^2}$$

Proof. In Lemma 3, make the substitutions

$$x = \sqrt{\frac{\pi}{\alpha}}\, u \qquad y = \sqrt{\frac{\alpha}{\pi}}\, v \qquad dx = \left(\frac{\pi}{\alpha}\right)^{-s/2} du$$ ∎

THEOREM 2. *If $\alpha > 0$, then the function $x \mapsto e^{-\alpha\|x\|^2}$ is radial and strictly positive definite on any real inner-product space.*

Proof. Select n distinct points $x_1, x_2, \ldots, x_n$ in the inner-product space. These n points span a finite-dimensional real inner-product space, which must be isomorphic to a Euclidean space $\mathbb{R}^s$. We may therefore assume that these points are in $\mathbb{R}^s$. In order to prove that the matrix $A_{ij} = \exp\left(-\alpha\|x_i - x_j\|^2\right)$ is positive definite, it suffices to prove that the function $f : \mathbb{R}^s \to \mathbb{R}$ defined by $f(x) = e^{-\alpha\|x\|^2}$ is strictly positive definite. To do this, we use Theorem 3 in Chapter 13 (page 92). According to that theorem, it is enough to verify that f is the Fourier transform of a measure on $\mathbb{R}^s$ whose carrier has positive Lebesgue measure. By Lemma 4, every Gaussian function is the Fourier transform of another. Specifically, $x \mapsto e^{-\alpha\|x\|^2}$ is the Fourier transform of the measure μ defined by

$$d\mu(x) = \left(\frac{\pi}{\alpha}\right)^{s/2} \exp\left(-\frac{\pi^2}{\alpha}\|x\|^2\right) dx$$

Since the Gaussian is positive, the carrier of μ is $\mathbb{R}^s$.   ∎

The theorem just proved is of independent interest in the interpolation problem. It asserts that Gaussian functions can be used for interpolation in $\mathbb{R}^s$. It can be proved directly, as suggested in Problem 13.

Example. Consider this interpolation problem:

nodes in $\mathbb{R}^2$	(3, 1)	(1, 2)	(4, 1)	(3, 3)	(1, 4)
function values	2	4	2	3	5

We solved this with the Gaussian function $e^{-\alpha\|x\|^2}$, using first $\alpha = 10$ and then $\alpha = 1$. The surfaces are shown in Figures 15.1 and 15.2. The effect of the parameter α is quite evident graphically. The computing and graphics were done in Mathematica.

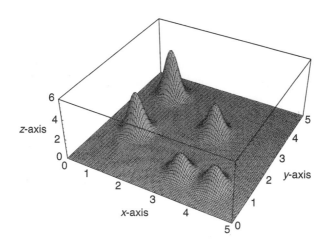

Figure 15.1 *Interpolation by Gaussian Functions, $\alpha = 10$.*

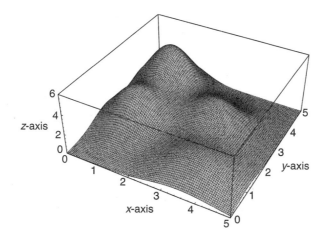

Figure 15.2 *Interpolation by Gaussian Functions, $\alpha = 1$.*

THEOREM 3. *If $x_1, x_2, \ldots, x_n$ are n distinct points in a real inner-product space, then the matrix having elements $\exp(x_i x_j)$ is positive definite.*

Proof. Let $A_{ij} = \exp\{-\|x_i - x_j\|^2\}$. Then

$$A_{ij} = \exp\{-\|x_i\|^2 + 2x_i x_j - \|x_j\|^2\}$$
$$= \exp\{-\|x_i\|^2\}\exp\{2x_i x_j\}\exp\{-\|x_j\|^2\}$$
$$= (D\,E\,D)_{ij}$$

where $D = \text{diag}\left(\exp(-\|x_i\|^2)\right)$ and $E_{ij} = \exp(2x_i x_j)$. Since A is positive definite (by the preceding theorem) and D is nonsingular, we see that E is positive definite. (Problem 8.) ∎

Proof of Theorem 1. Now we can prove the theorem of Schoenberg cited at the beginning of this chapter. Since f is completely monotone, we can invoke the Bernstein-Widder Theorem of Chapter 14 to conclude that there exists a bounded Borel measure μ such that

$$f(t) = \int_0^\infty e^{-st}\, d\mu(s) \qquad (0 \le t < \infty)$$

Since f is not constant, $\mu((0, \infty)) > 0$. Hence, the measure $d\mu$ is not concentrated at 0. If $c = (c_1, c_2, \ldots, c_n) \ne 0$, then with $A_{ij} = f(\|x_i - x_j\|^2)$ we have

$$c^* A c = \sum_{i=1}^n \sum_{j=1}^n \bar{c}_i c_j f(\|x_i - x_j\|^2)$$
$$= \sum_{i=1}^n \sum_{j=1}^n \bar{c}_i c_j \int_0^\infty \exp\{-s\|x_i - x_j\|^2\}\, d\mu(s)$$
$$= \int_0^\infty \sum_{i=1}^n \sum_{j=1}^n \bar{c}_i c_j \exp\{-\|\sqrt{s}x_i - \sqrt{s}x_j\|^2\}\, d\mu(s)$$

For $s > 0$, the integrand is positive, by Theorem 2. Since the measure μ is not concentrated at 0, the integral is positive. ∎

Examples. The results of this chapter indicate that if $x_1, x_2, \ldots, x_n$ are any n distinct points in $\mathbb{R}^s$, then we can interpolate arbitrary data at these nodes by functions of the form

$$1.\ \sum_{j=1}^n \frac{c_j}{\sqrt{1 + \|x - x_j\|^2}} \qquad 3.\ \sum_{j=1}^n c_j e^{xx_j}$$

$$2.\ \sum_{j=1}^n c_j e^{-\|x - x_j\|^2} \qquad 4.\ \sum_{j=1}^n \frac{c_j}{1 + \|x - x_j\|^2}$$

Problems

1. Let $\lambda_1 = 0$ and $\lambda_2 = 2\pi i$. According to Lemma 1 of Chapter 13, the functions $t \mapsto e^{\lambda_k t}$ $(k = 1, 2)$ form a linearly independent set. Reconcile this with the equation
$$e^{\lambda_2 t} = e^{2\pi i t} = (e^{2\pi i})^t = 1^t = 1 = e^{\lambda_1 t}$$

2. Prove that the Fourier transform of a radial function on $\mathbb{R}^s$ is radial. It may be easier to prove a more general statement: If f has the property $f(x) = f(Ax)$ for every unitary matrix, then $\hat{f}$ has the same property.

3. Prove that arbitrary data at points $x_1, x_2, \ldots, x_n$ in $\mathbb{R}^s$ can be interpolated by an "inverse multiquadric function" $f(x) = \sum_{j=1}^{n} c_j (1 + \|x - x_j\|^2)^{-1/2}$.

4. Repeat Problem 3 using $f(x) = \sum_{j=1}^{n} c_j \log\left[(2 + \|x - x_j\|^2)/(1 + \|x - x_j\|^2)\right]$.

5. Repeat Problem 3 using $f(x) = \sum_{j=1}^{n} c_j e^{xx_j}$.

6. Find some other fixed points of the Fourier transform.

7. Prove that the rank of the Gram matrix of $\{x_1, x_2, \ldots, x_n\}$ is the dimension of the linear span of that set.

8. Prove that if A is positive definite and B is nonsingular, then B^*AB is positive definite.

9. Let f be a function on $\mathbb{R}^s$ that is a product of univariate functions:
$$f(\xi_1, \ldots, \xi_s) = f_1(\xi_1) \cdots f_s(\xi_s)$$
Prove that $\hat{f}(\xi_1, \ldots, \xi_s) = \hat{f}_1(\xi_1) \cdots \hat{f}_s(\xi_s)$.

10. Let $f(t) = \sum_{j=1}^{n} c_j e^{\lambda_j t}$ where $\lambda_1 < \lambda_2 < \cdots < \lambda_n$. What are necessary and sufficient conditions for the complete monotonicity of f on $[0, \infty)$?

11. Prove that every Gram matrix is nonnegative definite.

12. Prove Lemma 3 without use of polar coordinates. In the formula for I^2, substitute $y = sx$ and obtain a double integral in x and s. (*Math. Mag.* 67 (1994), page 47.)

13. Prove Theorem 3 by a direct calculation of the quadratic form c^*Ac.

14. Don't make the mistake of believing that $\overline{a + ib} = a - ib$. (Take $a = 0$ and $b = i$.) What is the relevance of this to the theorems in this chapter?

15. Prove that if $x_1, x_2, \ldots, x_n$ are n distinct nodes in an inner-product space, then the functions $x \mapsto e^{-\alpha_j \|x - x_j\|^2}$ can be used for interpolation at the nodes, provided that the coefficients α_j are large enough.

16. Carry out the following proof of Lemma 2. Denote the left and right sides of the equation by $f(y)$ and $g(y)$. Prove that $f(0) = g(0)$, that $f'(y) = -2\pi y f(y)$, and that $g'(y) = -2\pi y g(y)$. It follows that $f'/f = g'/g$. Complete the reasoning.

17. If $f : [0, b] \times \mathbb{R} \to \mathbb{R}$ is continuous and if it satisfies a Lipschitz condition in its second argument, then the initial-value problem $x'(s) = f(s, x(s))$, $x(0) = \beta$ has a unique solution on $[0, b]$. Show that this result applies in the proof of Lemma 2.

18. Is Theorem 2 valid in a complex inner-product space?

19. *Edict*: Every mathematician, on attaining the age of 21, must elect one of the following definitions of the Fourier transform on $\mathbb{R}^s$, and swear eternal fealty to it for the remainder of his or her professional life.

$$\int f(y)e^{-2\pi ixy}\, dy, \quad \frac{1}{(2\pi)^{s/2}} \int f(y)e^{-ixy}\, dy, \quad \int f(y)e^{-ixy}\, dy, \quad \int f(y)e^{ixy}\, dy$$

What are the relative merits of these definitions?

References

[BS]   Baxter, B. J. C., and N. Sivakumar. "On shifted cardinal interpolation by Gaussians and multiquadrics." *J. Approx. Theory 87* (1996), 36–59.

[DMc]   Dym, H., and H. P. McKean. *Fourier Series and Integrals*. Academic Press, New York, 1972.

[Gold] Goldberg, R. R. *Fourier Transforms*. Cambridge University Press, Cambridge, 1970.

[HJ1]   Horn, R. A., and C. R. Johnson. *Matrix Analysis*. Cambridge University Press, Cambridge, 1985.

[Kor]   Körner, T. W. *Fourier Analysis*. Cambridge University Press, 1989.

[RS]   Reed, M., and B. Simon. *Methods of Modern Mathematical Physics,* Vol. I, *Functional Analysis*. 2nd ed., Academic Press, New York, 1980.

[Ru2]   Rudin, W. *Functional Analysis*. McGraw-Hill, New York, 1973.

[S1]   Schoenberg, I. J. "Metric spaces and completely monotone functions." *Ann. Math. 39* (1938), 811–841.

[SW]   Stein, E. M., and G. Weiss. *Introduction to Fourier Analysis on Euclidean Spaces*. Princeton University Press, Princeton, NJ, 1971.

16

The Micchelli Interpolation Theorem

The theorem of Schoenberg, in Chapter 15, provides an abundant assortment of functions f for which interpolation is possible by expressions of the form

$$\sum_{j=1}^{n} c_j f(\|x - x_j\|^2)$$

The following are examples of suitable functions, justified by Schoenberg's Theorem:

    1. $f(t) = (t + 1)^{-1}$
    2. $f(t) = e^{-t}$
    3. $f(t) = (t + 1)^{-1/2}$

Example 3 is especially important as it leads to "inverse multiquadric interpolation," which has found favor in geophysics.

    Functions that are *not* included in Schoenberg's Theorem but are nevertheless useful in this context are, for example,

    4. $f(t) = \sqrt{t}$
    5. $f(t) = \sqrt{1 + t}$
    6. $f(t) = \log(1 + t)$

Among these, we find that Example 5 is especially useful since it leads to "multiquadric interpolation," also employed in geophysics (see Hardy [Ha], [Ha2]).

    Examples 4–6 arise from a theorem of Micchelli [M1]. Following Sun [Su1] we define the appropriate function class as follows:

Definition. We let $\mathcal{DM}$ denote the class of all functions $f : [0, \infty) \to [0, \infty)$ such that

    1. $f \in C[0, \infty)$ and $f \in C^{\infty}(0, \infty)$
    2. f' is completely monotone but not constant on $(0, \infty)$.

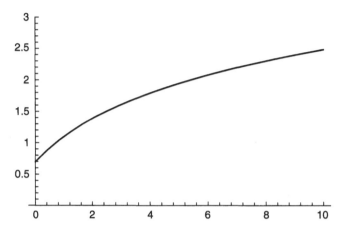

Figure 16.1   *Typical Function in the Class* $\mathcal{DM}$

Recall that the complete monotonicity of f' means that $(-1)^k f^{(k+1)}(t) > 0$ for $k = 0, 1, 2, \ldots$ and $t > 0$. One can verify easily that the functions f in Examples 4–6 belong to the class $\mathcal{DM}$. Figure 16.1 shows the shape of a typical function in this class. It is $t \mapsto \log(2 + t)$.

Sun, in [Su1], has given a characterization of the class $\mathcal{DM}$. His theorem parallels the Bernstein-Widder Theorem.

THEOREM 1. *In order that f be a member of $\mathcal{DM}$, it is necessary and sufficient that f have the form*

(1)
$$f(t) = c^2 + \int_{(0,\infty)} \frac{1 - e^{-st}}{s} \, d\mu(s)$$

in which μ is a nontrivial Borel measure on $(0, \infty)$ satisfying the inequalities

(2)
$$\int_{(1,\infty)} \frac{d\mu(s)}{s} < \infty \quad \text{and} \quad \mu\big((0, t)\big) < \infty \text{ for all } t$$

Remark. The class $\mathcal{DM}$ is almost the same as Schoenberg's class T in [S1], pages 824–825. For membership in T it is required that $f(0) = 0$, but f need not be continuous at 0. The integral representation in Equation (1) is adapted from Equation (4.4) in [S4]. Notice that γ need not be bounded, and the second integral in (2) is permitted to be infinite.

Proof. If f has the form given in Equation (1), then we can easily verify the properties required in the definition of $\mathcal{DM}$. First of all, $f(t)$ is well-defined because the first inequality in (2) implies the convergence of the improper integral that defines $f(t)$. Secondly, $f(t) \geq 0$ because $1 \geq e^{-st}$. Differentiation gives us

(3)
$$f'(t) = \int_{(0,\infty)} e^{-st} \, d\mu(s)$$

from which we easily see that f' is completely monotone on $(0, \infty)$. For the continuity of f at 0, let $t \downarrow 0$ in Equation (1). Then $1 - e^{-st} \downarrow 0$, and (by the monotone convergence theorem) the right hand side of Equation (1) approaches $c^2 = f(0)$. Since μ is nontrivial on $(0, \infty)$, Equation (3) shows that f' is not constant.

For the second half of the proof, assume that $f \in \mathcal{DM}$. By the Bernstein-Widder Theorem for completely monotone functions on $(0, \infty)$, (quoted below) f' must have a representation as in Equation (3), where we initially know nothing about μ. For $0 < \varepsilon < t$, we have

$$\int_\varepsilon^t f'(\sigma)\, d\sigma = f(t) - f(\varepsilon)$$

Since f is continuous at 0, we can let $\varepsilon \downarrow 0$ to obtain

$$f(t) - f(0) = \int_0^t f'(\sigma)\, d\sigma = \int_0^t \int_0^\infty e^{-s\sigma}\, d\mu(s)\, d\sigma$$

$$= \int_0^\infty \int_0^t e^{-s\sigma}\, d\sigma\, d\mu(s)$$

$$= \int_0^\infty \frac{1 - e^{-st}}{s}\, d\mu(s)$$

The change in the order of integration in this calculation is justified by Tonelli's Theorem (quoted below). (The important hypothesis for this theorem is the nonnegativity of the integrand.) Since the last integral must be finite, we conclude that the first part of Inequality (2) is true. In fact, we can write

$$f(1) = \int_0^\infty \frac{1 - e^{-s}}{s}\, d\mu(s) \geq \int_1^\infty \frac{1 - e^{-s}}{s}\, d\mu(s) \geq \int_1^\infty \frac{1/2}{s}\, d\mu(s)$$

The second part of Inequality (2) must be true because f' is not constant. ∎

The second version of the Bernstein-Widder Theorem, needed in the preceding proof, is as follows. See [W5], page 161.

THEOREM 2. *In order that f be completely monotone on $(0, \infty)$, i.e., $(-1)^k f^{(k)}(t) \geq 0$ for $t > 0$ and $k = 0, 1, 2, \ldots$, it is necessary and sufficient that there exist a Borel measure on $(0, \infty)$ such for each $t > 0$, $\mu\big((0, t)\big) < \infty$ and*

$$f(t) = \int_{(0,\infty)} e^{-st}\, d\mu(s)$$

The Tonelli Theorem, also required in the proof of Theorem 1, is as follows. See [HewS], pages 384–385 or [R], page 270.

THEOREM 3. *Let (X, μ) and (Y, v) be two σ-finite measure spaces. If f is nonnegative and measurable on $X \times Y$, then*

$$\int_X \int_Y f(x, y)\, dv(y)\, d\mu(x) = \int_Y \int_X f(x, y)\, d\mu(x)\, dv(y)$$

From matrix theory we require the celebrated Courant-Fischer Theorem, also known as the "min-max" theorem. In the statement of the theorem, V represents linear subspaces in $\mathbb{C}^n$.

THEOREM 4. *Let A be a complex Hermitian $n \times n$ matrix whose eigen-values are labelled $\lambda_1 \geq \lambda_2 \geq \cdots \geq \lambda_n$. Then for $k = 1, 2, \ldots, n$ we have*

$$\lambda_k = \max_{\dim V = k} \min_{0 \neq v \in V} \frac{v^* A v}{v^* v} = \min_{\dim V = n-k+1} \max_{0 \neq v \in V} \frac{v^* A v}{v^* v}$$

Convenient references for this theorem are [GvL], page 269 and [HJ1], page 179.

LEMMA 1. *Let A be an $n \times n$, real, symmetric matrix having nonnegative trace (that is, $\sum A_{ii} \geq 0$). If $v^T A v < 0$ whenever $\sum |v_i| > \sum v_i = 0$, then A has $n - 1$ negative eigenvalues and one positive eigenvalue.*

Proof. Arrange the eigenvalues of A in the order $\lambda_1 \geq \lambda_2 \geq \cdots \geq \lambda_n$. By the preceding theorem (Courant-Fischer Theorem), we have, with $e = (1, 1, \ldots, 1)$,

$$\lambda_2 = \min_{\substack{V \subset \mathbb{C}^n \\ \dim V = n-1}} \max_{0 \neq v \in V} \frac{v^* A v}{v^* v} \leq \min_{\substack{V \subset \mathbb{R}^n \\ \dim V = n-1}} \max_{0 \neq v \in V} \frac{v^T A v}{v^T v} = \min_{u \neq 0} \max_{0 \neq v \perp u} \frac{v^T A v}{v^T v}$$

$$\leq \max_{0 \neq v \perp e} \frac{v^T A v}{v^T v} = \max_{\|v\| = 1, v \perp e} v^T A v < 0$$

Recall now that the trace of a matrix equals the sum of its eigenvalues, [HJ1], page 42. Hence, in this lemma, $\sum \lambda_i = \text{trace}(A) \geq 0$. We conclude that $\lambda_1 > 0$, since all the other eigenvalues are negative. ∎

THEOREM 5. (Micchelli [M1]) *Let $f : [0, \infty) \to [0, \infty)$. Assume that f is continuous on $[0, \infty)$ and that f' is completely monotone but not constant on $(0, \infty)$. Then for any n distinct points $x_1, x_2, \ldots, x_n$ in a real inner-product space the matrix $A_{ij} = f(\|x_i - x_j\|^2)$ is nonsingular.*

Proof. By Theorem 1, f has a representation

$$f(t) = f(0) + \int_{(0,\infty)} \frac{1 - e^{-st}}{s} \, d\mu(s)$$

If $c \in \mathbb{R}^n$, $c \neq 0$, and $\sum c_i = 0$, then

$$c^T A c = \sum \sum c_i c_j f(\|x_i - x_j\|^2)$$

$$= \sum \sum c_i c_j \left[f(0) + \int_{(0,\infty)} \frac{1 - e^{-s\|x_i - x_j\|^2}}{s} \, d\mu(s) \right]$$

$$= -\int_{(0,\infty)} \sum \sum c_i c_j \frac{e^{-s\|x_i - x_j\|^2}}{s} \, d\mu(s) < 0$$

In the last step of this calculation we employ Theorem 2 from Chapter 15, page 104 according to which the matrix $\left(e^{-s\|x_i - x_j\|^2}\right)$ is positive definite. Notice that the trace of A is $nf(0) \geq 0$. Our result now follows from Lemma 1.   ∎

One way of constructing a measure μ on $(0, \infty)$ that has the properties needed in Theorem 1 is to start with a continuous nonnegative function g on $[0, \infty)$ such that

$$\int_1^\infty \frac{g(s)}{s}\, ds < \infty$$

Then we define $\mu(A)$ for any Borel set A by the equation

$$\mu(A) = \int_A g(s)\, d\mu(s)$$

The equation for f in Theorem 1 now becomes

$$f(t) = c^2 + \int_0^\infty \frac{1 - e^{-st}}{s} g(s)\, ds$$

For example,

$$t^\alpha = c \int_0^\infty (1 - e^{st}) s^{-\alpha - 1}\, ds \qquad (0 < \alpha < 1)$$

Thus, the function $t \mapsto t^\alpha$ is in the class $\mathcal{DM}$.

Example. We illustrate the interpolation theory of this chapter by using the data of the Example in Chapter 15, page 105. Here, however, we employ the function $x \mapsto \log(\alpha + \|x\|^2)$. The interpolant is shown for $\alpha = 0.1$ in Figure 16.3. In Figure 16.2 we used the modified function $x \mapsto 10 - \log(0.00005 + 200\|x\|^2)$.   ∎

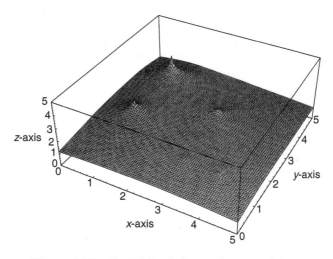

Figure 16.2   *Radial Basis Interpolant, $\alpha = 1.0$*

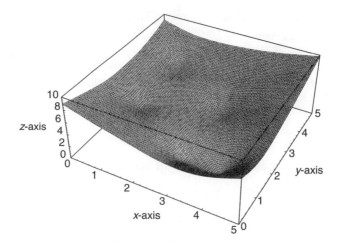

Figure 16.3   *Radial Basis Interpolant, $\alpha = 0.1$*

The remainder of this chapter explores the topic of embedding metric spaces into Hilbert space. Questions about such embedding motivated the work of Schoenberg and others in the period 1930–1940.

LEMMA 2. *For a real, symmetric, $n \times n$ matrix A these two properties are equivalent*

    a. *A is nonnegative definite;*
    b. *There exist points, $x_i \in \mathbb{R}^n$ such that $A_{ij} = \langle x_i, x_j \rangle$. (That is, A is a Gram matrix.)*

Proof. If **b** is true and if $c \in \mathbb{R}^n$, then we have

$$c^T A c = \sum \sum c_i c_j \langle x_i, x_j \rangle = \left\langle \sum c_i x_i, \sum c_j x_j \right\rangle \geq 0$$

If **a** is true, then the eigenvalues of A are nonnegative. By the finite-dimensional spectral theorem, there exists an orthogonal matrix U such that

$$UAU^T = D = \text{diag}(\lambda_1, \lambda_2, \dots, \lambda_n)$$

Here $\lambda_1, \dots, \lambda_n$ are the eigenvalues of A. Putting

$$D^{1/2} = \text{diag}\left(\sqrt{\lambda_1}, \sqrt{\lambda_2}, \dots, \sqrt{\lambda_n}\right)$$

we have

$$A = U^T D U = (U^T D^{1/2})(D^{1/2} U) = B^T B$$

where $B = D^{1/2} U$. If the columns of B are $x_1, x_2, \dots, x_n$, we will have $A_{ij} = \langle x_i, x_j \rangle$. ∎

Definition.   A **quasi-metric space** is a pair (X, d), where X is a set and d is a mapping $d : X \times X \to \mathbb{R}$ that has these properties:

1. $d(x, x) = 0$
2. $d(x, y) = d(y, x) \geq 0$

Observe that we do not require the positivity of $d(x, y)$ when $x \neq y$, nor do we require the triangle inequality. We say that the quasi-metric space (X, d) can be **embedded** in a Hilbert space H if there is a mapping $\phi : X \longrightarrow H$ such that $d(x, y) = \| \phi(x) - \phi(y) \|$ for all x and y in X.

THEOREM 6. (Schoenberg [S5], [S4]) *Let (X, d) be a finite quasi-metric space whose points are $x_0, x_1, \ldots, x_n$. Let $A_{ij} = d(x_i, x_0)^2 + d(x_j, x_0)^2 - d(x_i, x_j)^2$, where $1 \leq i, j \leq n$. In order that (X, d) be embeddable in $\mathbb{R}^n$, it is necessary and sufficient that the $n \times n$ matrix A be nonnegative definite.*

Proof. Suppose that an embedding exists. Then there exist points $y_0, y_1, \ldots, y_n$ in $\mathbb{R}^n$ such that $\| y_i - y_j \| = d(x_i, x_j)$ for $0 \leq i, j \leq n$. Putting $z_i = y_i - y_0$, we have (for $1 \leq i, j \leq n$)

$$
\begin{aligned}
A_{ij} &= d(x_i, x_0)^2 + d(x_j, x_0)^2 - d(x_i, x_j)^2 \\
&= \| y_i - y_0 \|^2 + \| y_j - y_0 \|^2 - \| y_i - y_j \|^2 \\
&= \| z_i \|^2 + \| z_j \|^2 - \| z_i - z_j \|^2 \\
&= \| z_i \|^2 + \| z_j \|^2 - \{ \| z_i \|^2 - 2\langle z_i, z_j \rangle + \| z_j \|^2 \} \\
&= 2\langle z_i, z_j \rangle
\end{aligned}
$$

Thus, A is a Gram matrix, and is nonnegative definite, by Lemma 2.

Now suppose that A is nonnegative definite. Notice that A is symmetric, because $d(x_i, x_j) = d(x_j, x_i)$. By Lemma 2, $A_{ij} = 2\langle v_i, v_j \rangle$ for suitable points $v_i \in \mathbb{R}^n$. Thus,

(1) $$d(x_i, x_0)^2 + d(x_j, x_0)^2 - d(x_i, x_j)^2 = 2\langle v_i, v_j \rangle$$

If we set $j = i$ in this equation, we obtain

(2) $$d(x_i, x_0)^2 = \langle v_i, v_j \rangle = \| v_i \|^2$$

Putting Equations (1) and (2) together, we have

$$
\begin{aligned}
d(x_i, x_j)^2 &= d(x_i, x_0)^2 + d(x_j, x_0)^2 - 2\langle v_i, v_j \rangle \\
&= \| v_i \|^2 + \| v_j \|^2 - 2\langle v_i, v_j \rangle \\
&= \| v_i - v_j \|^2 \qquad (1 \leq i, j \leq n)
\end{aligned}
$$

Now set $v_0 = 0$, so that

$$d(x_i, x_0)^2 = \| v_i \|^2 = \| v_i - v_0 \|^2$$

An embedding is therefore $x_i \mapsto v_i, 0 \leq i \leq n$.   ∎

We state without proof a stronger result due essentially to Schoenberg. See [WeW], page 5 for the proof. (Actually, Theorem 6 establishes half of Theorem 7.)

THEOREM 7. *A quasi-metric space (X, d) can be embedded in a Hilbert space if and only if for each finite set of points $x_0, x_1, \ldots, x_n$ in X the $n \times n$ matrix*

$$A_{ij} = d(x_i, x_0)^2 + d(x_j, x_0)^2 - d(x_i, x_j)^2$$

is nonnegative definite.

Example. Let X be this set of 4 points in $\mathbb{R}^2$:

$$x_0 = (-1, 0), \quad x_1 = (0, 1), \quad x_2 = (1, 0), \quad x_3 = (0, -1)$$

We give X the metric $d(x, y) = \|x - y\|_\infty$. Can (X, d) be embedded in Hilbert space? To answer this, one can use Theorem 6. The matrix A defined in that theorem will be

$$A = \begin{bmatrix} 2 & 4 & -2 \\ 4 & 8 & 4 \\ -2 & 4 & 2 \end{bmatrix}$$

in this example. It is *not* nonnegative definite. In fact, if $c = (1, -1, 1)$, then $c^T A c = -8$. Hence (X, d) cannot be embedded.   ∎

LEMMA 3. *Let A be an $(n + 1) \times (n + 1)$ symmetric matrix having all diagonal elements equal to zero. These properties of A are equivalent:*

a. *The $n \times n$ matrix $B_{ij} = A_{i0} + A_{j0} - A_{ij}$ is nonnegative definite*

b. *$v^T A v \leq 0$ whenever $v \in \mathbb{R}^{n+1}$ and $\sum_{i=0}^{n} v_i = 0$*

Proof. Let u be any vector in $\mathbb{R}^n$ having components $u_1, \ldots, u_n$. Put $u_0 = -\sum_{i=1}^{n} u_i$ and $\tilde{u} = (u_0, u_1, \ldots, u_n)$. Then we have

$$\tilde{u}^T A \tilde{u} = \sum_{i=0}^{n} u_i \sum_{j=0}^{n} u_j A_{ij} = \sum_{i=0}^{n} u_i \left[u_0 A_{i0} + \sum_{j=1}^{n} u_j A_{ij} \right]$$

$$= u_0 \left[u_0 A_{00} + \sum_{j=1}^{n} u_j A_{0j} \right] + \sum_{i=1}^{n} u_i \left[u_0 A_{i0} + \sum_{j=1}^{n} u_j A_{ij} \right]$$

$$= u_0 \sum_{j=1}^{n} u_j A_{0j} + u_0 \sum_{i=1}^{n} u_i A_{i0} + \sum_{i=1}^{n} \sum_{j=1}^{n} u_i u_j A_{ij}$$

$$= \left(-\sum_{i=1}^{n} u_i \right) \sum_{j=1}^{n} u_j A_{0j} + \left(-\sum_{j=1}^{n} u_j \right) \sum_{i=1}^{n} u_i A_{i0} + \sum_{i=1}^{n} \sum_{j=1}^{n} u_i u_j A_{ij}$$

$$= \sum_{i=1}^{n} \sum_{j=1}^{n} u_i u_j (-A_{0j} - A_{i0} + A_{ij}) = -u^T B u$$

If **b** is true, then $\tilde{u}^T A \tilde{u} \leq 0$ and $u^T B u \geq 0$. Hence **a** is true. On the other hand, if **a** is true then $u^T B u \geq 0$ and $\tilde{u}^T A \tilde{u} \leq 0$. Since $\tilde{u}$ can be *any* vector satisfying the hypotheses in **b**, we see that **b** is true.   ∎

The following theorem is close to Schoenberg's Theorem 6 in [S4].

THEOREM 8. *Let $f \in \mathcal{DM}$ (defined at the beginning of this chapter) and satisfy $f(0) = 0$. Let X be a real inner-product space. For x and y in X, define $d(x, y) = \sqrt{f(\|x - y\|^2)}$. Then (X, d) is embeddable in Hilbert space.*

Proof. Note that d is a quasi-metric. Hence Theorem 7 is applicable. Select any $n + 1$ distinct points $x_0, x_1, \ldots, x_n$ in X, and define

$$A_{ij} = f(\|x_i - x_j\|^2) \qquad (0 \le i, j \le n)$$

$$B_{ij} = A_{i0} + A_{j0} - A_{ij} \qquad (1 \le i, j \le n)$$

By the proof of Theorem 3, A has the property

$$\left[v \in \mathbb{R}^{n+1}, \, v \ne 0, \, \sum v_i = 0 \right] \implies v^T A v < 0$$

Hence, $v^T A v \le 0$ whenever $\sum v_i = 0$. By Lemma 3, B is nonnegative definite. Since

$$B_{ij} = d(x_i, x_0)^2 + d(x_j, x_0)^2 - d(x_i, x_j)^2$$

Theorem 7 shows that (X, d) embeds in Hilbert space. ∎

Problems

1. Prove that for an $n \times n$ symmetric real matrix A, the smallest and largest eigenvalues are respectively the minimum and the maximum of $v^T A v / v^T v$.

2. Give an example of a quasi-metric space consisting of three points that cannot be embedded in the Cartesian plane.

3. Prove that any three-point metric space can be embedded in the Cartesian plane.

4. Let four points be given in $\mathbb{R}^2$, namely $(0, 0)$, $(0, 1)$, $(1, 0)$, and $(1, 1)$. Use the ℓ^1-norm and determine whether the resulting four-point metric space can be embedded in Euclidean $\mathbb{R}^3$.

5. Prove that if $f \in \mathcal{DM}$, then $f > 0$ on $(0, \infty)$.

6. Find the weakest hypotheses so that the function $f(t) = a + be^{ct}$ will belong to $\mathcal{DM}$.

7. Prove that if A is a real, symmetric matrix such that $x^T A x > 0$ for all x in $\mathbb{R}^n \backslash 0$, then $z^* A z > 0$ for all $z \in \mathbb{C}^n \backslash 0$.

8. One version of the Courant-Fischer Theorem states that

$$\lambda_k = \max_{w_1, w_2, \ldots, w_{n-k}} \min_{\substack{\|v\|=1 \\ v \perp w_1, w_2, \ldots, w_{n-k}}} v^* A v$$

Prove the equivalence of this with the version in Theorem 2.

9. Prove that a real, symmetric matrix having at least one positive diagonal element must have a positive eigenvalue.

10. Prove that a real, symmetric, nonnegative, nontrivial matrix must have a positive eigenvalue.

11. In the proof of Theorem 1, we needed to know that $s \mapsto s^{-1}(1 - e^{-st})$ is bounded on $(0, 1)$. Prove that it is no greater than $e^t - 1$.

12. In the proof of Theorem 1, there is an interchange of the operations differentiation and integration. Using Theorem 5 in Chapter 14 (page 97), show that this interchange is valid.

13. Is there a function f in the class $\mathcal{DM}$ such that $f(1) = 0$?

14. For a symmetric, positive definite matrix A, the Cholesky factorization provides a stable and efficient method for solving the equation $Ax = b$. Is there a similar procedure that applies when A has the properties in Lemma 1?

15. In creating Figure 16.2 we used a function of the form $x \mapsto c - f(\|x\|^2)$, where f satisfies the hypotheses of Micchelli's Theorem. Prove that the interpolation matrix will be nonsingular for any choice of nodes, if $c \leq f(0)$.

References

[GvL]   Golub, G. H., and C. van Loan. *Matrix Computations.* Johns Hopkins University Press, Baltimore, 1983.

[Ha]   Hardy, R. "Multiquadric equations of topography and other irregular surfaces." *J. Geophysics Res. 76* (1971), 1905–1915.

[Ha2]   Hardy, R. L. "Theory and applications of the multiquadric biharmonic method." *Computers Math. Applic. 19* (1990), 163–208. (This paper has a bibliography of 109 items.)

[HewS]   Hewitt, E., and K. Stromberg. *Real and Abstract Analysis.* Springer-Verlag, New York, 1965.

[HJ1]   Horn, R. A., and C. R. Johnson. *Matrix Analysis.* Cambridge University Press, Cambridge, 1985.

[M1]   Micchelli, C. A. "Interpolation of scattered data: Distance matrices and conditionally positive definite functions." *Constructive Approximation 2* (1986), 11–22.

[R]   Royden, H. L. *Real Analysis.* 2nd ed., Macmillan, New York, 1968.

[S1]   Schoenberg, I. J. "Metric spaces and completely monotone functions." *Ann. Math. 39* (1938), 811–841.

[S4]   Schoenberg, I. J. "Metric spaces and positive definite functions." *Trans. Amer. Math. Soc. 44* (1938), 522–536.

[S5]   Schoenberg, I. J. "Remarks to Maurice Frechet's article." *Ann. Math. 36* (1935), 724–732.

[Su1]   Sun, Xingping. *Multivariate Interpolation Using Ridge or Related Functions.* Unpublished doctoral dissertation, University of Texas at Austin, August 1990.

[WeW]   Wells, J. H., and L. R. Williams. *Embeddings and Extensions in Analysis.* Springer-Verlag, New York, 1975.

[W5]   Widder, D. V. *The Laplace Transform.* Princeton University Press, Princeton, NJ, 1946.

17
Positive Definite Functions on Spheres

For interpolation of some data, such as geodetic, geological, and meteorological information gathered over the Earth's surface, we require suitable functions defined on spherical surfaces. Here we shall give some of the classical theory, much of which originated with I. J. Schoenberg.

The m-dimensional sphere, S^m, is the unit sphere in $\mathbb{R}^{m+1}$. A typical element of S^m is a point

$$x = (\xi_1, \xi_2, \ldots, \xi_{m+1}) \qquad (\xi_1^2 + \xi_2^2 + \cdots + \xi_{m+1}^2 = 1)$$

The sphere S^∞ is the unit sphere in ℓ^2. Its points are

$$x = (\xi_1, \xi_2, \ldots) \qquad \left(\sum_{k=1}^\infty \xi_k^2 = 1 \right)$$

The "sphere" S^1 is actually a circle of radius 1. Thus, a generic point of S^1 is of the form $x = (\cos \phi, \sin \phi)$. The distance between this point and another, say $y = (\cos \psi, \sin \psi)$, is the length of the shorter of the two arcs that join x and y. See Figure 17.1. We define this distance function as follows,

(1) $$d(x, y) = \min \{ |\phi - \psi|, 2\pi - |\phi - \psi| \}$$

An alternative formula arises from the observation that the arc in question is the angle between the points x and y. Denoting this angle by θ, we have

$$xy = \|x\| \, \|y\| \cos \theta = \cos \theta$$

whence

(2) $$d(x, y) = \theta = \text{Arccos } xy$$

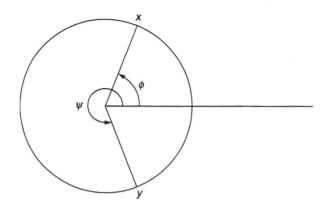

Figure 17.1   *Distance Function on the Circle*

The Arccosine function used here gives values in the interval $[0, \pi]$. This formula for d can be used in any S^m or S^∞, since the measurement of distance between two points x and y of S^m or S^∞ takes place on a circle that is the intersection of S^m or S^∞ with the two-dimensional plane determined by 0, x, and y.

LEMMA 1. *The function d in Equation (2) is a metric on the spheres S^m and S^∞.*

Proof. Only the triangle inequality is questionable. We want to prove that

$$d(x, y) \leq d(x, z) + d(z, y)$$

for any three points on S^m. Observe that d is invariant under rotations. In fact, for any orthogonal matrix, U,

$$d(Ux, Uy) = \text{Arccos } UxUy = \text{Arccos}(U^T Ux)y = \text{Arccos } xy = d(x, y)$$

We choose our coordinate system so that $x = (0, 0, 1, 0, 0, \ldots, 0)$, $y = (\sin \alpha, 0, \cos \alpha, 0, 0, \ldots, 0)$, and $z = (\sin \gamma \cos \beta, \sin \gamma \sin \beta, \cos \gamma, 0, 0, \ldots, 0)$. Here, α and γ are in $[0, \pi]$, but β can be in $[0, 2\pi)$. We have arranged that all but the first three coordinates of our points are zero, since they certainly lie in a three-dimensional subspace of $\mathbb{R}^{m+1}$. The points x, y, and z are illustrated in Figure 17.2.

We want to prove that

$$\alpha \leq \gamma + \text{Arccos } zy$$

or

(3) $$\alpha - \gamma \leq \text{Arccos}(\sin \alpha \sin \gamma \cos \beta + \cos \alpha \cos \gamma)$$

If $\alpha - \gamma \leq 0$, then Inequality (3) is trivially true. Hence we may assume that $\alpha - \gamma > 0$. Both sides of Inequality (3) are in the interval $(0, \pi]$, where the cosine is decreasing. Hence an equivalent inequality to prove is

$$\cos(\alpha - \gamma) \geq \sin \alpha \sin \gamma \cos \beta + \cos \alpha \cos \gamma$$

or

$$\cos \alpha \cos \gamma + \sin \alpha \sin \gamma \geq \sin \alpha \sin \gamma \cos \beta + \cos \alpha \cos \gamma$$

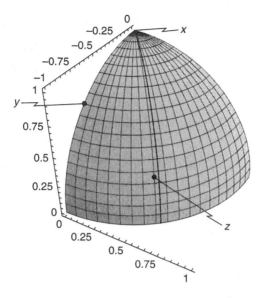

Figure 17.2 *Points on a Sphere in* $\mathbb{R}^3$

or

$$\sin \alpha \sin \gamma \geq \sin \alpha \sin \gamma \cos \beta$$

This is certainly correct, for $\sin \alpha \sin \gamma \geq 0$ and $\cos \beta \leq 1$. ∎

The next step in understanding the metric space $\mathcal{S}^m$ is to describe its polar coordinate system. The familiar system on $\mathcal{S}^2$ is shown in Figure 17.3.

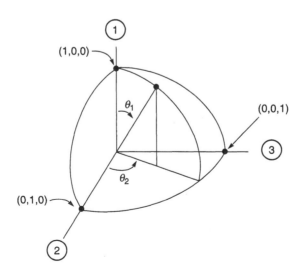

Figure 17.3 *Coordinates on* $\mathcal{S}^2$

We have taken the "north pole" to be the point $(1, 0, 0)$. Notice the preferred numbering of the axes. Also observe that the range of θ_1 is $[0, \pi]$, while the range of θ_2 is $[0, 2\pi)$. We follow this model as we proceed to the higher-dimensional spheres. The polar coordinates of a point $(\xi_1, \xi_2, \ldots, \xi_{m+1})$ on $\mathbf{S}^m$ are $(\theta_1, \theta_2, \ldots, \theta_m)$, and the correspondence between the "rectangular" coordinates ξ_j and the polar coordinates θ_j is described as follows.

$$\xi_1 = \cos\theta_1 \qquad\qquad (0 \le \theta_1 \le \pi)$$
$$\xi_2 = \sin\theta_1 \cos\theta_2 \qquad\qquad (0 \le \theta_2 \le \pi)$$
$$\xi_3 = \sin\theta_1 \sin\theta_2 \cos\theta_3 \qquad\qquad (0 \le \theta_3 \le \pi)$$
$$\cdots$$
$$\xi_j = (\sin\theta_1 \sin\theta_2 \cdots \sin\theta_{j-1})\cos\theta_j \qquad\qquad (0 \le \theta_j \le \pi)$$
$$\xi_{j+1} = (\sin\theta_1 \sin\theta_2 \cdots \sin\theta_j)\cos\theta_{j+1} \qquad\qquad (0 \le \theta_{j+1} \le \pi)$$
$$\cdots$$
$$\xi_{m-1} = (\sin\theta_1 \sin\theta_2 \cdots \sin\theta_{m-2})\cos\theta_{m-1} \qquad (0 \le \theta_{m-1} \le \pi)$$
$$\xi_m = (\sin\theta_1 \sin\theta_2 \cdots \sin\theta_{m-1})\cos\theta_m \qquad (0 \le \theta_m < 2\pi)$$
$$\xi_{m+1} = (\sin\theta_1 \sin\theta_2 \cdots \sin\theta_{m-1})\sin\theta_m$$

Notice that if $x = (\xi_1, \ldots, \xi_{m+1}) \in \mathbf{S}^m$ then

$$x = \left(\cos\theta_1, \ (\sin\theta_1)y\right)$$

where y is a point in $\mathbf{S}^{m-1}$ given by

$$y = (\cos\theta_2, \ \sin\theta_2 \cos\theta_3, \ \sin\theta_2 \sin\theta_3 \cos\theta_4, \ldots)$$

Another topic that enters into the theory of positive definite functions on spheres is that of **Gegenbauer polynomials,** also known as **ultraspherical polynomials.** For any $\lambda > -1/2$, there is an orthogonal family $P_0^{(\lambda)}, P_1^{(\lambda)}, \ldots$. Each member of the family is a polynomial, and its subscript indicates its exact degree. The orthogonality condition is

(4)
$$\int_{-1}^{1} P_k^{(\lambda)}(x) P_j^{(\lambda)}(x)(1 - x^2)^{\lambda - 1/2} = 0 \quad \text{if} \quad k \ne j$$

The reader familiar with Jacobi polynomials will recognize the Gegenbauer polynomials as a special case, namely, the case when $\alpha = \beta = \lambda - 1/2$, where $x \mapsto (1 - x)^\alpha (1 + x)^\beta$ is the weight function of the Jacobi polynomials.

Equation (4) does not specify the polynomials $P_k^{(\lambda)}$ fully because the equation is homogeneous. In Szegő's book [Sz, Section 4.7] and in the Tables of Abramowitz and Stegun [AS], page 778, the normalization is

(5)
$$P_n^{(\lambda)}(1) = \binom{n + 2\lambda - 1}{n}$$

(This is not consistent with the standard normalization of the Jacobi polynomials.) For $\lambda = 0$, we follow Schoenberg and adopt the normalization $P_n^{(0)}(1) = 1$.

A generating function of the Gegenbauer polynomials is

(6)
$$(1 - 2rw + r^2)^{-\lambda} = \sum_{n=0}^{\infty} r^n P_n^{(\lambda)}(w) \qquad (\lambda > 0).$$

Some special cases of these polynomials are noteworthy. They correspond to $\lambda = 0$, $\lambda = 1/2$, and $\lambda = 1$. For these we have

$P_n^{(0)}(x) = T_n(x)$ Chebyshev polynomial of the first kind, used on S^1

$P_n^{(1/2)}(x) = P_n(x)$ Legendre polynomial, used on S^2

$P_n^{(1)}(x) = U_n(x)$ Chebyshev polynomial of the second kind, used on S^3

A real-valued function $f : [0, \pi] \rightarrow \mathbb{R}$ is said to be **positive definite** on S^m if for any n and any n points $x_1, x_2, \ldots, x_n$ on S^m the $n \times n$ matrix with elements $f(d(x_i, x_j))$ is nonnegative definite. Schoenberg's Theorem [S3] is as follows.

THEOREM 1. *In order that a continuous function f be positive definite on S^m, it is necessary and sufficient that f be expressible in the form*

$$f(t) = \sum_{k=0}^{\infty} a_k P_k^{(\lambda)}(\cos t) \quad \left(\lambda = (m-1)/2, \quad a_k \geq 0, \quad \sum a_k P_k^{(\lambda)}(1) < \infty \right)$$

We will prove several parts of this theorem and add some information about strictly positive definite functions. First, let us examine the case of S^1.

THEOREM 2. *Let $f(t) = \sum_{k=0}^{\infty} a_k \cos kt$, where $a_k \geq 0$ and $\sum a_k < \infty$. Then f is continuous and positive definite on S^1.*

Proof. The continuity is a consequence of the Weierstrass M-test for uniform convergence. We just observe that $\sum |a_k \cos kt| \leq \sum |a_k| < \infty$. Let $x_1, x_2, \ldots, x_n$ be points in S^1. We can write

$$x_i = (\cos \theta_i, \sin \theta_i) \qquad (0 \leq \theta_i < 2\pi)$$

Then the distance between x_i and x_j is

$$\begin{aligned}
d(x_i, x_j) &= \text{Arccos } x_i x_j \\
&= \text{Arccos}[\cos \theta_i \cos \theta_j + \sin \theta_i \sin \theta_j] \\
&= \text{Arccos}[\cos(\theta_i - \theta_j)] \\
&= \begin{cases} |\theta_i - \theta_j| & \text{if } |\theta_i - \theta_j| \leq \pi \\ 2\pi - |\theta_i - \theta_j| & \text{if } |\theta_i - \theta_j| > \pi \end{cases}
\end{aligned}$$

In the ensuing calculation, we use the formula just obtained and the fact that $\cos(2\pi - \theta) = \cos\theta$. Now we have

$$\sum_{i=1}^{n}\sum_{j=1}^{n} c_i c_j f(x_i, x_j)$$

$$= \sum_{i=1}^{n}\sum_{j=1}^{n} c_i c_j \sum_{k=0}^{\infty} a_k \cos\left[kd(x_i, x_j)\right]$$

$$= \sum_{i=1}^{n}\sum_{j=1}^{n}\sum_{k=0}^{\infty} a_k c_i c_j \cos\left[k|\theta_i - \theta_j|\right]$$

$$= \sum_{k=0}^{\infty} a_k \sum_{i=1}^{n}\sum_{j=1}^{n} c_i c_j (\cos k\theta_i \cos k\theta_j + \sin k\theta_i \sin k\theta_j)$$

$$= \sum_{k=0}^{\infty} a_k \left(\sum_{i=1}^{n} c_i \cos k\theta_i \sum_{j=1}^{n} c_j \cos k\theta_j + \sum_{i=1}^{n} c_i \sin k\theta_i \sum_{j=1}^{n} c_j \sin k\theta_j \right)$$

$$= \sum_{k=0}^{\infty} a_k \left[\left(\sum_{i=1}^{n} c_i \cos k\theta_i \right)^2 + \left(\sum_{i=1}^{n} c_i \sin k\theta_i \right)^2 \right] \geq 0 \qquad \blacksquare$$

Using the preceding proof, we can now give a sufficient condition for strict positive definiteness. See [XC2].

THEOREM 3. *If, in Theorem 2, we have $a_k \geq 0$ for all k and $a_k > 0$ for $0 \leq k \leq n/2$, then for distinct points $x_1, x_2, \ldots, x_n$ in $\mathbf{S}^1$, the $n \times n$ matrix having elements $f(d(x_i, x_j))$ is positive definite.*

Proof. Suppose the quadratic form in the preceding proof is 0. Then

$$\sum_{i=1}^{n} c_i \cos k\theta_i = \sum_{i=1}^{n} c_i \sin k\theta_i = 0 \qquad (0 \leq k \leq n/2)$$

Define a linear functional $\mathcal{L}$ whose action on a function g is

$$\mathcal{L}(g) = \sum_{i=1}^{n} c_i g(\theta_i)$$

Letting $r = [n/2]$, we see that $\mathcal{L}(g) = 0$ if $g(\theta) = \cos k\theta$ or if $g(\theta) = \sin k\theta$, where $0 \leq k \leq r$. Thus $\mathcal{L}$ annihilates each trigonometric polynomial of degree at most r. By the interpolating property ("Haar property") of the trigonometric polynomials, there exists a trigonometric polynomial g of degree at most r such that $g(\theta_i) = c_i$ for $1 \leq i \leq n$. (Here we note that $n \leq 2r + 1$, and $2r + 1$ is the dimension of the trigonometric polynomial subspace involved here.) Now we see that all c_i are zero because

$$0 = \mathcal{L}(g) = \sum_{i=1}^{n} c_i g(\theta_i) = \sum_{i=1}^{n} c_i^2 \qquad \blacksquare$$

COROLLARY. *If all a_k are positive in Theorem 3, then f is strictly positive definite on $\mathbb{S}^1$.*

THEOREM 4. *Let m be a positive integer, and put $\lambda = \frac{1}{2}(m-1)$. Let*

$$f(t) = \sum_{k=0}^{\infty} a_k P_k^{(\lambda)}(\cos t), \qquad a_k \geq 0, \qquad \sum_{k=0}^{\infty} a_k P_k^{(\lambda)}(1) < \infty$$

Let $x_1, x_2, \ldots, x_n$ be points in $\mathbb{S}^m$. Then the matrix $A_{ij} = f(d(x_i, x_j))$ is nonnegative definite. If the points x_i are distinct and $a_k > 0$ for $0 \leq k < n$, then A is positive definite.

Proof. The proof is by induction on m. When $m = 1$, we have $\lambda = 0$, and $P_k^{(0)}(\cos t) = T_k(\cos t) = \cos[k \operatorname{Arccos}(\cos t)] = \cos kt$. Thus Theorem 3 disposes of this case. Let $m \geq 2$. A polar coordinate system is introduced on $\mathbb{S}^m$, with the "north pole" chosen to be any point p for which the inner products $p x_i$ are all different and all in the open interval $(-1, 1)$. We represent the points x_i in the form of ordered pairs

(7) $$x_i = (\cos \phi_i, (\sin \phi_i) y_i), \qquad y_i \in \mathbb{S}^{m-1}, \qquad \cos \phi_i = p x_i$$

(Think of the point y_i as a point on the equator lying on the same meridian as x_i.) From Equation (7) we see easily that

(8) $$\cos(d(x_i, x_j)) = \cos \phi_i \cos \phi_j + \sin \phi_i \sin \phi_j \cos(d(y_i, y_j))$$

Presently, we shall substitute this in the formula for f. We shall then require Gegenbauer's addition formula for ultraspherical polynomials:

(9) $$P_k^{(\lambda)}(\cos \alpha \cos \beta + \sin \alpha \sin \beta \cos \gamma)$$
$$= P_k^{(\lambda)}(\cos \alpha) P_k^{(\lambda)}(\cos \beta) + \sum_{s=1}^{k} b_k^{(s)} Q_k^{(s)}(\alpha) Q_k^{(s)}(\beta) P_s^{(\lambda - 1/2)}(\cos \gamma)$$

The functions $Q_k^{(s)}$ are not required in explicit form for our proof, but are

$$Q_k^{(s)}(\theta) = (\sin \theta)^s P_{k-s}^{(\lambda + s)}(\cos \theta)$$

Our notation does not indicate their dependence on λ, and the same is true of $b_k^{(s)}$, which are nonnegative. Using Equations (8) and (9), we obtain

$$P_k^{(\lambda)}\big(\cos(d(x_i, x_j))\big) = P_k^{(\lambda)}\big(\cos \phi_i \cos \phi_j + \sin \phi_i \sin \phi_j \cos(d(y_i, y_j))\big)$$

$$= P_k^{(\lambda)}(\cos \phi_i) P_k^{(\lambda)}(\cos \phi_j) + \sum_{s=1}^{k} b_k^{(s)} Q_k^{(s)}(\phi_i) Q_k^{(s)}(\phi_j) P_s^{(\lambda - 1/2)}\big(\cos(d(y_i, y_j))\big)$$

This important equation appears in [S3], page 101. With it in hand we can now compute in detail the matrix A:

$$A_{ij} = \sum_{k=0}^{\infty} a_k P_k^{(\lambda)}\big(\cos(d(x_i, x_j))\big) = \sum_{k=0}^{\infty} a_k P_k^{(\lambda)}(\cos \phi_i) P_k^{(\lambda)}(\cos \phi_j)$$

$$+ \sum_{k=0}^{\infty} a_k \sum_{s=1}^{k} b_k^{(s)} Q_k^{(s)}(\phi_i) Q_k^{(s)}(\phi_j) P_s^{(\lambda - 1/2)}\big(\cos(d(y_i, y_j))\big)$$

This equation shows that A is of the form $E + F$, where

$$E_{ij} = \sum_{k=0}^{\infty} a_k P_k^{(\lambda)}(\cos \phi_i) P_k^{(\lambda)}(\cos \phi_j)$$

$$F_{ij} = \sum_{k=0}^{\infty} a_k \sum_{s=1}^{k} b_k^{(s)} Q_k^{(s)}(\phi_i) Q_k^{(s)}(\phi_j) B_{ij}^{(s)}$$

$$B_{ij}^{(s)} = P_s^{(\lambda-1/2)}\Big(\cos\big(d(y_i, y_j)\big)\Big)$$

First observe that each matrix $B^{(s)}$ is nonnegative definite by Schoenberg's Theorem applied in S^{m-1}. The row and column factors $Q_k^{(s)}(\phi_i)$ do not affect the nonnegative definite property, nor do the nonnegative scalars a_k and $b_k^{(s)}$. (See Problem 10.) Thus F is nonnegative definite. The matrix E is also nonnegative definite, being a sum of nonnegative definite matrices of the form uu^T. Hence A is nonnegative definite. This proves the first conclusion in the theorem. To prove the second, suppose that for some $c \in \mathbb{R}^n$ we have $c^T A c = 0$. Then $c^T E c = c^T F c = 0$. Only the first of these equations seems to be useful. It gives us

$$c^T E c = \sum_{i=1}^{n} \sum_{j=1}^{n} c_i c_j \sum_{k=0}^{\infty} a_k P_k^{(\lambda)}(\cos \phi_i) P_k^{(\lambda)}(\cos \phi_j)$$

$$= \sum_{k=0}^{\infty} a_k \left[\sum_{i=1}^{n} c_i P_k^{(\lambda)}(\cos \phi_i) \right]^2 = 0$$

Assuming $a_0, a_1, \ldots, a_{n-1}$ to be positive, we have

(10) $$\sum_{i=1}^{n} c_i P_k^{(\lambda)}(\cos \phi_i) = 0 \qquad (0 \le k < n)$$

Now $\{P_k^{(\lambda)} : 0 \le k < n\}$ generates the polynomial space Π_{n-1}. We can find a polynomial p in this space such that $p(\cos \phi_i) = c_i$. Here we must note that the numbers $\cos \phi_i$ are all different because we chose the pole of our coordinate system to ensure this. Equation (10) implies that $c = 0$ by the argument used in the proof of Theorem 3.   ∎

Recent work of Menegatto has added to our knowledge of the functions that can be used for interpolation on S^m. We shall quote several of his theorems without proof. The first is a necessary condition for strict positive definiteness. See [Men3], Corollary 3.5.

THEOREM 5. *Let f be a positive definite continuous function on S^m having the representation $f(t) = \sum_{k=0}^{\infty} a_k P_k^{(\lambda)}(\cos t)$, where $a_k \ge 0$, $\lambda = (m-1)/2$, and $\sum_{k=0}^{\infty} a_k P_k^{(\lambda)}(1) < \infty$. If f is strictly positive definite, then the set $\{k : a_k > 0\}$ must contain infinitely many odd and infinitely many even integers.*

Another class of functions studied by Menegatto is the class of conditionally negative definite functions on S^m. This class, denoted by $CND(S^m)$, consists of continuous functions $f : [0, \pi] \to \mathbb{R}$ such that each matrix of the form $A_{ij} = f(d(x_i, x_j))$ has the property

$$\sum_{i=1}^{n} c_i = 0 \quad \Longrightarrow \quad c^T A c \leq 0$$

Matrices of this type arose in Chapter 16. Menegatto characterized the class $CND(S^m)$ as follows. See [Men3], Theorem 2.4.

THEOREM 6. *In order that f be conditionally negative definite on S^m, it is necessary and sufficient that it be representable in the form*

$$f(t) = f(0) + \sum_{k=0}^{\infty} a_k [1 - p_k^{(\lambda)}(\cos t)]$$

in which $a_k \geq 0$, $\lambda = (m-1)/2$, $\sum_{k=0}^{\infty} a_k < \infty$, and $p_k^{(\lambda)}(x) = P_k^{(\lambda)}(x)/P_k^{(\lambda)}(1)$.

In order to use functions in $CND(S^m)$ for interpolation, we require a strengthening of their properties, so that the matrices A in the definition will be nonsingular (when the points x_i are distinct). Here there is only a sufficient condition.

THEOREM 7. *Let f be in $CND(S^m)$ and be represented as in the preceding theorem. If, in addition, $f(0) \geq 0$ and $a_k > 0$ for infinitely many odd and infinitely many even integers, then each matrix $A_{ij} = f(d(x_i, x_j))$ composed from n distinct points x_i is nonsingular. Indeed, it has one positive and $n - 1$ negative eigenvalues.*

The last topic in this chapter concerns the positive definite functions on S^∞. Since

$$S^1 \subset S^2 \subset \cdots \subset S^\infty$$

we see at once that any function that is positive definite on S^∞ must be positive definite on *each* S^m. As one might expect, it is necessary to start with the work of Schoenberg from 1942.

THEOREM 8. *[S3]. If $f \in C[0, \pi]$ and is positive definite on S^∞, then f must have the form $f(t) = \sum_{k=0}^{\infty} a_k \cos^k t$, in which $a_k \geq 0$ and $\sum_{k=0}^{\infty} a_k < \infty$.*

The converse of this was proved by Bingham [Bi]. Thus the continuous positive definite functions on S^∞ are completely characterized. As for the *strictly* positive definite functions, we have this theorem of Menegatto, which improved on an earlier result in [CX3]. See [Men4].

THEOREM 9. *Let $f(t) = \sum_{k=0}^{\infty} a_k \cos^k t$, in which $a_k \geq 0$ for all k and $\sum a_k < \infty$. Let $x_1, x_2, \ldots, x_n$ be distinct points on S^∞, and put $A_{ij} = f(d(x_i, x_j))$. In order that A be positive definite, it is sufficient that $a_\mu > 0$ for some even index $\mu \geq n - 1$ and that $a_\nu > 0$ for some odd index $\nu \geq n - 1$.*

Proof. Assume all the hypotheses. With the choice of a suitable coordinate system, we have $x_i \in S^m$, where m can be $n - 1$. Let $\lambda = (m - 1)/2$ and $p_j(t) = P_j^{(\lambda)}(t)/P_j^{(\lambda)}(1)$. It is known [Bi] that $t^k = \sum_{j=0}^{k} b_{kj}p_j(t)$, where $b_{kj} \geq 0$ in all cases and $b_{kj} > 0$ when $k - j$ is even. Let $\beta = \max\{\mu, \nu\}$, and define a partial sum of f:

$$g(t) = \sum_{k=0}^{\beta} a_k \cos^k t = \sum_{k=0}^{\beta} a_k \sum_{j=0}^{k} b_{kj}p_j(\cos t)$$

$$= \sum_{j=0}^{\beta} \left(\sum_{k=j}^{\beta} a_k b_{kj} \right) p_j(\cos t) = \sum_{j=0}^{\beta} c_j p_j(\cos t)$$

Put $h(t) = \sum_{j=0}^{n-1} c_j p_j(\cos t)$ and $B_{ij} = h(d(x_i, x_j))$. To prove that B is positive definite, we shall use Theorem 4. Thus we should verify that $c_j > 0$ for $0 \leq j \leq n - 1$. We have

$$c_j = \sum_{k=j}^{\beta} a_k b_{kj} \geq a_\mu b_{\mu j} + a_\nu b_{\nu j} > 0$$

Here we note that $j \leq n - 1 \leq \mu$, $\nu \leq \beta$. Also either $\mu - j$ or $\nu - j$ is even, so either $b_{\mu j} > 0$ or $b_{\nu j} > 0$. We complete the proof by noting that $g - h$ is positive definite on S^m by Theorem 4, whereas $f - g$ is positive definite on S^∞ by Bingham's converse of Schoenberg's Theorem. Since

$$A_{ij} = B_{ij} + (g - h)(d(x_i, x_j)) + (f - g)(d(x_i, x_j))$$

we see that A is positive definite.   ∎

Menegatto then obtains the following complete characterization of strictly positive definite functions on S^∞.

THEOREM 10. *In order that a continuous function f be strictly positive definite on S^∞, it is necessary and sufficient that $f(t) = \sum_{k=0}^{\infty} a_k \cos^k t$, in which all $a_k \geq 0$, $\sum_{k=0}^{\infty} a_k < \infty$, and $a_k > 0$ for infinitely many odd and infinitely many even values of k.*

A proof of the triangle inequality for the spherical distance function can be found in [Ry]. A discussion of n-dimensional spherical coordinates is given in [Bls]. The addition formula for Gegenbauer polynomials is in [As], page 30. A large bibliography accompanies the survey by Fasshauer and Schumaker [FaSc].

Problems

1. Use the generating function to prove that $P_n^{(\lambda)}(1) = \dbinom{n + 2\lambda - 1}{n}$. Of course, use the formula

$$\binom{x}{n} = \frac{x}{n} \frac{x - 1}{n - 1} \frac{x - 2}{n - 2} \cdots \frac{x - n + 1}{1}$$

2. Give a formal proof (using induction) that the polar coordinate equations imply $\sum_{j=1}^{m+1} \xi_j^2 = 1$.

3. Let A be a real $n \times n$ matrix that is symmetric. Suppose that $u^T A u > 0$ when $0 \neq u \in \mathbb{R}^n$. Prove that $v^T A v > 0$ when $0 \neq v \in \mathbb{C}^n$.

4. Prove that the Euclidean distance and the geodesic distance are related by

$$\|x - y\| = 2 \sin \left[\frac{1}{2} d(x, y) \right] \qquad (x, y \in \mathcal{S}^m)$$

5. Let $x_1, x_2, \ldots, x_n$ be n distinct points on $\mathcal{S}^m$. Put $A_{ij} = 1 - x_i x_j$, and prove that A is conditionally negative definite. (That means $c^T A c \leq 0$ when $\sum_{i=1}^n c_i = 0$.)

6. Use Theorem 10 to prove that these functions are strictly positive definite on all $\mathcal{S}^m$ $(m = 1, 2, \ldots, \infty) : f(t) = e^{\cos t}$ and $f(t) = (1 - \theta \cos t)^{-1}, 0 < \theta < 1$.

7. Repeat the instructions of Problem 6 for the function $f(t) = \log(1 - \beta \cos t)^{-1}$, $0 < \beta < 1$.

8. Use the generating function for Legendre polynomials:

$$\sum_{k=0}^{\infty} P_k(x) z^k = (1 - 2xz + z^2)^{-1/2}$$

to prove that the function $(3/2 - \cos t)^{-1/2}$ is strictly positive definite on $\mathcal{S}^2$.

9. The weight function used in the definition of the ultraspherical polynomials is $(1 - x^2)^{\lambda - 1/2}$. Prove that this is integrable on $[-1, 1]$, using the hypothesis $\lambda > -1/2$.

10. Let A be an $n \times n$ nonnegative definite matrix. Let $\beta_1, \ldots, \beta_n$ be arbitrary complex numbers. Define

$$B_{ij} = \beta_i \bar{\beta}_j A_{ij}$$

Prove that B is nonnegative definite. Prove a similar theorem about positive definiteness.

11. Prove that the topology on $\mathcal{S}^m$ engendered by the geodesic metric is the same as the relative topology that $\mathcal{S}^m$ receives as a subset of $\mathbb{R}^{m+1}$.

12. Show that the correspondence between rectangular and polar coordinates on $\mathbb{R}^m$ is bijective if we exclude points x whose last two coordinates are 0 and restrict $\theta_1, \ldots, \theta_{m-1}$ to $(0, \pi)$.

References

[AS]   Abramowitz, M., and I. A. Stegun. *Handbook of Mathematical Functions*. National Bureau of Standards, Washington, DC, 1964. Reprint, Dover, New York.

[As]   Askey, R. *Orthogonal Polynomials and Special Functions*. Regional Conference Series in Applied Mathematics, vol. 21. SIAM, Philadelphia, 1975.

[Bi]   Bingham, N. H. "Positive definite functions on spheres." *Proc. Camb. Phil. Soc. 73* (1973), 145–156.

[Bls]   Blumenson, L. E. "A derivation of n-dimensional spherical coordinates." *Amer. Math. Monthly* (1960), 63–66.

[CX3]   Cheney, E. W., and Yuan Xu. "A set of research problems in approximation theory." In *Topics in Polynomials of One and Several Variables and Their Applications,* ed. by T. M. Rassias, H. M. Srivastava, and A. Yanushauskas. World Scientific Publishers, 1992. pp. 109–123.

[FaSc]   Fasshauer, G. E., and L. L. Schumaker. "Scattered data fitting on the sphere." In *Mathematical Methods for Curves and Surfaces II,* ed. by M. Dæhlen, T. Lyche, and L. L. Schumaker. Vanderbilt University Press, Nashville, TN, 1998.

[Men1]   Menegatto, V. A. *Interpolation on Spherical Spaces.* Unpublished doctoral dissertation, University of Texas at Austin, August, 1992.

[Men2]   Menegatto, V. A. "Interpolation on spherical domains." *Analysis 14* (1994), 415–424.

[Men6]   Menegatto, V.A. "Strictly positive definite kernels on the circle." *Rocky Mt. J. Math. 25* (1995), 1149–1163.

[Ry]   Ryan, P. J. *Euclidean and Non–Euclidean Geometry.* Cambridge University Press, Cambridge, 1986.

[S3]   Schoenberg, I. J. "Positive definite functions on spheres." *Duke Math. J. 9* (1942), 96–108.

[Sz]   Szegő, G. *Orthogonal Polynomials.* Amer. Math. Soc. Colloquium Publ. vol. 23, New York, 1959.

[XC2]   Xu, Yuan, and E. W. Cheney. "Strictly positive definite functions on spheres," *Proc. Amer. Math. Soc. 116* (1992), 977–981.

18
Approximation by Positive Definite Functions

In previous chapters, notably Chapters 15 and 16, we considered the problem of interpolation on $\mathbb{R}^s$ by functions of the form

$$F(x) = \sum_{j=1}^{n} c_j f(x - y_j)$$

In this chapter we turn our attention to more general types of interpolation and to the question of approximating a given function by one of the type F above. Specifically, we ask: For what functions f is it true that for each compact set Q in $\mathbb{R}^s$ the set

$$\{x \longmapsto f(x - y) : y \in Q\}$$

is fundamental in $C(Q)$? Recall that a set V in a normed linear space E is said to be **fundamental** if the closure of the span of V is E. In other words, the set of all linear combinations of elements of V is dense in E. Another way to express it is to say that for each $g \in E$ and for each $\varepsilon > 0$, there is a linear combination $\sum c_j v_j$, where $v_j \in V$ and $\|g - \sum c_j v_j\| < \varepsilon$. (The sum here is finite.)

We answer the question posed above by proving that the functions f considered in Chapters 12 to 16 have this remarkable approximation property.

The fundamentality theorems here originated in the work of A. L. Brown [Bro]. He proved Lemma 3 (below) on $C(Q)$, where Q is any compact subset of $\mathbb{R}^s$. He established Theorem 2 for $f(x) = e^{-\|x\|^2}, \|x\|, \sqrt{\|x\|}$ in the same setting. W. A. Light in [L3] proved Lemma 3 and Theorems 1 and 2 in the same setting. The observation that the proofs are valid in $C_0(\mathbb{R}^s)$ was made by Dr. Junjiang Lei in the Approximation Theory Seminar (Austin, Texas, February 1993).

We begin with an easy result in functional analysis that provides a useful technique for proving fundamentality theorems.

131

LEMMA 1. *For a subset V in a normed linear space E, the following two properties are equivalent:*

> **a.** *V is fundamental (that is, its linear span is dense in E).*
> **b.** $V^\perp = 0$ *(that is, 0 is the only element of E^* that annihilates V).*

Proof. Assume **a** and let $\phi \in V^\perp$. Then $\phi(v) = 0$ for all $v \in V$. By the linearity of ϕ, $\phi(x) = 0$ for all x in the linear span of V. By the continuity of ϕ, $\phi(x) = 0$ for all x in closure (span(V)), which is E. Hence $\phi = 0$.

Now assume that **a** is false. Find a point $y \in E\backslash$closure (span(V)). By the Hahn-Banach Theorem, there is a continuous linear functional ϕ such that $\phi(y) = 1$ and $\phi(v) = 0$ for all $v \in V$. Hence **b** is false. Incidentally, the 0 in the equation of part **b** in Lemma 1 is the 0-subspace in E^*. ∎

We will have frequent need for the Riesz Representation Theorem for spaces of continuous functions. A basic version is as follows. See [Ru0], page 139 or [Ru2], page 310; [DS], page 265; and [HewS], page 365.

RIESZ REPRESENTATION THEOREM. *Let X be a compact Hausdorff space, and let C(X) be the Banach space of all continuous real-valued functions on X, normed with the sup-norm. If ϕ is a continuous linear functional on C(X), then there is a signed Borel measure v on X such that, for all $f \in C(X)$,*

$$\phi(f) = \int_X f(x) \, dv(x)$$

Among the signed Borel measures we find the Dirac measures, δ_w, defined for $w \in \mathbb{R}^s$ by declaring $\delta_w(A) = 1$ if $w \in A$, and $\delta_w(A) = 0$ otherwise. Then $\int f(x) \, d\delta_w(x) = f(w)$.

LEMMA 2. *Let X be a compact Hausdorff space, and let $F \in C(X \times X)$. If, for every nonzero signed Borel measure μ on X, we have*

(1)
$$\int_X \int_X F(x, y) \, d\mu(x) \, d\mu(y) \neq 0$$

then the set of functions $\{F^y : y \in X\}$ is fundamental in C(X).

Proof. If the given set is not fundamental, then by Lemma 1 there exists a nonzero functional $\phi \in C(X)^*$ such that $\phi(F^y) = 0$ for all $y \in X$. By the Riesz Representation Theorem, ϕ corresponds to a nonzero signed Borel measure μ in such a way that

$$\phi(g) = \int_X g(x) \, d\mu(x) \qquad (g \in C(X))$$

Consequently,

$$\int_X F^y(x) \, d\mu(x) = \int_X F(x, y) \, d\mu(x) = 0 \qquad (y \in X)$$

It follows that the double integral in Equation (1) is zero. This contradiction concludes the proof. ∎

In view of Lemma 2, it is natural to consider this property of a function F in $C(X \times X)$:

(2)
$$\int_X \int_X F(x, y) \, d\mu(x) \, d\mu(y) > 0$$

whenever μ is a nonzero signed Borel measure on X. We can regard the integral in Inequality (2) as a quadratic form in the variable μ, and we can say that "F is strictly positive definite in the measure sense."

THEOREM 1. *Let μ be a nonnegative finitely valued Borel measure on $\mathbb{R}^s$, and assume that $\mu(O) > 0$ for every nonempty open set O. Let Q be a compact set in $\mathbb{R}^s$. Then the Fourier transform of μ has the property that the set of functions*

$$\{x \longmapsto \hat{\mu}(x - y) : y \in Q\}$$

is fundamental in $C(Q)$.

Proof. The proof is effected by showing that $\hat{\mu}$ is strictly positive definite in the measure sense, and by appealing to Lemma 2. Let v be a nonzero signed Borel measure on Q. We can think of v being defined for all Borel sets in $\mathbb{R}^s$ by writing $v(A) = v(A \cap Q)$. With the aid of Fubini's Theorem for general signed measures, as in [DS], page 193, we have

$$\int \int \hat{u}(x - y) \, dv(x) \, dv(y) = \int \int \int e^{-2\pi i (x-y)z} \, d\mu(z) \, dv(x) \, dv(y)$$

$$= \int \left[\int e^{-2\pi i x z} \, dv(x) \right] \left[\int e^{2\pi i y z} \, dv(y) \right] d\mu(z)$$

$$= \int \left| \int e^{-2\pi i x z} \, dv(x) \right|^2 d\mu(z) = \int |\hat{v}(z)|^2 \, d\mu(z)$$

Since v is a nonzero, finite, signed measure, its Fourier transform is a nonzero continuous function. Hence $|\hat{v}(z)|$ is positive on an open set, O. Since $\mu(O) > 0$, the final expression in the above equation is positive. Our argument uses the fact that the Fourier transform is one-to-one on the space of tempered distributions. For this, refer to Theorem 7.15 and Problem 7.10 in [Ru2]. The measure v can be interpreted as a tempered distribution. ∎

THEOREM 2. *For any $\alpha > 0$ and for any compact set Q in $\mathbb{R}^s$, the set of Gaussian radial functions*

$$\{x \longmapsto e^{-\alpha \|x-y\|^2} : y \in Q\}$$

is fundamental in $C(Q)$.

Proof. This follows from Theorem 1 and the fact that the Gaussian function $e^{-\alpha\|x\|^2}$ is the Fourier transform of another Gaussian function, which is positive everywhere. See Lemma 4 in Chapter 15, page 104. ∎

LEMMA 3. *Let f be completely monotone but not constant on* $[0, \infty)$. *Let* μ *be a nontrivial signed Borel measure on* $\mathbb{R}^s$. *Then*

$$\int \int f(\|x - y\|^2) \, d\mu(x) \, d\mu(y) > 0$$

Proof. By the Bernstein-Widder Theorem (Chapter 14, page 95), f can be represented in the form

$$f(\sigma) = \int_0^\infty e^{-t\sigma} \, d\nu(t)$$

where ν is a finite-valued nonnegative Borel measure. Proceeding as in the proof of Theorem 1, we have

$$I = \int \int \int_0^\infty e^{-t\|x-y\|^2} \, d\nu(t) \, d\mu(x) \, d\mu(y)$$

Again the Fubini Theorem is applicable, since the integrand is in $L^1(\nu \times |\mu| \times |\mu|)$. Hence

$$I = \int_0^\infty \left[\int \int e^{-t\|x-y\|^2} \, d\mu(x) \, d\mu(y) \right] d\nu(t)$$

The bracketed expression here is positive for $t > 0$, by the Problem 3. Since f is not constant, the measure ν is not concentrated at 0. Hence $I > 0$. ∎

THEOREM 3. *Let f be completely monotone but not constant on* $[0, \infty)$. *Let Q be a compact subset (containing at least two points) in* $\mathbb{R}^s$. *Then the set of functions*

$$\{x \longmapsto f(\|x - y\|^2) : y \in Q\}$$

is fundamental in $C(Q)$.

Proof. This follows at once from Lemmas 2 and 3. ∎

In the next results we refer to the space $C_0(\mathbb{R}^s)$. This space consists of all the continuous functions $g : \mathbb{R}^s \to \mathbb{R}$ that "vanish at infinity." This means that for each $\varepsilon > 0$ the set $\{x : |g(x)| \geq x\varepsilon\}$ is compact. The norm in $C_0(\mathbb{R}^s)$ is the usual one:

$$\|g\|_\infty = \sup_{x \in \mathbb{R}^s} |g(x)|$$

Thus normed, the space $C_0(\mathbb{R}^s)$ is a Banach space.

Another version of the Riesz Representation Theorem asserts that if Φ is a continuous linear functional on $C_0(\mathbb{R}^s)$ then there exists a (unique regular) real-valued Borel measure μ on $\mathbb{R}^s$ such that

$$\Phi(g) = \int_{\mathbb{R}^s} g(x) \, d\mu(x) \qquad (g \in C_0(\mathbb{R}^s))$$

For this theorem, consult [Ru0], page 139.

THEOREM 4. *Let f be completely monotone but not constant on $[0, \infty)$. Assume also that $f(t) \to 0$ as $t \to \infty$. Then the set $\{x \mapsto f(\|x - y\|^2) : y \in \mathbb{R}^s\}$ is fundamental in $C_0(\mathbb{R}^s)$.*

Proof. Since $f(t) \to 0$ as $t \to \infty$, there corresponds to any $\varepsilon > 0$ an $r > 0$ such that $f(t^2) < \varepsilon$ when $t > r$. For fixed y and $\|x\| > r + \|y\|$, we have $\|x - y\| > r$. Hence $f(\|x - y\|^2) < \varepsilon$. This indicates that the function $x \mapsto f(\|x - y\|^2)$ is in $C_0(\mathbb{R}^s)$. Now proceed as in the proof of Theorem 2. The measure μ will be on $\mathbb{R}^s$ instead of Q, but we establish that $\mu = 0$ in the same manner as before. ∎

LEMMA 4. *Let f be completely monotone on $[0, \infty)$ and given by $f(u) = \int e^{-tu} \, d\nu(t)$, where ν is a finite-valued Borel measure on $[0, \infty)$. In order that f vanish at ∞, it is necessary and sufficient that $\nu(\{0\}) = 0$.*

Proof. Let $c = \nu(\{0\})$. Then

$$f(u) = \int_{[0, \infty)} e^{-tu} \, d\nu(t) = c + \int_{(0, \infty)} e^{-tu} \, d\nu(t)$$

By the Monotone Convergence Theorem,

$$\lim_{u \to \infty} f(u) = c$$
∎

LEMMA 5. *Let f be a continuous function on $[0, \infty)$ whose derivative is completely monotone but not constant on $(0, \infty)$. Assume that $f'(\infty) = 0$. Let μ be a signed Borel measure on $\mathbb{R}^s$ such that $\mu(\mathbb{R}^s) = 0$. If $\mu \neq 0$ then*

$$(3) \qquad \int_{\mathbb{R}^s} \int_{\mathbb{R}^s} f(\|x - y\|^2) \, d\mu(x) \, d\mu(y) < 0$$

Proof. By a slight modification of Theorem 1 in Chapter 16, page 110, f has the form

$$f(\sigma) = c + \int_0^\infty t^{-1}(1 - e^{-t\sigma}) \, d\nu(t)$$

where c is any real number and ν is nonnegative and satisfies

$$\int_1^\infty t^{-1} \, d\nu(t) < \infty \qquad \int_{0^+}^\infty d\nu(t) > 0$$

Notice that in the present circumstances, f is not required to be nonnegative. Hence c is not necessarily nonnegative. The integral in question is therefore

$$I = \int \int \left[c + \int_0^\infty t^{-1}(1 - e^{-t\|x-y\|^2}) \, dv(t) \right] d\mu(x) \, d\mu(y)$$

Since $\mu(\mathbb{R}^s) = 0$, the c-term drops out in the integration. By L'Hôpital's rule the value of $t^{-1}(1 - e^{-t\sigma})$ at $t = 0$ is σ. Letting J denote $v(\{0\})$, we have

$$I = \int \int \left[J\|x - y\|^2 + \int_{(0,\infty)} t^{-1}\left(1 - \exp(-t\|x - y\|^2)\right) dv(t) \right] d\mu(x) \, d\mu(y)$$

From the proof of Theorem 1 in Chapter 16, page 110, and from Lemma 4 above, we see that $J = f'(\infty)$. Hence by our hypothesis $J = 0$. An application of Fubini's Theorem is now called for. A version applying to signed measures is required ([DS], page 193). Since the signed measure μ necessarily satisfies $|\mu|(\mathbb{R}^s) < \infty$, the integrand is absolutely integrable with respect to $dv \times d|\mu| \times d|\mu|$. Thus we arrive at

$$I = \int_{(0,\infty)} t^{-1}\left[\int \int (1 - e^{-t\|x-y\|^2}) \, d\mu(x) \, d\mu(y) \right] dv(t)$$

$$= -\int_{(0,\infty)} t^{-1}\left[\int \int e^{-t\|x-y\|^2} \, d\mu(x) \, d\mu(y) \right] dv(t)$$

By Lemma 3, the bracketed term is a positive function of t when $t > 0$. Since f' is not constant, v is not concentrated at 0. Hence $I < 0$. ∎

> **THEOREM 5.** *Let $f \in C[0, \infty)$, and assume that f' is completely monotone but not constant on $(0, \infty)$. Assume also that $f(0) \geq 0$ and that $f'(\infty) = 0$. Let Q be a compact set, containing at least two points, in $\mathbb{R}^s$. Then the set of functions*
>
> $$\{x \longmapsto f(\|x - y\|^2) : y \in Q\}$$
>
> *is fundamental in $C(Q)$.*

Proof. Assume all the hypotheses and that the given set is not fundamental. Then there exists a nontrivial continuous linear functional that annihilates all the functions in that set. By the Riesz Representation Theorem, there is a nontrivial signed Borel measure μ on Q such that

$$(4) \qquad \int_Q f(\|x - y\|^2) \, d\mu(x) = 0 \qquad (y \in Q)$$

This leads to

$$\int_Q \int_Q f(\|x - y\|^2) \, d\mu(x) \, d\mu(y) = 0$$

By Lemma 5, we conclude that $\mu(Q) \neq 0$. We therefore construct a second measure having the property $\nu(Q) = 0$:

$$\nu = \mu + \lambda \delta_z$$

Here, δ_z is the Dirac measure concentrated at a point $z \in Q$, and λ is a scalar chosen so that $\nu(Q) = 0$. Thus $\lambda = -\mu(Q)$. Computing the new quadratic form, we have

$$\int \int f(\|x - y\|^2) \, d\nu(x) \, d\nu(y) = \int \left[\int f(\|x - y\|^2) \, d\mu(x) + \lambda f(\|z - y\|^2) \right] d\nu(y)$$
$$= \lambda^2 f(0) \geq 0$$

By Lemma 6, we would expect this to be negative. Hence we conclude that $\nu = 0$, or $\mu = -\lambda \delta_z = \mu(Q) \delta_z$. Armed with this information, we return to Equation (4), and infer that $f(\|z - y\|^2) = 0$ for all $y \in Q$. Since Q has at least two points, we can select y so that $t_0 = \|z - y\|^2 > 0$. We only have to prove that $f(t_0) > 0$ to reach a contradiction. Since $f'' \leq 0$, f' is nonincreasing on $(0, \infty)$. Hence $f'(t) \geq f'(\infty) = 0$ for $t \in (0, \infty)$. Since f' is not constant, there is a point t_1 where $f'(t_1) > 0$. If $0 < \xi \leq t_1$, then $f'(\xi) > 0$ because f' is nonincreasing. If $0 < t \leq t_1$, then by the Mean Value Theorem, $f(t) - f(0) = f'(\xi) t > 0$ for some $\xi \in (0, t)$. Hence $f(t) > f(0) \geq 0$. Since f is nondecreasing, $f(t) > 0$ for all t in $(0, \infty)$, in particular, $f(t_0) > 0$. ∎

An example of a function satisfying the hypotheses of Theorem 4 is $f(t) = \log(1 + t)$.

The techniques used in this chapter can be applied to general positive definite functions, as in Chapters 20 and 21. We shall give several such results.

In Lemma 5, another property is suggested, namely that Inequality (3) is true for nonzero signed measures that satisfy the equation $\mu(X) = 0$. Such measures annihilate constant functions, or we can say $\mu \in \Pi_0^\perp = \{1\}^\perp$. A function having this property is said to be **strictly conditionally negative definite in the measure sense.**

In general, if G is a subset of $C(X)$ we shall say that F is strictly negative definite on $G^\perp$ if the inequality (3) holds for all nonzero $\mu \in G^\perp$.

THEOREM 6. *Let X be a compact Hausdorff space, $F \in C(X \times X)$, and $G \subset C(X)$. If F is strictly negative definite on $G^\perp$, then*

$$G \cup \{F^y : y \in X\}$$

is a fundamental subset of $C(X)$.

Proof. If the given set is not fundamental, then (as in previous proofs) we can find a nontrivial signed Borel measure μ on X such that

$$\int_X g(x) \, d\mu(x) = 0 \qquad (g \in G)$$

and

$$\int_X F^y(x) \, d\mu(x) = 0 \qquad (y \in X)$$

Then it follows that $\mu \in G^{\perp}$ and that

$$\int_X \int_X F(x, y) \, d\mu(x) \, d\mu(y) = 0$$

This contradicts the hypothesis that F is strictly negative definite on $G^{\perp}$.   ∎

If we apply this theorem to $F(x, y) = f(\|x - y\|^2)$, where f is as in Lemma 5, we can only conclude that

$$\{1\} \cup \{F^y : y \in Q\}$$

is fundamental in $C(Q)$. In fact, however, the constant functions need not be adjoined to the set of sections F^y, as we see from Theorem 5.

Now we want to discuss a more general interpolation problem in which point evaluation functionals are replaced by arbitrary linear functionals. (This topic was discussed briefly in Chapters 6 and 9.) A suitable setting for this discussion is a space $C(X)$, where X is compact Hausdorff.

Suppose that $\{\phi_1, \phi_2, \ldots, \phi_n\}$ is a linearly independent set in $(C(X))^*$. We want to find a convenient function g such that

$$(5) \qquad\qquad \phi_i(g) = \lambda_i \qquad (1 \le i \le n)$$

where the λ_i are prescribed real numbers. Can positive definite functions be used to solve this problem?

Assume that the functionals correspond to signed Borel measures v_i as in the Riesz Theorem. Thus for $f \in C(X)$,

$$\phi_i(f) = \int_X f(x) \, dv_i(x) \qquad (1 \le i \le n)$$

Next we select a function $F \in C(X \times X)$ that is strictly positive definite in the measure sense. Thus, for any nontrivial signed Borel measure μ, the following inequality holds:

$$\int \int F(x, y) \, d\mu(x) \, d\mu(y) > 0$$

It is possible to solve the interpolation problem with a set of base functions g_j given by

$$g_j(x) = \phi_j(F_x) = \int F(x, y) \, dv_j(y) \qquad (1 \le j \le n)$$

Thus, we set $g = \sum_{j=1}^n c_j g_j$, and write out the interpolation conditions:

$$\lambda_i = \phi_i(g) = \phi_i\left(\sum_{j=1}^n c_j g_j \right) = \sum_{j=1}^n c_j \phi_i(g_j) = \sum_{j=1}^n A_{ij} c_j$$

in which

$$A_{ij} = \phi_i(g_j) = \int g_j(x) \, dv_i(x) = \int \int F(x, y) \, dv_j(y) \, dv_i(x)$$

Of course we want this matrix to be nonsingular. In fact, it is positive definite, as we now verify. Let $b \in \mathbb{R}^n$, $b \neq 0$. Then

$$b^T A b = \sum_{i=1}^{n} \sum_{j=1}^{n} b_i b_j A_{ij} = \sum_{i=1}^{n} \sum_{j=1}^{n} b_i b_j \int \int F(x, y) \, dv_j(y) \, dv_i(x)$$

$$= \int \int F(x, y) \left[\sum_{j=1}^{n} b_j \, dv_j(y) \right] \left[\sum_{i=1}^{n} b_i \, dv_i(x) \right]$$

If we define the measure μ by putting

$$\mu = \sum_{j=1}^{n} b_j v_j$$

then $\mu \neq 0$, because $\{\phi_1, \phi_2, \ldots, \phi_n\}$ is linearly independent, and so is $\{v_1, v_2, \ldots, v_n\}$. The quadratic form is then

$$b^T A b = \int \int F(x, y) \, d\mu(y) \, d\mu(x) > 0$$

The foregoing analysis establishes this theorem:

THEOREM 7. *Let X be a compact Hausdorff space, and let $\{\phi_1, \phi_2, \ldots, \phi_n\}$ be a linearly independent set of continuous linear functionals on $C(X)$. If $F \in C(X \times X)$ and if F is strictly positive definite in the measure sense, then the interpolation problem $\phi_i(g) = \lambda_i$ $(1 \leq i \leq n)$ has a solution of the form $g(x) = \sum_{j=1}^{n} c_j \phi_j(F_x)$.*

Some references to interpolation with arbitrary linear functionals are [Da2], especially pages 26–31, and [Sard].

Problems

1. Let $x, y, z \in \mathbb{R}^s$. Prove that
$$\left| e^{ixz} - e^{iyz} \right| \leq \|x - y\| \, \|z\|$$

2. Suppose that F satisfies the hypothesis in Lemma 2. Prove that the double integral $\int \int F(x, y) \, dv(x) \, dv(y)$ is positive for all $v \neq 0$ or negative for all $v \neq 0$.

3. Prove that if $\alpha > 0$ then the function $(x, y) \mapsto \exp(-\alpha\|x - y\|^2)$ is strictly positive definite in the measure sense.

4. Give an example of a subset of $C[0, 1]$ that is not fundamental and is not annihilated by any nontrivial functional of the form $\sum_{i=1}^{n} c_i x_i^*$.

References

[Bro]   Brown, A. L. "Uniform approximation by radial basis functions." In *Advances in Numerical Analysis*. vol. 2, ed. by W. A. Light. Oxford University Press, Oxford, 1992. pp. 203–206.

[Da2]   Davis, P. J. *Interpolation and Approximation.* Blaisdell, New York, 1963. Reprint, Dover, New York.

[Do2]   Donoghue, W. F. *Distributions and Fourier Transforms.* Pure and Applied Mathematics Series, vol. 32. Academic Press, New York, 1969.

[DS]   Dunford, N., and J. T. Schwartz. *Linear Operators, Part I, General Theory.* Interscience, New York, 1958. Reprint, Wiley.

[L3]   Light, W. A. "Some aspects of radial basis function approximation." In *Approximation Theory, Spline Functions and Applications,* ed. by S. P. Singh. Kluwer Academic, Boston, 1992, 163–190.

[Ru0]   Rudin, W. *Real and Complex Analysis.* 2nd ed., McGraw-Hill, New York, 1974.

[Ru2]   Rudin, W. *Functional Analysis.* McGraw-Hill, New York, 1973.

[Sard] Sard, A. *Linear Approximation.* Amer. Math. Soc. Math. Surveys, No. 9, Providence, RI, 1963.

19
Approximate Reconstruction of Functions and Tomography

A very general question in applied mathematics is how to reconstruct a function from incomplete information about it. Interpolation processes attempt to reconstruct a function from a finite table of its values. More generally, we can attempt to reconstruct a function f from known values of some linear functionals, say $\phi_1(f)$, $\phi_2(f)$, ..., $\phi_n(f)$. Classical interpolation of the Lagrange type addresses this problem when each ϕ_i is a point-evaluation functional: $\phi_i(f) = f(x_i)$.

Now we want to consider the reconstruction problem when each $\phi_i(f)$ is a line integral of f. This problem arises in "computed tomography." In tomography, one seeks to locate hidden lesions or tumors inside the human body by means of radiographs (x-ray images). When an x-ray beam of known intensity passes through a body, its energy is attenuated in different degrees by different types of tissue. The quantities that *can* be measured directly (or can be easily computed) are the integrals of the density function along lines.

The same principles are involved in "nondestructive testing" of materials. X-rays, sound waves, or other types of waves can be sent through an object in order to detect cracks or other nonuniform characteristics.

Consider a beam of x-rays following a line L that passes through an object. We treat this as a one-dimensional problem for the moment and use t as a coordinate measured along the line L. Let $f(t)$ be the attenuation factor associated with the point t, and let $I(t)$ be the intensity of the beam at t. In traversing a small distance Δt, the intensity is reduced by an amount ΔI according to the formula

$$\Delta I = -I(t) f(t) \, \Delta t$$

This leads to the differential equation

$$\frac{dI}{dt} = -I(t) f(t)$$

which we solve by writing it in the form

$$\frac{dI}{I} = -f(t)\, dt$$

and integrating between t_0 and t_1. The result is

$$\log I(t_1) - \log I(t_0) = -\int_{t_0}^{t_1} f(t)\, dt$$

We take t_0 and t_1 to be points outside the object. In fact t_0 can be the source of the beam, and $I(t_0)$ will be the intensity of the beam at the source. The point t_1 is located where the intensity can be measured after the beam has passed through the object. Hence $I(t_0)$ and $I(t_1)$ are known or measured quantities. These measurements then provide the value of $\int_L f(t)\, dt$. This description is from Natterer's book [Na]. We are led to the question of how to recover f from a knowledge of all its line integrals.

In this problem, the practical dimensions of our variables are two or three, but we can just as easily include the case of any dimension, m. In this discussion we follow Smith [Sm] and Natterer [Na].

Let $f : \mathbb{R}^m \to \mathbb{R}$. This is the *density* function, that is, the function to be reconstructed from various line integrals. In describing these integrals we often specify a direction by selecting a vector v of norm 1; that is, $v \in S^{m-1}$. Two types of x-ray scanners are in use. They produce these data:

$$(P_v f)(x) = \int_{-\infty}^{\infty} f(x + tv)\, dt \qquad \text{Parallel beam transform}$$

$$(D_x f)(v) = \int_0^{\infty} f(x + tv)\, dt \qquad \text{Divergent beam transform}$$

In both equations, x is an arbitrary point in $\mathbb{R}^m$. Note that P_v and D_x are linear transformations. A basic problem is now to reconstruct f. If $P_v f$ is available, then f is to be reconstructed from $P_v f$, for $v \in S^{m-1}$. If $D_x f$ is available, then f is to be reconstructed

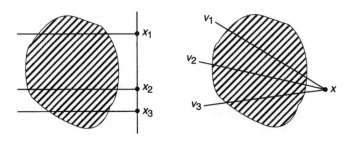

Figure 19.1   *Parallel Beams and Divergent Beams*

from $D_x f$, for $x \in v^\perp$. This is a theoretical question. In the practical version, only a finite number of these integrals will be available, and an approximate reconstruction is sought. In current medical practice the parallel beam scanner has yielded to the divergent beam scanner, but in other applications, such as in nondestructive testing of materials, both methods may be in use. First, we discuss the divergent beam scan. We denote the Euclidean norm by $\| \cdot \|$.

LEMMA 1. *Let* $r(x) = \|x\|$, *where* $x \in \mathbb{R}^m$. *Then*

$$\int_{S^{m-1}} (D_x f)(v) \, dS^{m-1}(v) = (r^{1-m} * f)(x)$$

Proof. The convolution on the right side of the equation can be expressed as follows (in which S_t^{m-1} is a sphere of radius t).

$$(r^{1-m} * f)(x) = \int_{\mathbb{R}^m} f(x - y)\|y\|^{1-m} \, dy$$

$$= \int_0^\infty \int_{S_t^{m-1}} f(x - y)\|y\|^{1-m} \, dS_t^{m-1}(y) \, dt$$

$$= \int_0^\infty \int_{S^{m-1}} f(x - tv) \, dS^{m-1}(v) \, dt$$

$$= \int_{S^{m-1}} (D_x f)(-v) \, dS^{m-1}(v)$$

$$= \int_{S^{m-1}} (D_x f)(v) \, dS^{m-1}(v) \qquad \blacksquare$$

LEMMA 2. *In the sense of distributions,* $\widehat{r^\alpha} = c \, r^{-m-\alpha}$. *(Here,* $r(x) = \|x\|$ *and* $x \in \mathbb{R}^m$.*)*

Proof. The equation is to be interpreted in the sense of distributions, or generalized functions. The proof can be found in Gelfand and Shilov [GS], page 192. Here is an outline. We find that $\widehat{r^\alpha}$ is positively homogeneous of degree $-m - \alpha$ because for $t > 0$,

$$\widehat{r^\alpha}(tx) = \int r^\alpha(y) e^{-2\pi i t x y} \, dy \qquad (ty = z)$$

$$= \int r^\alpha(z/t) e^{-2\pi i x z} (t^{-m} \, dz)$$

$$= t^{-m-\alpha} \int r^\alpha(z) e^{-2\pi i x z} \, dz = t^{-m-\alpha} \widehat{r^\alpha}(x)$$

Next we observe that $\widehat{r^\alpha}$ is "spherically symmetric" or "rotationally invariant." This is an instance of a general principle that the Fourier transform preserves such symmetry. To see that this is so, let f be rotationally invariant. Let U be a rotation; that is, any orthogonal $m \times m$ matrix with $\det(U) = +1$. Then, with the substitution $y = Uz$, we have

$$\hat{f}(Ux) = \int f(y)e^{-2\pi iUxy}\,dy$$

$$= \int f(y)e^{-2\pi ixU^T y}\,dy$$

$$= \int f(Uz)e^{-2\pi ixz}(\det U)\,dz$$

$$= \int f(z)e^{-2\pi ixz}\,dz$$

$$= \hat{f}(x)$$

Since $\widehat{r^\alpha}$ has the two properties mentioned, we conclude that $\widehat{r^\alpha} = cr^{-m-\alpha}$. The constant c depends on m and α. It turns out to be ([GS] page 193)

$$c = 2^{\alpha+m}\,\Gamma\!\left(\frac{\alpha+m}{2}\right)\!\Big/\Gamma(-\alpha/2) \qquad \blacksquare$$

Definition. Define an operator L by the equation

$$(Lf)^\wedge = r\hat{f} \qquad (r(x) = \|x\|)$$

This operator is one of the **Riesz potentials**, defined for $\alpha < m$ by the equation $(L_\alpha f)^\wedge = r^\alpha \hat{f}$. At the end of this chapter, a direct formula for L is given.

THEOREM 1. *The reconstruction formula for the divergent beam x-ray transform is $f = c^{-1}Lg$, where*

$$g(x) = \int_{S^{m-1}} (D_x f)(v)\,dS^{m-1}(v) \qquad (x \in \mathbb{R}^m)$$

Proof. First we note that

$$L(r^{1-m} * f) = cf$$

where c is the constant in Lemma 2. To prove this, we take a Fourier transform of the left-hand side, and then use the definition of L and Lemma 2:

$$[L(r^{1-m} * f)]^\wedge = r(r^{1-m} * f)^\wedge$$

$$= r(r^{1-m})^\wedge \hat{f}$$

$$= crr^{-1}\hat{f} = c\hat{f}$$

Now by Lemma 1, $Lg = cf$. $\qquad \blacksquare$

THEOREM 2. *The reconstruction formula for the parallel beam x-ray transform is*

$$f = (2c)^{-1}L\left[\int_{S^{m-1}} (P_v f)(\bullet)\,dS^{m-1}(v)\right]$$

Proof.

$$(P_v f)(x) = \int_{-\infty}^{\infty} f(x + tv)\, dt$$

(1)
$$= \int_{-\infty}^{0} f(x + tv) + \int_{0}^{\infty} f(x + tv)\, dt$$

$$= \int_{0}^{\infty} f(x - tv)\, dt + \int_{0}^{\infty} f(x + tv)\, dt$$

$$= (D_x f)(-v) + (D_x f)(v)$$

By Theorem 1, $f = c^{-1} L g$. As v runs over $\mathbb{S}^{m-1}$, so does $-v$. Hence

$$f = c^{-1} L\left[\int_{\mathbb{S}^{m-1}} (D_\bullet f)(-v)\, d\mathbb{S}^{m-1}(v) \right]$$

Hence we have, on adding these two equations,

$$2f = c^{-1} L\left[\int_{\mathbb{S}^{m-1}} \{(D_\bullet f)(v) + (D_\bullet f)(-v)\}\, d\mathbb{S}^{m-1}(v) \right]$$

$$= c^{-1} L\left[\int_{\mathbb{S}^{m-1}} (P_v f)(\bullet)\, d\mathbb{S}^{m-1}(v) \right] \qquad \blacksquare$$

THEOREM 3. *("Fourier Slice Theorem") If $v \in \mathbb{S}^{m-1}$ and $x \perp v$, then $\hat{f}(x) = (P_v f)^{\wedge}(x)$. (We regard $P_v f$ as defined on $v^{\perp}$.)*

Proof. An integral over $\mathbb{R}^m$ can be expressed as an iterated integral:

$$\int_{\mathbb{R}^m} f(y)\, dy = \int_{v^{\perp}} \int_{-\infty}^{\infty} f(z + tv)\, dt\, dz$$

Applying this fact to $f(y)e^{-2\pi i x y}$ gives

$$\hat{f}(x) = \int_{\mathbb{R}^m} f(y)e^{-2\pi i x y}\, dy = \int_{v^{\perp}} \int_{-\infty}^{\infty} f(z + tv)e^{-2\pi i (z + tv)x}\, dt\, dz$$

Assume that $x \perp v$. Then the preceding equation asserts that

$$\hat{f}(x) = \int_{v^{\perp}} \int_{-\infty}^{\infty} f(z + tv)\, dt\, e^{-2\pi i z x}\, dz$$

$$= \int_{v^{\perp}} (P_v f)(z)e^{-2\pi i z x}\, dz$$

$$= (P_v f)^{\wedge}(x) \qquad \blacksquare$$

Theorem 3 shows that the density function is fully determined by its radiographs.

The operator L can be given in a direct formula. We need the functions

$$H_j(x) = \xi_j \|x\|^{-m-1} \qquad x = (\xi_1, \xi_2, \cdots, \xi_m)$$

This can be interpreted as a principal-value distribution whose action on a test function ϕ is defined by

$$\langle H_j, \phi \rangle = \lim_{\varepsilon \downarrow 0} \int_{\|x\| > \varepsilon} H_j(x)\phi(x)\, dx$$

This eventually leads to the formula

$$Lf = \sum_{j=1}^{m} \frac{\partial}{\partial \xi_j} H_j * f$$

where the convolution must be interpreted in the principal-value sense. Thus

$$(Lf)(x) = \sum_{j=1}^{m} \frac{\partial}{\partial \xi_j} \lim_{\varepsilon \downarrow 0} \int_{\|y\| > \varepsilon} H_j(y)f(x-y)\, dy$$

Problems

1. Prove that the Riesz potentials have this property: $L_\alpha^{-1} = L_{-\alpha}$.

2. A function $f : \mathbb{R}^m \to \mathbb{R}$ is said to be "radial" if it is of the form $f(x) = g(\|x\|)$ for some function g. We say that f is "rotationally invariant" if for all orthogonal matrices U, we have $f(x) = f(Ux)$. Prove that these two properties are equivalent.

3. Let $r(x) = \|x\|$. Prove that if the radial function $g \circ r$ is positively homogeneous of degree λ, then $g(t) = ct^\lambda$.

References

[CEH]   Censor, Y., T. Elfving, and G. Herman, eds. "Special issue on linear algebra in image reconstruction from projections." *Linear Alg. Appl. 130* (1990), 1–305.

[Dav]   Davison, M. E. "A singular value decomposition for the Radon transform in n-dimensional Euclidean space." *Numer. Func. Anal. Optim. 3* (1981), 321–340.

[De]   Deans, S. R. *The Radon Transform and Some of Its Applications.* Wiley, New York, 1983.

[GS]   Gelfand, I. M., and G. E. Shilov. *Generalized Functions,* vol. 1. Academic Press, New York, 1964.

[HSSW]   Hamaker, C., K. T. Smith, D. C. Solmon, and S. L. Wagner. "The divergent beam x-ray transform." *Rocky Mt. J. Math. 10* (1980), 253–283.

[HaSo]   Hamaker, C., and D. C. Solmon. "The angles between the null spaces of x-rays." *J. Math. Anal. Appl. 62* (1978), 1–23.

[Helg2]   Helgason, S. *The Radon Transform.* Birkhäuser, Basel, 1980.

[HazS]   Hazou, I., and D. C. Solmon. "Approximate inversion formulas and convolution kernels for the exponential x-ray transform." *Zeit. Angew. Math. Mech. 66* (1986), T370–T372.

[Her]   Hertle, A. "A characterization of Fourier and Radon transforms on Euclidean space." *Trans. Amer. Math. Soc. 273* (1982), 595–609.

[HN]   Herman, G., and F. Natterer. *Mathematical Aspects of Computerized Tomography.* Springer-Verlag, Berlin, 1981.

[HLN]   Herman, G., A. K. Louis, and F. Natterer. *Mathematical Methods in Tomography.* Lecture Notes in Math., vol. 1497. Springer-Verlag, New York, 1991.

[Herm]   Herman, G. T. (ed.) *Image Reconstruction from Projections.* Topics in Applied Physics, vol. 32. Springer-Verlag, New York, 1979.

[Herm2]   Herman, G. T. *Image Reconstruction from Projections.* Academic Press, New York, 1980.

[Lei1]   Lei, Junjiang. "The multivariate Radon transform." *Approx. Theory Appl. 3* (1987), 30–49.

[Log]   Logan, B. F. "The uncertainty principle in reconstructing functions from projections." *Duke Math. J. 42* (1975), 661–706.

[LoS]   Logan, B. F., and L. A. Shepp. "Optimal reconstruction of a function from its projections." *Duke Math. J. 42* (1975), 645–659.

[Ma1]   Madych, W. R. "Summability and approximate reconstruction from Radon transform data." *Contemp. Math. 113* (1990), 189–219.

[Ma2]   Madych, W. R. "Degree of approximation in computerized tomography," in *Approximation Theory III,* ed. by E. W. Cheney. Academic Press, New York, 1980, 615–621.

[MN7]   Madych, W. R., and S. A. Nelson. "Polynomial based algorithms for computed tomography." *SIAM J. Applied Math. 43* (1983), 157–185. Part II, ibid., *44* (1984), 193–208.

[MN8]   Madych, W. R., and S. A. Nelson. "Approximate inversion formulas for Radon transform data." In *Approximation Theory IV,* ed. by C. K. Chui, L. L. Schumaker, and J. D. Ward. Academic Press, New York, 1983, pp. 599–604.

[MN9]   Madych, W. R., and S. A. Nelson. "Characterization of tomographic reconstructions which commute with rigid motions." *J. Functional Analysis 46* (1982), 258–263.

[Na]   Natterer, F. *The Mathematics of Computerized Tomography.* Wiley, New York, 1986.

[PSS]   Petersen, B. E., K. T. Smith, and D. C. Solmon. "Sums of plane waves, and the range of the Radon transform." *Math. Ann. 243* (1979), 153–161.

[Shepp]   Shepp, L. A. (ed.) *Computed Tomography.* Proc. Symp. in Applied Math., vol. 27. Amer. Math. Soc., Providence, RI, 1983.

[SK]   Shepp, L. A., and J. B. Kruskal. "Computerized tomography: The new medical x-ray technology." *Amer. Math. Monthly 85* (1978), 420–439.

[Sm]   Smith, K. T. "Reconstruction formulas in computed tomography." In *Computed Tomography,* ed. by L. A. Shepp. Proc. Symp. in Applied Math, vol. 27. Amer. Math. Soc., Providence, RI, 1983.

[SSW]   Smith, K. T., D. C. Solmon, and S. L. Wagner. "Practical and mathematical aspects of reconstructing objects from radiographs." *Bull. Amer. Math. Soc. 83* (1977), 1227–1270.

[Sol]   Solmon, D. C. "The x-ray transform." *J. Math. Anal. Appl. 71* (1976), 61–83.

[Stri]   Strichartz, R. S. "Radon inversion—variations on a theme." *Amer. Math. Monthly 89* (1982), 377–384.

[Tay2]   Taylor, M. E. *Noncommutative Harmonic Analysis.* Amer. Math. Soc. Mathematical Surveys and Monographs, vol. 22. Amer. Math. Soc., Providence, RI, 1986.

20
Approximation by Convolution

Many of the linear methods used to produce approximations are based on the convolution of a "kernel function" with the function being approximated. Weierstrass himself used such a method to prove his famous approximation theorem in 1888. For two functions defined on $\mathbb{R}^s$, their convolution is defined formally by the equation

$$(f * g)(x) = \int_{\mathbb{R}^s} f(y)g(x - y)\, dy$$

Since the integral is improper, a question of its convergence arises. If, for example, $f \in L^1(\mathbb{R}^s)$ and $g \in L^\infty(\mathbb{R}^s)$, the integral exists and defines $f * g$ as an L^∞-function. This is easily verified (Problem 1). A more interesting theorem is given in Problem 2. For functions in L^1, we have

> **THEOREM 1.** *If f and g belong to $L^1(\mathbb{R}^s)$, then so does $f * g$, and $\|f * g\|_1 \leq \|f\|_1 \|g\|_1$.*

Proof. (From [Ru0], page 156.) The functions f and g, being Lebesgue measurable, are equal almost everywhere to Borel functions. Since the integrals that arise here are not affected by substituting equivalent functions, we may assume that f and g are Borel functions.

Put $F(x, y) = f(y)g(x - y)$. To prove that F is a Borel function, write

$$F(x, y) = f(y)g(x - y) = (f \circ \psi)(x, y)(g \circ \phi)(x, y)$$

where $\psi(x, y) = y$ and $\phi(x, y) = x - y$. Since f, g, ϕ, and ψ are Borel functions, so are $f \circ \psi$, $g \circ \phi$ and their product, F. Here we can use, for example, Theorems 1.9(c) and 1.12(d) in [Ru0].

Now write

$$\int \int |F(x, y)|\, dx\, dy = \int |f(y)| \int |g(x - y)|\, dx\, dy = \|f\|_1 \|g\|_1 < \infty$$

Hence F is integrable. By the Fubini Theorem, [Ru0, Theorem 7.8] $(f * g)(x)$ is well-defined almost everywhere and $f * g$ is an L^1-function. Furthermore,

$$\|f * g\|_1 = \int \left| \int F(x, y)\, dy \right| dx \le \int \int |F(x, y)|\, dy\, dx$$

$$= \int \int |F(x, y)|\, dx\, dy = \|f\|_1 \|g\|_1 \qquad \blacksquare$$

The basic theorem showing how convolutions can be used to approximate functions is as follows.

THEOREM 2. *Let $H_1, H_2, \ldots$ be a sequence in $L^1(\mathbb{R}^s)$ such that $\sup_k \|H_k\|_1 < \infty$ and $\int_{\mathbb{R}^s} H_k(x)\, dx = 1$ for all k. Assume further that for any $\delta > 0$ the following equation is true:*

$$\lim_{k \to \infty} \int_{\|x\| > \delta} |H_k(x)|\, dx = 0$$

Then, for each bounded function f in $C(\mathbb{R}^s)$, we have $H_k * f \to f$, the convergence being uniform on compact sets in $\mathbb{R}^s$.

Proof. Since the integral of H_k is 1, we have

(1)
$$\begin{aligned}
|(H_k * f)(x) - f(x)| &= \left| \int_{\mathbb{R}^s} H_k(y)[f(x - y) - f(x)]\, dy \right| \\
&\le \int_{\mathbb{R}^s} |H_k(y)|\, |f(x - y) - f(x)|\, dy
\end{aligned}$$

For any positive r, the r-ball at 0 is defined by

$$B_r = \{x \in \mathbb{R}^s : \|x\| \le r\}$$

Let us now fix r and establish the desired uniform convergence on B_r. Define

$$\omega(f; \delta) = \sup\{|f(x - y) - f(x)| : \|x\| \le r,\ \|y\| \le \delta\}$$

This is a "modulus of continuity" of f. Since a continuous real-valued function defined on a compact set is necessarily uniformly continuous, we have $\omega(f; \delta) \downarrow 0$ as $\delta \downarrow 0$. Let $\varepsilon > 0$, and select δ so small that $\omega(f; \delta) < \varepsilon/(2M)$, where $M = \sup_k \|H_k\|_1$. By the hypotheses on H_k, one can select m so that

$$\int_{\|y\| > \delta} |H_k(y)|\, dy < \frac{\varepsilon}{4\|f\|_\infty} \qquad (k \ge m)$$

Now let $x \in B_r$ and $k \geq m$. The final integral in Equation (1) can be split into two parts, which will be estimated separately as follows.

$$\int_{\|y\|>\delta} |H_k(y)| \, |f(x-y) - f(x)| \, dy \leq 2\|f\|_\infty \int_{\|y\|>\delta} |H_k(y)| \, dy$$

$$\leq 2\|f\|_\infty \frac{\varepsilon}{4\|f\|_\infty} = \frac{\varepsilon}{2}$$

and

$$\int_{\|y\|\leq\delta} |H_k(y)| \, |f(x-y) - f(x)| \, dy \leq \omega(f; \delta) \int_{\|y\|\leq\delta} |H_k(y)| \, dy$$

$$\leq \frac{\varepsilon}{2M} \|H_k\|_1 \leq \frac{\varepsilon}{2} \qquad \blacksquare$$

Example 1. Theorem 2 is illustrated by the sequence

$$H_k(x) = \begin{cases} c_k(1 - \|x\|^2)^k & \|x\| \leq 1 \\ 0 & \|x\| > 1 \end{cases}$$

In this definition, $x \in \mathbb{R}^s$, $\|x\|$ is the Euclidean norm of x, and c_k is a constant chosen so that $\int_{\mathbb{R}^s} H_k(x) \, dx = 1$. This sequence of kernels will be used to prove the Weierstrass Approximation Theorem for compact sets in $\mathbb{R}^s$. (See Theorem 3 below.) The only hypothesis in Theorem 2 that is troublesome for this example is the equation

$$\lim_{k\to\infty} \int_{\|x\|>\delta} |H_k(x)| \, dx = 0 \qquad (\delta > 0)$$

Let δ be given. If $\delta \geq 1$, then the integrand $H_k(x)$ is zero for $\|x\| > \delta$. Hence, we may assume that $0 < \delta < 1$. Using polar coordinates, we have, for any ρ in $(0, 1)$,

$$c_k^{-1} = \int_{\|x\|\leq 1} (1 - \|x\|^2)^k \, dx = \omega_{s-1} \int_0^1 (1 - r^2)^k r^{s-1} \, dr$$

$$\geq \omega_{s-1} \int_0^\rho (1 - r^2)^k r^{s-1} \, dr \geq (1 - \rho^2)^k \omega_{s-1} \int_0^\rho r^{s-1} \, dr$$

$$= \omega_{s-1}(1 - \rho^2)^k \rho^s / s$$

(The coefficients ω_{s-1} are explained in Chapter 23, page 177.) This establishes that

$$c_k \leq s(1 - \rho^2)^{-k} \rho^{-s} \omega_{s-1}^{-1}$$

Similarly, we have

$$\int_{\|x\|>\delta} |H_k(x)| \, dx = \omega_{s-1} \int_\delta^1 c_k(1 - r^2)^k r^{s-1} \, dr$$

$$\leq \omega_{s-1} c_k (1 - \delta^2)^k \int_\delta^1 r^{s-1} \, dr$$

$$\leq \omega_{s-1} c_k (1 - \delta^2)^k \int_0^1 r^{s-1} \, dr$$

$$\leq s(1 - \rho^2)^{-k} \rho^{-s} (1 - \delta^2)^k s^{-1}$$

$$= \left(\frac{1 - \delta^2}{1 - \rho^2} \right)^k \rho^{-s}$$

If we choose ρ so that $0 < \rho < \delta$, then this last expression converges to 0 as k tends to ∞. ∎

Example 1 and Theorem 3, below, are adapted from Smith [Sm2], Section 6.8. Figure 20.1 shows the graphs of kernels H_1 and H_{10} (when $s = 2$, of course).

THEOREM 3 (The Weierstrass Approximation Theorem). *Let K be a compact set in $\mathbb{R}^s$. The polynomials in s variables form a dense set in $C(K)$.*

Proof. Let $f \in C(K)$ and select r so that $K \subset B_r$. Define $K' = (1/3r)K$ and $g(x) = f(3rx)$, for $x \in K'$. By the Tietze extension theorem, we may assume that $g \in C(\mathbb{R}^s)$, $\|g\|_\infty < \infty$, and $g(x) = 0$ on $\mathbb{R}^s \setminus B_{1/2}$. Let H_n be the kernel in Example 1. Applying Theorem 2, we have $H_n * g \to g$, uniformly on K'. It remains to be seen whether these are *polynomial* approximations. For $x \in K'$

$$(H_n * g)(x) = (g * H_n)(x) = \int_{B_{1/2}} g(y) H_n(x - y) \, dy$$

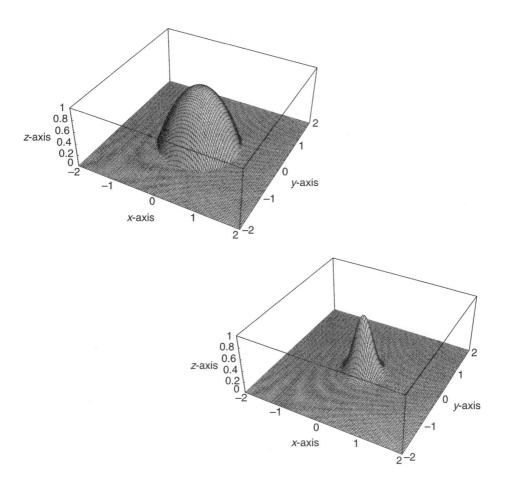

Figure 20.1   *The Kernels H_1 and H_{10}*

In this integral $\|x\| \leq 1/3$ and $\|y\| \leq 1/2$. Hence $\|x - y\| < 1$, and $H_n(x - y) = c_n(1 - \|x - y\|^2)^n$. Thus $H_n * g$ is a polynomial, p_n. Put $q_n(x) = p_n(x/3r)$. For $x \in K$, let $z = x/3r$. Then

$$\begin{aligned} |f(x) - q_n(x)| &= |f(3rz) - q_n(3rz)| \\ &= |g(z) - p_n(z)| \leq \|g - p_n\|_{K'} \to 0 \end{aligned}$$ ∎

It often occurs that a sequence of kernels like the ones in Theorem 2 can be obtained by "dilating" a single kernel. This simple process is described as follows.

Given a function H in $L^1(\mathbb{R}^s)$, we define

$$(2) \qquad\qquad H_k(x) = k^s H(kx) \qquad (k \in \mathbb{N}, x \in \mathbb{R}^s)$$

We call these *dilations* or *dilates* of H.

> **THEOREM 4.** *Let H be a function in $L^1(\mathbb{R}^s)$ such that $\int H(x)\, dx = 1$. Let H_k denote the dilates of H. For each bounded function $f \in C(\mathbb{R}^s)$, we have (in the topology of uniform convergence on compact sets)*
>
> $$H_k * f \to f \qquad (\text{as } k \to \infty)$$

Proof. The substitution $z = ky$ gives us

$$\int_{\mathbb{R}^s} H_k(y)\, dy = \int_{\mathbb{R}^s} H(ky)k^s\, dy = \int_{\mathbb{R}^s} H(z)\, dz = 1$$

The same substitution shows that $\|H_k\|_1 = \|H\|_1$. For any $\delta > 0$, we have

$$\int_{\|y\|>\delta} |H_k(y)|\, dy = \int_{\|y\|>\delta} |H(ky)|k^s\, dy = \int_{\|z\|>k\delta} |H(z)|\, dz \to 0$$

The hypotheses of Theorem 2 have now been verified, and the conclusion may be drawn that $H_k * f \to f$ for every bounded continuous function f on $\mathbb{R}^s$.   ∎

This theorem is the prototype of many similar results to be found in the literature. See [Ru2], page 157; [SW], page 10; [Shap3]; [Shap2]; [J]; [But1]; and [Dev2]. A theorem similar to Theorem 4 but for periodic functions will be found in Chapter 21 (page 158).

Example 2. In any space $\mathbb{R}^s$, let H be the function

$$H(x) = \begin{cases} c & \|x\| \leq 1 \\ 0 & \|x\| > 1 \end{cases}$$

where c is determined so that $\int_{\mathbb{R}^s} H = 1$. Then, as in the proof of Theorem 2, we have

$$|(H_k * f - f)(x)| \leq c \int_{\|z\|\leq 1} \left| f\left(x - \frac{z}{k}\right) - f(x) \right|\, dz \leq \omega\left(f, r, \frac{1}{k}\right)$$

Here $\omega(f, r, \delta)$ is the modulus of continuity of f on the sets B_r and $x \in B_r$.   ∎

For any kernel H that is Riemann integrable, the convolution $f * H_k$ can be approximated by a Riemann sum:

(3)
$$(f * H_k)(x) = \int f(y)H_k(x - y)\, dy$$
$$\approx \sum_i c_i f(y_i) H_k(x - y_i)$$

It follows that $f * H_k$ is a limit of linear combinations of *translates* of H_k. Thus, when we take into account the dilation due to the parameter k, we see that f is being represented as a limit of linear combinations of dilations and translates of H. Theorem 2 therefore has the following corollary:

THEOREM 5. *If* $H \in C(\mathbb{R}^s) \bigcap L^1(\mathbb{R}^s)$ *and* $\int_{\mathbb{R}^s} H = 1$, *then the set*

$$\{x \longmapsto H(kx + v) : k \in \mathbb{N},\ v \in \mathbb{R}^s\}$$

is fundamental in $C(\mathbb{R}^s)$.

A formal proof of Theorem 5 is not presented here because a similar result (Theorem 1 in Chapter 24, page 185) occurs later, accompanied by a full proof.

In Theorem 5, the continuity of H is assumed only to ensure that the approximating Riemann sums in Equation (3) are in the space $C(\mathbb{R}^s)$. If H is not continuous, then we are, in effect, approximating a continuous f by linear combinations of discontinuous functions in Equation (3).

Example 3. The Weierstrass kernel is

$$W(x) = \pi^{-s/2} e^{-\|x\|^2} \qquad (x \in \mathbb{R}^s)$$

It is known that $W \in L^1(\mathbb{R}^s)$ and $\int_{\mathbb{R}^s} W(x)\, dx = 1$. (See Chapter 15 for the proofs.) Furthermore, if $f \in L^p(\mathbb{R}^s)$, where $1 \le p \le \infty$, then $(W_k * f)(x) \to f(x)$ for almost all x. See [SW], page 13.

Example 4. The Cauchy kernel is $x \mapsto c(1 + \|x\|^2)^{-1}$, where $x \in \mathbb{R}^s$.

Example 5. For $s = 1$, the Jackson kernel is $x \mapsto 3(x^{-1} \sin x)^4/\pi$, where $x \in \mathbb{R}$.

LEMMA 1. *If f is a Lebesgue measurable function on $\mathbb{R}^s$, then the function* $(x, y) \mapsto f(x - y)$ *is measurable on* $\mathbb{R}^s \times \mathbb{R}^s$.

Proof. [Sm2], page 308. Let $F(x, y) = f(x - y)$ and $U(x, y) = (x - y, x + y)$, where x and y are points in $\mathbb{R}^s$. We shall verify first that for any set $A \subset \mathbb{R}$

(4)
$$F^{-1}(A) = U^{-1}[f^{-1}(A) \times \mathbb{R}^s]$$

This is done by confirming the following equivalences:

$$(x, y) \in F^{-1}(A) \iff F(x, y) \in A \iff f(x - y) \in A$$
$$\iff x - y \in f^{-1}(A) \iff (x - y, x + y) \in f^{-1}(A) \times \mathbb{R}^s$$
$$\iff U(x, y) \in f^{-1}(A) \times \mathbb{R}^s \iff (x, y) \in U^{-1}[f^{-1}(A) \times \mathbb{R}^s]$$

Thus, Equation (4) is correct. If A is open, then $f^{-1}(A)$ is measurable (because f is measurable). Then the set $f^{-1}(A) \times \mathbb{R}^s$ is measurable (by the theory of product measures). Since U^{-1} is a C^1-function, $U^{-1}[f^{-1}(A) \times \mathbb{R}^s]$ is measurable [Sm2, Theorem 4.5]. Hence $F^{-1}(A)$ is measurable by Equation (4), and F is measurable. ∎

The following theorem and proof are from [Shap3].

THEOREM 6. *Let $H \in L^1(\mathbb{R}^s)$ and satisfy $\int_{\mathbb{R}^s} H(x)\, dx = 1$. For each $f \in L^1(\mathbb{R}^s)$ we have*

$$\lim_{k \to \infty} \|H_k * f - f\|_1 = 0$$

Proof. Proceed as in the proof of Theorem 1, obtaining

$$\|H_k * f - f\|_1 = \int_{\mathbb{R}^s} |(H_k * f - f)(x)|\, dx$$

(5)
$$= \int_{\mathbb{R}^s} \left| \int_{\mathbb{R}^s} H_k(y)[f(x-y) - f(x)]\, dy \right| dx$$

$$\leq \int_{\mathbb{R}^s} \int_{\mathbb{R}^s} |H_k(y)|\,|f(x-y) - f(x)|\, dy\, dx$$

By the preceding lemma, $(x, y) \mapsto f(x - y)$ is measurable. So is $(x, y) \mapsto H_k(y)$. Their product is also measurable. The integrand in the last integral above is thus a measurable function of two variables. The Fubini Theorem [Ru0, Theorem 7.8] may therefore be applied. The order of integration can be reversed, and we arrive at

(6)
$$\|H_k * f - f\|_1 \leq \int_{\mathbb{R}^s} |H_k(y)| g(y)\, dy$$

in which the function g is given by

$$g(y) = \int_{\mathbb{R}^s} |f(x-y) - f(x)|\, dx$$

By appealing to the following lemma, we find that g is continuous at 0. Hence, if $\varepsilon > 0$, we can find $r > 0$ such that $g(y) < \varepsilon/(2\|H\|_1)$ when $\|y\| < r$. The integral in Equation (6) is divided into two parts. The first integral is over the ball B_r and the second over $\mathbb{R}^s \backslash B_r$. For the first of these,

$$\int_{B_r} |H_k(y)| g(y)\, dy \leq \frac{\varepsilon}{2\|H\|_1} \int_{\mathbb{R}^s} |H_k(y)|\, dy = \varepsilon/2$$

For the second, we now assume that k is chosen so large that

$$\int_{\|z\| > kr} |H(z)|\, dz < \frac{\varepsilon}{4\|f\|_1}$$

Then

$$\int_{\mathbb{R}^s \backslash B_r} |H_k(y)| g(y)\, dy \leq 2\|f\|_1 \int_{\mathbb{R}^s \backslash B_r} |H_k(y)|\, dy$$

$$= 2\|f\|_1 \int_{\|z\| \geq kr} |H(z)|\, dz < \varepsilon/2 \qquad ∎$$

LEMMA 2. *Let T_y denote the shift operator on $L^1(\mathbb{R}^s)$, and let $f \in L^1(\mathbb{R}^s)$. Then $\lim\limits_{y \to 0} \|T_y f - f\|_1 = 0$.*

Proof. [HewS], page 199. Let $\varepsilon > 0$. Recall that the set of continuous functions having compact support is dense in $L^1(\mathbb{R}^s)$. Select such a function, g, for which $\|f - g\|_1 < \varepsilon/3$. Since g has compact support, there is an $r > 0$ such that $g(x) = 0$ when $\|x\|_\infty \geq r$. Since g is uniformly continuous, there is a $\delta \in (0, 1)$ such that

$$\|y\|_\infty \leq \delta \implies |g(x + y) - g(x)| < \frac{\varepsilon}{4(2r + 2)^s}$$

Note that if μ is Lebesgue measure on $\mathbb{R}^s$, then $\mu\{x : \|x\|_\infty \leq r + 1\} = (2r + 2)^s$. If $\|y\|_\infty \leq \delta$, then we have

$$
\begin{aligned}
\|T_y g - g\|_1 &= \int |g(x + y) - g(x)| \, dx \\
&= \int_{\|x\|_\infty \leq r+1} |g(x + y) - g(x)| \, dx \\
&< \frac{\varepsilon}{4(2r + 2)^s} \mu\{x : \|x\|_\infty \leq r + 1\} \\
&= \frac{\varepsilon}{4}
\end{aligned}
$$

It now follows that

$$\|T_y f - f\|_1 \leq \|T_y f - T_y g\|_1 + \|T_y g - g\|_1 + \|g - f\|_1 \leq \frac{\varepsilon}{3} + \frac{\varepsilon}{4} + \frac{\varepsilon}{3} < \varepsilon \quad \blacksquare$$

With the help of Lemma 1, an alternative proof of Theorem 1 can be given. By Lemma 1, $|F|$ as defined in the proof of Theorem 1 is Lebesgue measurable, even if we do not first replace f and g by equivalent Borel functions. Then the Fubini Theorem applies and the calculations in the proof of Theorem 1 can proceed with the original f and g. The technicalities in both proofs of Theorem 1 are required because the composition of two Lebesgue measurable functions need not be measurable. See, for example, [HewS], page 151.

Problems

1. Prove that if $g \in L^1(\mathbb{R}^s)$ and $f \in L^\infty(\mathbb{R}^s)$, then $g * f \in L^\infty(\mathbb{R}^s)$.

2. (Continuation) Prove that under the hypotheses of Problem 1, $g * f$ is, in fact, a bounded and uniformly continuous function. (This is harder. Lusin's Theorem is useful.)

3. Prove that $f * g = g * f$.

4. Let $H \in L^1(\mathbb{R}^s)$, $H \geq 0$, and $\|H\|_1 = 1$. Prove that if f is uniformly continuous on $\mathbb{R}^s$, then the modulus of continuity satisfies $\omega(H * f; \delta) \leq \omega(f; \delta)$.

5. Prove that if H has the properties in Theorem 2, and f is a periodic function of period p, then $H_k * f$ is also periodic with period p.

6. Prove that if H has the properties in Theorem 2, and if f is a trigonometric polynomial of order n then so is $H_k * f$.

7. Let H be as in Theorem 2. Prove that if H has compact support and if f is uniformly continuous then $H_k * f$ converges uniformly to f on $\mathbb{R}^s$.

8. Prove that for f and g in $L^1(\mathbb{R}^s)$, we have

$$\| |f| * |g| \|_1 = \|f\|_1 \|g\|_1$$

9. (Continuation)   Find the exact conditions for the equality

$$\|f * g\|_1 = \|f\|_1 \|g\|_1$$

10. Prove that if f and g belong to $L^1(\mathbb{R}^s)$, then $\int f * g = \int f \int g$.

11. Let $H_1, H_2, \ldots$ be a sequence satisfying the hypotheses of Theorem 2. Define operators L_k by writing $L_k f = H_k * f$. Prove that each L_k is a bounded linear operator on the space $C_b(\mathbb{R}^s)$ consisting of bounded continuous functions. Use the sup-norm on $\mathbb{R}^s$. Compute the norm of L_k, and determine conditions under which L_k is a projection.

12. Prove that if f is Borel measurable and if g is Lebesgue measurable, then $f \circ g$ is Lebesgue measurable.

References

[But1]   Butzer, P. L. "Representation and approximation of functions by general singular integrals." *Neder. Akad. Wetensch. Proc. Ser. 63A (Indag. Math. 22)*, (1960), 1–24.

[Dev2]   DeVore, R. A. *The Approximation of Continuous Functions by Positive Linear Operators.* Lecture Notes in Math., vol. 293. Springer-Verlag, New York, 1972.

[HewS]   Hewitt, E., and K. Stromberg. *Real and Abstract Analysis.* Springer-Verlag, New York, 1965.

[J]   Jackson, D. *The Theory of Approximation.* Amer. Math. Soc., Providence, RI, 1930.

[Ru0]   Rudin, W. *Real and Complex Analysis.* 2nd ed., McGraw-Hill, New York, 1974.

[Ru2]   Rudin, W. *Functional Analysis.* McGraw-Hill, New York, 1973.

[Shap2]   Shapiro, H. S. "Convergence almost everywhere of convolution integrals with a dilation parameter." In *Constructive Theory of Functions of Several Variables.* Lecture Notes in Math., vol. 571. Springer-Verlag, New York, 1977, pp. 250–266.

[Shap3]   Shapiro, H. S. *Smoothing and Approximation of Functions.* Van Nostrand, New York, 1969.

[Sm2]   Smith, K. T. *Primer of Modern Analysis.* Bogden and Quigley, Boston, 1971. Reprinted by Springer-Verlag, New York.

[SW]   Stein, E. M., and G. Weiss. *Introduction to Fourier Analysis on Euclidean Spaces.* Princeton University Press, Princeton, NJ, 1971.

21
The Good Kernels

In the preceding chapter we saw that any kernel $H \in L^1(\mathbb{R}^s)$ whose integral was unity could be used to provide approximations by convolution:

$$(1) \qquad (H_k * f)(x) = \int H_k(y)f(x-y)\, dy = \int k^s H(ky)f(x-y)\, dy$$

Now we wish to address the question of whether further assumptions on H can lead to enhanced approximation properties. The sort of assumption we might make about H is illustrated in this incomplete list:

1. $H(x)$ decays rapidly to 0 as $x \to \infty$.
2. $\hat{H}(x)$ decays rapidly or has compact support.
3. H has some extra smoothness.
4. $H * p = p$ for all $p \in \Pi_m(\mathbb{R}^s)$.

Some theorems to illustrate this theme are given next. In the first of these, adapted from [Shap3], we require the class $\text{Lip}_\alpha(\mathbb{R}^s)$, which consists of all functions f on $\mathbb{R}^s$ for which a constant c exists satisfying

$$|f(x) - f(y)| \le c\|x - y\|_\infty^\alpha \qquad (x, y \in \mathbb{R}^s)$$

THEOREM 1. *Let $H \in L^1(\mathbb{R}^s)$, $\int_{\mathbb{R}^s} H = 1$, and have the property that the function $x \mapsto \|x\|^\alpha H(x)$ is integrable, for some $\alpha \in (0, 1]$. For each $f \in \text{Lip}_\alpha(\mathbb{R}^s)$ there exists a constant A such that*

$$\|H_k * f - f\|_\infty \le A/k^\alpha \qquad (k = 1, 2, 3, \ldots)$$

Proof. Using the change of variables $z = ky$, we have

$$|(H_k * f)(x) - f(x)| \leq \int |H_k(y)|\ |f(x - y) - f(x)|\ dy$$

$$\leq \int |H(z)|\ \left|f\left(x - \frac{z}{k}\right) - f(x)\right|\ dz$$

$$\leq \int |H(z)|c\left\|\frac{z}{k}\right\|^\alpha dz$$

$$\leq ck^{-\alpha} \int \|z\|^\alpha |H(z)|\ dz$$

$$= Ak^{-\alpha} \qquad\qquad \blacksquare$$

In the next theorem we require the definition of an "integer-periodic" function. This means that

$$f(x + m) = f(x) \qquad (x \in \mathbb{R}^s,\ m \in \mathbb{Z}^s)$$

Also required is the Fourier transform, defined for a function f by the formula

(2) $$\hat{f}(x) = \int_{\mathbb{R}^s} f(y)e^{-2\pi ixy}\ dy \qquad (x \in \mathbb{R}^s)$$

Here xy denotes the ordinary inner product (or "dot" product) of the points x and y in $\mathbb{R}^s$. Finally, we require the definition of a multivariate trigonometric polynomial of degree at most n. Such a function has the form

(3) $$g(x) = \sum_{\substack{m \in \mathbb{Z}^s \\ \|m\|_1 \leq n}} c_m e^{2\pi imx}$$

THEOREM 2. [Shap3] *Let $H \in L^1(\mathbb{R}^s)$ and $\int_{\mathbb{R}^s} H = 1$. Assume that $\hat{H}(x) = 0$ when $\|x\|_1 \geq 1$. If f is in $C(\mathbb{R}^s)$ and is integer-periodic, then $H_k * f$ is a multivariate trigonometric polynomial of degree at most $k - 1$.*

Proof. For any multi-integer m, let $e_m(x) = e^{2\pi imx}$. We shall prove that

(4) $$H_k * e_m = \hat{H}(m/k)e_m$$

(Thus, e_m is an eigenvector of the operator H_k*, and $\hat{H}(m/k)$ is the corresponding eigenvalue.) To establish Equation (4), let $x \in \mathbb{R}^s$. Then

$$(H_k * e_m)(x) = \int H_k(y)e^{2\pi im(x-y)}\ dy$$

$$= e^{2\pi imx} \int H_k(y)e^{-2\pi imy}\ dy$$

$$= e_m(x) \int H(z)e^{-2\pi imz/k}\ dz$$

$$= e_m(x)\hat{H}(m/k)$$

It follows that if g is any trigonometric polynomial, then $H_k * g$ is a trigonometric polynomial of degree at most $k - 1$, because if $\|m\|_1 \geq k$, then $\hat{H}(m/k) = 0$. By the Stone-Weierstrass Theorem, each integer-periodic continuous function f is expressible as $\lim_{n\to\infty} g_n$, where g_n is a trigonometric polynomial and the convergence is uniform. By the continuity of H_k*, $H_k * f = \lim_{n\to\infty} H_k * g_n$. The latter is a trigonometric polynomial of degree at most $k - 1$. ∎

The preceding theorem, used in conjunction with Theorem 4 of Chapter 20 (page 152), provides a specific sequence of trigonometric polynomials converging to a given function f; namely $f = \lim_{k\to\infty} H_k * f$. In order to obtain an H for which $\text{supp}(\hat{H}) \subset \{x : \|x\|_1 \leq 1\}$, we can start with a suitable $\hat{H}$ and take its inverse Fourier transform. This inverse is given by the formula

$$H(x) = \int_{\mathbb{R}^s} \hat{H}(y) e^{2\pi i x y} \, dy \qquad (x \in \mathbb{R}^s)$$

An example in the one-variable case is given by $H(x) = \text{sinc}^2(x)$. This function is discussed in Chapter 29. Its Fourier transform has as its support the interval $[-1, 1]$ and $\int \text{sinc}^2 = 1$. Consequently, Theorem 2 applies to it. This kernel is one form of **Fejér's kernel.** Fejér used it to prove his famous theorem that the Fourier series of a continuous periodic function is uniformly $(C, 1)$-summable to the function. See, for example, [C1], page 122. Our Theorem 2 shows that the function

$$p_k(x) = \int \text{sinc}^2(kx - ky) f(y) k \, dy$$

is a trigonometric polynomial of degree at most $k - 1$, and that $p_k \to f$ uniformly, provided that f is integer-periodic and continuous.

The next two theorems come from [CLX]. We shall use this notation:

$$V_\alpha(x) = x^\alpha/\alpha! \qquad\qquad (x \in \mathbb{R}^s \,, \alpha \in \mathbb{Z}^s_+)$$

$$H_\lambda(x) = \lambda^s H(\lambda x) \qquad\qquad (x \in \mathbb{R}^s \,, \lambda > 0)$$

$$(T_v f)(x) = f(x - v) \qquad\qquad (x, v \in \mathbb{R}^s)$$

$$|f|_j = \max_{|\alpha| = j} \, \sup_{x \in \mathbb{R}^s} |D^\alpha f(x)|$$

$$\|f\|_k = \sum_{j=0}^{k} |f|_j$$

The symbol $D^\alpha f$ denotes a partial derivative of f given by

$$(D^\alpha f)(x) = \frac{\partial^{\alpha_1 + \cdots + \alpha_s}}{\partial x_1^{\alpha_1} \cdots \partial x_s^{\alpha_s}} \qquad (\alpha = (\alpha_1, \alpha_2, \ldots, \alpha_s))$$

THEOREM 3. *Let $H \in L^1(\mathbb{R}^s)$ and satisfy, for some k,*

    a. $V_\alpha H \in L^1(\mathbb{R}^s)$ *when* $|\alpha| \leq k$
    b. $|H_\lambda * V_\alpha - V_\alpha| \leq \lambda^{-k} A$ *when* $|\alpha| < k$

(Here A is a function on $\mathbb{R}^s$.) Then there is a constant c such that for $f \in C^k(\mathbb{R}^s)$ we have

(5) $$|H_\lambda * f - f| \le \lambda^{-k}[c\,|f|_k + A(0)\|f\|_k]$$

Proof. Fix $f \in C^k(\mathbb{R}^s)$ and $x \in \mathbb{R}^s$. Let p be the Taylor polynomial of degree $k-1$ for the function f, centered at x:

$$p(y) = \sum_{|\alpha| < k} (D^\alpha f)(x) V_\alpha(y - x)$$

If we set $c_\alpha = (D^\alpha f)(x)$, we can write

(6) $$p = \sum_{|\alpha| < k} c_\alpha T_x V_\alpha$$

By Taylor's Theorem, the remainder $r = f - p$ satisfies the equation

$$r(y) = \sum_{|\alpha|=k} (D^\alpha f)(\xi_y) V_\alpha(y - x)$$

Now we can write

$$H_\lambda * f - f = H_\lambda * (f - p) + (H_\lambda * p - p) + (p - f)$$
$$= H_\lambda * r + (H_\lambda * p - p) - r$$

Evaluate at x and use the fact that $r(x) = 0$ to get

(7) $$|(H_\lambda * f - f)(x)| \le |(H_\lambda * r)(x)| + |(H_\lambda * p - p)(x)|$$

The two terms on the right will be analyzed separately.

$$|(H_\lambda * r)(x)| = \left| \int H_\lambda(x - y) r(y)\, dy \right|$$

$$\le \int |H_\lambda(x - y)| \sum_{|\alpha|=k} |(D^\alpha f)(\xi_y)|\,|V_\alpha(y - x)|\, dy$$

$$\le |f|_k \int |H(\lambda x - \lambda y)| \sum_{|\alpha|=k} |V_\alpha(y - x)| \lambda^s\, dy$$

$$= |f|_k \int |H(z)| \sum_{|\alpha|=k} \left| V_\alpha\!\left(\frac{-z}{\lambda}\right) \right|\, dz$$

$$= \lambda^{-k} |f|_k \sum_{|\alpha|=k} \int |H(z) V_\alpha(z)|\, dz$$

By Hypothesis (a), the integrals in this last expression are finite, and their sum can be denoted by c. This justifies the first term on the right in Inequality (5). For the second term on the right of Inequality (7), we have

$$H_\lambda * p - p = H_\lambda * \sum_{|\alpha|<k} c_\alpha T_x V_\alpha - \sum_{|\alpha|<k} c_\alpha T_x V_\alpha$$

$$= \sum c_\alpha (H_\lambda * T_x V_\alpha - T_x V_\alpha)$$

$$= \sum c_\alpha T_x (H_\lambda * V_\alpha - V_\alpha)$$

Using Hypothesis (b), we have

$$|H_\lambda * p - p| \le \sum |c_\alpha| T_x |H_\lambda * V_\alpha - V_\alpha|$$
$$\le \sum |c_\alpha| T_x (\lambda^{-k} A)$$
$$= \lambda^{-k} \sum |c_\alpha| T_x A$$

Evaluating at x gives us

$$|(H_\lambda * p - p)(x)| \le \lambda^{-k} A(0) \sum |c_\alpha|$$
$$= \lambda^{-k} A(0) \|f\|_k \qquad \blacksquare$$

The function A in Theorem 3 can be defined as

$$A(x) = \sup_{\lambda > 0} \max_{|\alpha| < k} \lambda^k |(H_\lambda * V_\alpha - V_\alpha)(x)|$$

Estimate (5) in Theorem 3 is of use only if $\|f\|_k < \infty$.

According to Theorem 3, if we want *rapid* convergence of our approximations $H_\lambda * f$ to f, we must require that the kernel H satisfy the extra conditions (a) and (b). The next theorem shows how to modify a given kernel h in a simple manner to obtain a new kernel H that has property (b). We must start with an h that already has property (a).

THEOREM 4. *Let $h \in L^1(\mathbb{R}^s)$ and $\int_{\mathbb{R}^s} h = 1$. Assume that for some k, $V_\alpha h \in L^1(\mathbb{R}^s)$ when $|\alpha| \le k$. Construct a new kernel*

(8) $$H(x) = \sum_{i=1}^m a_i h(x/t_i) t_i^{-s} \qquad (x \in \mathbb{R}^s)$$

where a_i and t_i are chosen so that $t_i \ne 0$ and

(9) $$p(0) = \sum_{i=1}^m a_i p(t_i) \qquad (p \in \Pi_{k-1}(\mathbb{R}))$$

Then $H * V_\alpha = V_\alpha$ for $|\alpha| < k$ and $H * V_\alpha - V_\alpha$ is a constant function for $|\alpha| = k$.

Proof. For any f such that $fh \in L^1(\mathbb{R}^s)$, we have

$$(H * f)(x) = \sum_{i=1}^m a_i \int h(y/t_i) f(x - y) t_i^{-s} \, dy$$
$$= \sum_{i=1}^m a_i \int h(z) f(x - t_i z) \, dz$$
$$= \int h(z) \sum_{i=1}^m a_i f(x - t_i z) \, dz$$

Let $f = V_\alpha$, with $|\alpha| < k$. The function $t \mapsto V_\alpha(x - tz)$ is in $\Pi_{k-1}(\mathbb{R})$, and thus by Equation (9)

$$\sum_{i=1}^m a_i V_\alpha(x - t_i z) = V_\alpha(x)$$

Hence,

$$(H * V_\alpha)(x) = \int h(z) V_\alpha(x) \, dz = V_\alpha(x)$$

If $|\alpha| = k$, then we use the binomial theorem to write

$$(H * V_\alpha)(x) = \int h(z) \sum_{i=1}^{m} a_i V_\alpha(x - t_i z) \, dz$$

$$= \int h(z) \sum_{i=1}^{m} \sum_{\beta \leq \alpha} a_i t_i^{|\beta|} V_\beta(-z) V_{\alpha-\beta}(x) \, dz$$

$$= \int h(z) \sum_{\beta \leq \alpha} A_\beta V_\beta(-z) V_{\alpha-\beta}(x) \, dz$$

where we have put $A_\beta = \sum_{i=1}^{m} a_i t_i^{|\beta|}$. By Equation (9), $A_\beta = 1$ if $\beta = 0$ and $A_\beta = 0$ if $0 < |\beta| < k$. Hence only the terms corresponding to $\beta = 0$ and $\beta = \alpha$ remain in the sum above, and we have

$$(H * V_\alpha)(x) = \int h(z) [V_\alpha(x) + A_\alpha V_\alpha(-z)] \, dz$$

$$= V_\alpha(x) + A_\alpha \int h(z) V_\alpha(-z) \, dz \qquad \blacksquare$$

Equation (9) can be valid with m as small as k. If $t_1, t_2, \ldots, t_k$ are k different nonzero points in $\mathbb{R}$, we can find the appropriate coefficients a_i by using the Lagrange interpolation formula, with t_i as nodes. That formula asserts that each p in Π_{k-1} obeys the equation

$$p = \sum_{i=1}^{k} p(t_i) \ell_i$$

This implies that

$$p(0) = \sum_{i=1}^{k} p(t_i) \ell_i(0)$$

and that in Equation (9) we should take $a_i = \ell_i(0)$, $i = 1, 2, \ldots, k$. Since

$$\ell_i(t) = \prod_{\substack{j=1 \\ j \neq i}}^{k} \frac{t - t_j}{t_i - t_j}$$

we have

$$a_i = \ell_i(0) = \prod_{\substack{j=1 \\ j \neq i}}^{k} \frac{-t_j}{t_i - t_j}$$

An even simpler way to obtain a formula such as Equation (9) is to use the "forward difference functional" defined by

$$\Delta^k f = \sum_{j=0}^{k} (-1)^{k-j} \binom{k}{j} f(j)$$

It is known that Δ^k annihilates Π_{k-1}. Hence, for $p \in \Pi_{k-1}$,

$$0 = \Delta^k p = (-1)^k p(0) + \sum_{j=1}^{k} (-1)^{k-j} \binom{k}{j} p(j)$$

Solving for $p(0)$, we obtain

$$p(0) = \sum_{j=1}^{k} (-1)^{j-1} \binom{k}{j} p(j)$$

For results about Δ^k, see [AS], page 877.

Problems

1. Prove that if $\text{supp}(\hat{H}) \subset \{x \in \mathbb{R}^s : \|x\|_\infty \leq r\}$ then for each continuous integer-periodic f and for each k, $H_k * f$ is a trigonometric polynomial of degree at most $s(rk - 1)$.

2. Prove these properties of convolution:
   a. $|f * g| \leq |f| * |g| \leq \|f\|_\infty \|g\|_1$
   b. $T_x(f * g) = (T_x f) * g = f * (T_x g)$
   c. $f * g = g * f$
   d. $|f * (gh)| \leq \|g\|_\infty |f| * |h|$

3. Prove that if $H * V_\alpha = V_\alpha$ then $H_\lambda * V_\alpha = V_\alpha$, and conversely.

4. Prove that in Equation (9), $m \geq k$.

5. The forward difference *operator* is defined by $(\Delta f)(x) = f(x + 1) - f(x)$. Prove that
   a. $(\Delta^n f)(x) = \sum_{j=0}^{n} (-1)^{n-j} \binom{n}{j} f(x + j)$
   b. $\Delta(\Pi_k) = \Pi_{k-1}$

6. Let $v_k(t) = t^k$. Find the formula for $\Delta^k v_k$.

7. The class $\text{Lip}_\alpha(\mathbb{R}^s)$ consists of all functions for which

$$\sup_{x \neq y} \frac{|f(x) - f(y)|}{\|x - y\|^\alpha} < \infty$$

Prove that this expression defines a seminorm on $\text{Lip}_\alpha(\mathbb{R}^s)$. Identify the functions whose seminorm is zero.

8. Prove that $\text{Lip}_\alpha(\mathbb{R})$ is not interesting when $\alpha > 1$.

9. Prove that the linear span of $\{e_m : m \in \mathbb{Z}^s\}$ is an algebra. Here $e_m(x) = e^{2\pi i m x}$. Prove that this algebra separates the points of $\mathbb{R}^s$ and contains the constant functions. Draw a suitable conclusion from the Stone-Weierstrass Theorem, as mentioned in the proof of Theorem 2.

10. Prove that if $0 < \alpha < \beta \le 1$, then $\text{Lip}_\beta(\mathbb{R}) \subset \text{Lip}_\alpha(\mathbb{R})$, and that this inclusion is proper.

11. Prove that the Binomial Theorem, in terms of the functions V_α, is given by

$$V_\alpha(x + y) = \sum_{0 \le \beta \le \alpha} V_\beta(x) V_{\alpha-\beta}(y)$$

12. Prove that the operator Δ^k annihilates Π_{k-1}.

References

[AS]   Abramowitz, M., and I. A. Stegun. *Handbook of Mathematical Functions.* National Bureau of Standards, Washington, 1964. Reprint, Dover, New York.

[C1]   Cheney, E. W. *Introduction to Approximation Theory.* McGraw-Hill, New York, 1966. 2nd ed., Chelsea, New York, 1982.

[CLX]   Cheney, E. W., W. A. Light, and Yuan Xu. "On kernels and approximation orders." In *Approximation Theory: Proceedings of the Sixth Southeastern International Approximation Conference,* ed. by G. Anastassiou. Dekker, New York, 1992, 227–242.

[Shap3] Shapiro, H. S. *Smoothing and Approximation of Functions.* Van Nostrand, New York, 1969.

[Stef] Steffensen, J. F. *Interpolation.* Chelsea, New York, 1950.

22

Ridge Functions

In multivariate approximation theory, special functions called "ridge" functions are useful. We begin with their definition.

Let X be a normed linear space. A function $f : X \to \mathbb{R}$ is called a **ridge function** if it can be represented in the form $f = g \circ \phi$, where $g : \mathbb{R} \to \mathbb{R}$ and $\phi \in X^*$. (Recall that X^* is the space of all continuous linear functionals on X.)

Example 1. We know that every continuous linear functional on $\mathbb{R}^s$ is of the form

$$\phi(x) = \alpha_1 \xi_1 + \alpha_2 \xi_2 + \cdots + \alpha_s \xi_s = ax$$

Here we have written $x = (\xi_1, \xi_2, \ldots, \xi_s)$ and $a = (\alpha_1, \alpha_2, \ldots, \alpha_s)$. A ridge function on $\mathbb{R}^s$ is then a function of the form

$$f(x) = g(\alpha_1 \xi_1 + \alpha_2 \xi_2 + \cdots + \alpha_s \xi_s) \qquad \blacksquare$$

Example 2. In the preceding example, let $s = 2$, and select a specific vector a, say $a = (1, 0)$. This leads to ridge functions of the form

$$f(x) = f(\xi_1, \xi_2) = g(ax) = g(\xi_1)$$

The graph of such a function is a ruled surface, for if (ξ_1, ξ_2, r) is a point on the graph then so is every point $(\xi_1, \xi_2 + t, r)$ for arbitrary t. Similar remarks can be made for any vector $a = (\alpha_1, \alpha_2)$. If $b = (-\alpha_2, \alpha_1)$, then we have $ab = 0$. Consequently

$$g(ax) = g\big(a(x + tb)\big) \qquad (t \in \mathbb{R})$$

and again the graph contains the line

$$(\xi_1, \xi_2, r) + t(-\alpha_2, \alpha_1, 0) \qquad r = g(ax) \qquad \blacksquare$$

Figure 22.1 shows the ridge function $z = p(x - y)$, where p is a certain polynomial of degree 4.

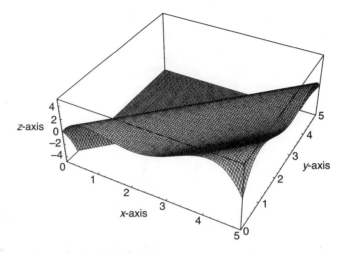

Figure 22.1   *A Typical Ridge Function*

We shall be interested almost exclusively in continuous ridge functions, and accordingly g must be chosen to be continuous; i.e., $g \in C(\mathbb{R})$. In this connection, see Problem 1.

A single ridge function is very limited in its capacity to approximate an arbitrary continuous function on X. In particular, the graph of a ridge function is a ruled surface. One therefore considers linear combinations of ridge functions. The general form would be

$$f = \sum_{i=1}^{m} c_i g_i \circ \phi_i \qquad (g_i \in C(\mathbb{R}) , \ \phi_i \in X^*)$$

At this level of generality, the coefficients c_i are not needed, for they can be absorbed by the function g_i. Not all continuous functions on X are linear combinations of ridge functions. (See Problem 2.) But every continuous function on X can be well approximated by such linear combinations, as we shall prove in Theorem 2, page 168. First, it is necessary to describe in more detail the space $C(X)$ of all continuous real-valued functions on the normed linear space X. The linear and multiplicative structures in $C(X)$ are given by the definitions

$$(\alpha f + \beta g)(x) = \alpha f(x) + \beta g(x)$$
$$(fg)(x) = f(x)g(x)$$

The topological structure in $C(X)$ can be defined by a suitable notion of convergence. Here we must consider generalized sequences or "nets." These are like ordinary sequences $[f_\nu]$ except that the index ν is allowed to run over an arbitrary directed set. (A **directed set** is a set V with an transitive order relation $\geq$ such that for any two elements p and q in V there is a third element ν satisfying $\nu \geq p$ and $\nu \geq q$.) Now we define convergence in $C(X)$. One writes $f_\nu \rightarrow f$ if for every compact set K in X,

$$\lim_\nu \sup_{x \in K} |f_\nu(x) - f(x)| = 0$$

This convergence is often called "compact convergence" or "uniform convergence on compacta." The topology itself is called "the topology of uniform convergence on compacta" or the "compact-open topology."

A convenient notation to assist in describing convergence in $C(X)$ is this:

$$\|f\|_K = \sup_{x \in K} |f(x)|$$

Then the statement $f_v \to f$ translates into

$$\|f_v - f\|_K \longrightarrow 0$$

for all compact sets K in X.

The topology of $C(X)$ can also be described by giving a base for the neighborhoods of 0. The elements of this base are sets depending on a positive number ε and a compact set K in X. We set

$$V(K, \varepsilon) = \{g \in C(X) : \|g\|_K < \varepsilon\}$$

These sets are "basic" neighborhoods of 0. Other neighborhoods of 0 are obtained by taking arbitrary unions of basic neighborhoods of 0. Neighborhoods of other points in $C(X)$ are obtained by translation. Thus a typical "basic" neighborhood of f is

$$\{g \in C(X) : \|f - g\|_K < \varepsilon\}$$

A subset D in $C(X)$ is **dense** if its closure is $C(X)$. Another way of saying this is that every nonempty open set intersects D. A third way of saying this is that for every f in $C(X)$ and for every neighborhood U of f, $D \cap U$ is nonempty. A fourth restatement is that for each $f \in C(X)$, for each compact $K \subset X$, and for each $\varepsilon > 0$ there is an element $g \in D$ for which $\|f - g\|_K < \varepsilon$. Finally, the density of D means that for any $f \in C(X)$ there is a net (or generalized sequence) $g_v \in D$ such that $g_v \to f$.

For further information on $C(X)$ and its topology, we refer the reader to [Ru2], page 24ff; [KN], pages 69, 81; and [Ke], page 229.

With the preceding definitions, $C(X)$ becomes a topological algebra. Since X is a normed linear space, it is *not* a compact topological space, and the usual Stone-Weierstrass Theorem does not apply to it. However, there is a suitable substitute as follows.

THEOREM 1. *Let X be a normed linear space (or, indeed, any Hausdorff topological space). If $\mathcal{A}$ is a subalgebra of $C(X)$ that contains constants and separates the points of X, then $\mathcal{A}$ is dense in $C(X)$.*

Proof. Let f be any element of $C(X)$. It is to be proved that each neighborhood of f contains an element of $\mathcal{A}$. It suffices to prove that each "basic" neighborhood of f contains a member of $\mathcal{A}$. Therefore, let K be any compact set in X and let ε be any positive number. By restricting f and all members of $\mathcal{A}$ to the compact set K, we make it possible to apply the classical version of the Stone-Weierstrass Theorem in $C(K)$. Its conclusion is that the set

$$\{g|K : g \in \mathcal{A}\}$$

is dense in $C(K)$. Hence there is an element g in $\mathcal{A}$ such that $\|f - g\|_K < \varepsilon$. ∎

The basic Stone-Weierstrass Theorem is proved in [Ru2], page 115. Other accounts are in [F], page 116; [RS], page 102; and [DS], page 272. See also [Bla2], [Pr], [GJ], [Sem], and [Sim].

A subset of $C(X)$ is said to be **fundamental** if its linear span is dense. For example, the monomial functions, $x_n(t) = t^n$, form a fundamental set in $C(\mathbb{R})$, by the version of the Stone-Weierstrass Theorem we have just proved. With this concept at our command, we can state a central problem in the study of ridge functions:

Problem. For a given normed linear space X, a given subset G in $C(\mathbb{R})$, and a given subset Φ in X^*, determine whether the set

$$\{g \circ \phi : g \in G \text{ and } \phi \in \Phi\}$$

is fundamental in $C(X)$.

We present next some basic results concerning this problem. The first is one essentially given by Diaconis and Shahshahani [DiS].

THEOREM 2. *Let X be a normed linear space. The set of ridge functions $\{\exp \circ \phi : \phi \in X^*\}$ is fundamental in $C(X)$.*

Proof. Let $\mathcal{A}$ be the linear space generated by the given set of ridge functions. We note that $\mathcal{A}$ is an algebra in $C(X)$ because

$$(\exp \circ \phi)(\exp \circ \psi) = \exp \circ (\phi + \psi)$$

Since we can take $\phi = 0$, we see that $\mathcal{A}$ contains constants. To see that $\mathcal{A}$ separates the points of X, let $x \neq y$. By the Hahn-Banach Theorem, there is an element ϕ in X^* such that $\phi(x - y) \neq 0$. Then $\exp\big(\phi(x)\big) \neq \exp\big(\phi(y)\big)$. By the Stone-Weierstrass Theorem (Theorem 1), $\mathcal{A}$ is dense in $C(X)$. ∎

THEOREM 3. *Let G be a fundamental set in $C(\mathbb{R})$, and let X be a normed linear space. Let Φ be a subset of X^* such that the set*

$$\{\phi/\|\phi\| : \phi \in \Phi , \phi \neq 0\}$$

is dense in the unit sphere of X^*. Then the following set of ridge functions is fundamental in $C(X)$:

$$\{g \circ \phi : g \in G , \phi \in \Phi\}$$

Proof. Let f be a member of $C(X)$, K a compact set in X, and $\varepsilon > 0$. By Theorem 2, there exist $h_i \in C(\mathbb{R})$ and $\psi_i \in X^*$ such that

$$\left\| f - \sum_{i=1}^{m} h_i \circ \psi_i \right\|_K < \varepsilon/3$$

By adjusting the functions h_i if necessary, we can assume that $\|\psi_i\| = 1$ for $1 \leq i \leq m$. (See Problem 7.) Put $M = \sup_{x \in K}\|x\|$. Select $\delta > 0$ so that when the inequalities $|s| \leq M$, $|t| \leq M$, and $|s - t| < \delta$ are satisfied, we shall have $|h_i(s) - h_i(t)| < \varepsilon/3m$ for $1 \leq i \leq m$. Select $\phi_i \in \Phi$ so that $\|\phi_i/\|\phi_i\| - \psi_i\| < \delta/M$ for $1 \leq i \leq m$. Put $\lambda_i = 1/\|\phi_i\|$ and $\mu = \max_i \|\phi_i\|$. Select coefficients a_{ij} and functions $g_{ij} \in G$ so that for $|t| \leq \mu M$ we have

$$\left| h_i(\lambda_i t) - \sum_{j=1}^{N} a_{ij} g_{ij}(t) \right| < \varepsilon/3m \qquad (1 \leq i \leq m)$$

Now a straightforward use of the triangle inequality shows that

$$\left| f(x) - \sum_{i=1}^{m} \sum_{j=1}^{N} a_{ij} g_{ij}(\phi_i(x)) \right| < \varepsilon \qquad (x \in K)$$

The details of this are as follows. Let $x \in K$. Then we have $\|x\| \leq M$, $|\psi_i(x)| \leq M$, $|\lambda_i \phi_i(x)| \leq M$, and

$$|\psi_i(x) - \lambda_i \phi_i(x)| \leq \|x\| \, \|\psi_i - \lambda_i \phi_i\| \leq M(\delta/M) = \delta$$

The definition of δ allows us to infer that

$$\left| \sum_{i=1}^{m} h_i(\psi_i(x)) - \sum_{i=1}^{m} h_i(\lambda_i \phi_i(x)) \right| \leq \sum_{i=1}^{m} \varepsilon/3m = \varepsilon/3$$

Since $|\phi_i(x)| \leq \|\phi_i\| \, \|x\| \leq \mu M$, the definition of a_{ij} and g_{ij} yields

$$\left| \sum_{i=1}^{m} h_i(\lambda_i \phi_i(x)) - \sum_{i=1}^{m} \sum_{j=1}^{N} a_{ij} g_{ij}(\phi_i(x)) \right| \leq \sum_{i=1}^{m} \varepsilon/3m = \varepsilon/3$$

Putting this together, we have

$$\left| f(x) - \sum_{i=1}^{m} \sum_{j=1}^{N} a_{ij} g_{ij}(\phi_i(x)) \right|$$

$$\leq \left| f(x) - \sum_{i=1}^{m} h_i(\psi_i(x)) \right| + \left| \sum_{i=1}^{m} h_i(\psi_i(x)) - \sum_{i=1}^{m} h_i(\lambda_i \phi_i(x)) \right|$$

$$+ \left| \sum_{i=1}^{m} h_i(\lambda_i \phi_i(x)) - \sum_{i=1}^{m} \sum_{j=1}^{N} a_{ij} g_{ij}(\phi_i(x)) \right| < \varepsilon \qquad \blacksquare$$

COROLLARY 1. *The set of ridge functions*

$$\{ \phi^n : n \in \mathbb{Z}_+, \phi \in X^*, \|\phi\| = 1 \}$$

is fundamental in $C(X)$, for any normed linear space X.

THEOREM 4. *If G is fundamental in $C(\mathbb{R})$ then the functions $x \mapsto g(zx)$, with $g \in G$ and $z \in \mathbb{Z}^s$, form a fundamental set in $C(\mathbb{R}^s)$. (Here $\mathbb{Z}^s$ is the set of s-tuples with integer coordinates.)*

Proof. In order to use Theorem 3, we must show that the set

$$\{ z/\|z\| : z \in \mathbb{Z}^s, z \neq 0 \}$$

is dense on the unit sphere of $\mathbb{R}^s$. To this end, let $\|x\| = 1$. Select a vector y having rational coordinates such that $\|x - y\| < \varepsilon$ and $\|y\| > 1$. (Every open set in $\mathbb{R}^s$ contains "rational" points.) By using a lowest common denominator, we can assume that the coordinates of y have the form p_i/q where p_i and $q \in \mathbb{Z}$. Putting $z = qy$, we have

$$\left\| x - \frac{z}{\|z\|} \right\| = \left\| x - \frac{y}{\|y\|} \right\| < \|x - y\| < \varepsilon$$

In this last step we have used a general result given in Problem 4. Since $z \in \mathbb{Z}^s$, we have established the required density. $\qquad \blacksquare$

THEOREM 5. *Let g be a continuous function on $\mathbb{R}$ such that the limits of $g(t)$ as $t \to \infty$ and $t \to -\infty$ exist and are different. Put $g_{ji}(t) = g(jt + i)$. Then $\{ g_{ji} : i \in \mathbb{Z}, j \in \mathbb{Z}_+ \}$ is fundamental in $C(\mathbb{R})$.*

Proof. Since $C(\mathbb{R})$ has been given the topology of uniform convergence on compact sets, it suffices to prove that for any compact set K in $\mathbb{R}$, the set $G = \{g_{ji}|K : i \in \mathbb{Z}, j \in \mathbb{Z}_+\}$ is fundamental in $C(K)$. We may take K to be an interval $[a, b]$ with rational endpoints. Fundamentality will be proved by establishing that $G^\perp = 0$. Let $\phi \in G^\perp$. By the Riesz Representation Theorem [Ru0], page 139, there exists a signed, regular, Borel measure μ on K such that

$$\phi(f) = \int_a^b f(t)\, d\mu(t) \qquad (f \in C(K))$$

Also, there is a Borel function h such that $|h(t)| = 1$ for all t and

$$\int_a^b f(t)\, d\mu(t) = \int_a^b f(t)h(t)\, d|\mu|(t) \qquad (f \in C(K))$$

Now fix two integers p and q such that $q > 0$ and $p/q \in K$. Put $\tilde{g}_{jk}(t) = g(jqt - jp + k)$, where $j \in \mathbb{Z}_+$ and $k \in \mathbb{Z}$. These functions belong to G. We find that $\lim_{j\to\infty} \tilde{g}_{jk}(t) = v_k(t)$, where

$$v_k(t) = \begin{cases} g(-\infty) & a \leq t < p/q \\ g(k) & t = p/q \\ g(\infty) & p/q < t \leq b \end{cases}$$

Our hypotheses on g imply that it is bounded. Therefore $h\tilde{g}_{jk} \in L^1(K, |\mu|)$, and the Dominated Convergence Theorem [Ru0], page 27 is applicable. Thus

$$0 = \lim_{j\to\infty} \phi(\tilde{g}_{jk}) = \lim_{j\to\infty} \int_a^b \tilde{g}_{jk}\, d\mu = \lim_{j\to\infty} \int_a^b \tilde{g}_{jk} h\, d|\mu|$$

$$(1) \qquad = \int_a^b v_k h\, d|\mu| = \int_a^b v_k\, d\mu$$

$$= g(-\infty)\mu\left[a, \frac{p}{q}\right) + g(k)\mu\left\{\frac{p}{q}\right\} + g(\infty)\mu\left(\frac{p}{q}, b\right]$$

In Equation (1), first let $k \to \infty$ and then let $k \to -\infty$. On comparing the results and using the hypothesis $g(\infty) \neq g(-\infty)$, we conclude that $\mu\{p/q\} = 0$ for all p/q. In Equation (1), let $p/q = a$ to get

$$0 = g(\infty)\mu(a, b] = g(\infty)\mu[a, b]$$

By letting $p/q = b$, we obtain the same equation with $g(-\infty)$. Hence $\mu[a, b] = 0$. By the additivity of μ,

$$(2) \qquad 0 = \mu[a, b] = \mu[a, r] + \mu(r, b]$$

for any rational r in $[a, b]$. From Equations (1) and (2),

$$0 = g(-\infty)\mu[a, r) + g(\infty)\mu(r, b]$$
$$= g(-\infty)\mu[a, r] + g(\infty)\mu(r, b]$$
$$= g(-\infty)\mu[a, r] - g(\infty)\mu[a, r]$$
$$= \{g(-\infty) - g(\infty)\}\mu[a, r]$$

Hence $\mu[a, r] = 0$ for all rational r. By Equation (2), $\mu(r, b] = 0$. Then for rational r and r',

$$0 = \mu[a, b] = \mu[a, r] + \mu(r, r') + \mu[r, b] = \mu(r, r')$$

Thus, μ assigns 0 measure to all intervals whose endpoints are rational. To show that $\phi = 0$, we take any $f \in C(K)$ and approximate it by a simple function $s = \sum_{j=1}^{n} c_j \chi_{I_j}$ in which each χ_{I_j} is the characteristic function of an interval having rational endpoints. (The following lemma shows how this can be done.) For the moment, we work in $L_{\infty}[a, b]$, because the function s is not continuous. One verifies easily that $\int_a^b s(t) \, d\mu(t) = 0$, since each I_j has rational endpoints. If $\|f - s\|_{\infty} < \varepsilon$, then

$$|\phi(f)| = \left| \int f(t) \, d\mu(t) \right| = \left| \int [f(t) - s(t)] \, d\mu(t) \right|$$

$$\leq \|f - s\|_{\infty} |\mu|([a, b]) < \varepsilon |\mu|([a, b])$$

Hence $|\phi(f)| < \varepsilon \|\phi\|$. Since ε is arbitrary, $\phi(f) = 0$. Since f is arbitrary, $\phi = 0$. ∎

LEMMA. *Let $f \in C[0, 1]$, and let $\varepsilon > 0$. Then there is a natural number n such that $\|f - \sum_{j=1}^{n} c_j \chi_{I_j}\|_{\infty} < \varepsilon$, where $I_j = \left[\frac{j-1}{n}, \frac{j}{n}\right)$ for $1 \leq j \leq n - 1$, $I_n = \left[\frac{n-1}{n}, 1\right]$, and $c_j = f\left(\frac{j-1}{n}\right)$.*

Proof. Select n so that $\omega(f; 1/n) < \varepsilon$. Here ω is the modulus of continuity. Let $t \in [0, 1]$. Then there is a unique k for which $t \in I_k$. Then

$$\left| f(t) - \sum_{j=1}^{n} c_j \chi_{I_j}(t) \right| = |f(t) - c_k \chi_{I_k}(t)| = |f(t) - c_k|$$

$$= \left| f(t) - f\left(\frac{k-1}{n}\right) \right| \leq \omega\left(f; t - \frac{k-1}{n}\right) \leq \omega\left(f; \frac{1}{n}\right) < \varepsilon \quad ∎$$

COROLLARY 2. *If g is as in Theorem 5, then for any normed linear space X, the set*

$$\{x \longmapsto g(j\phi(x) + k) : \phi \in X^*, \|\phi\| = 1, k \in \mathbb{Z}, j \in \mathbb{Z}_+\}$$

is fundamental in $C(X)$.

Proof. Theorem 5 indicates that

$$\{t \longmapsto g(jt + k) : k \in \mathbb{Z}, j \in \mathbb{Z}_+\}$$

is fundamental in $C(\mathbb{R})$. Use Theorem 3 to complete the proof. ∎

COROLLARY 3. *If g is as in Theorem 5, then the set*

$$\{x \longmapsto g(zx + k) : z \in \mathbb{Z}^s, k \in \mathbb{Z}\}$$

is fundamental in $C(\mathbb{R}^s)$.

Proof. Use the previous proof and Theorem 4. ∎

For some applications, such as in neural networks, it is very desirable to employ a single function g in the ridge functions. Theorems 1 and 5, above, concern this special case. Let us standardize the sort of function occurring in Theorem 5.

Definition. A sigmoid function is a function $\sigma : \mathbb{R} \to \mathbb{R}$ such that

$$\lim_{t \to \infty} \sigma(t) = 1 \quad \text{and} \quad \lim_{t \to -\infty} \sigma(t) = 0$$

Theorem 5 and its corollaries show that these functions can be used in ridge-function approximation. In fact, as we shall see, a sigmoid function need not be continuous to be useful. The Heaviside function is an example:

$$H(x) = \begin{cases} 1 & \text{if } x \geq 0 \\ 0 & \text{if } x < 0 \end{cases}$$

It models the action of an electrical switch that is turned on at the instant $x = 0$, or the action of a neuron in the brain when it is suddenly activated.

Fixing a bounded sigmoid function σ and an integer n, we consider the class of functions of the form

(3) $$g(x) = \sum_{i=1}^{n} a_i \sigma(mx + k_i) \quad (x \in \mathbb{R})$$

By varying the parameters $a_i \in \mathbb{R}$, $m \in \mathbb{N}$, and $k_i \in \mathbb{Z}$, we obtain the class being considered, and we denote it by $\Phi(\sigma, n)$. A question about "degree" of approximation can now be posed: How well can a function f in $C[0, 1]$ be approximated by elements of $\Phi(\sigma, n)$?

To make the question precise, define

$$\text{dist}(f, \Phi(\sigma, n)) = \inf\{\|f - g\|_\infty : g \in \Phi(\sigma, n)\}$$

The norm here is the supremum norm on the interval $[0, 1]$. The answer to our question is given by an elegant theorem of Debao Chen [DC1]. It is expressed in terms of the modulus of continuity of f, defined by the equation

$$\omega(f; \delta) = \sup\{|f(x) - f(y)| : 0 \leq x \leq 1 , 0 \leq y \leq 1 , |x - y| \leq \delta\}$$

THEOREM 6. (Debao Chen). *For each f in $C[0, 1]$*

$$\text{dist}(f, \Phi(\sigma, n)) \leq \|\sigma\| \omega\left(f; \frac{1}{n}\right)$$

Here, $\|\sigma\| = \sup|\sigma(x)|$ as x ranges over $(-\infty, \infty)$.

Proof. The proof will be constructive and, in fact, effected with the aid of a family of linear operators (these being of some interest in themselves). Having fixed n, we define

(4) $$x_i = i/n \qquad f_i = f(x_i) \qquad (0 \leq i \leq n)$$

(5) $$(L_m f)(x) = f_1 \sigma(mx + m) + \sum_{i=1}^{n-1} (f_{i+1} - f_i)\sigma(mx - mx_i)$$

Observe that if m is a multiple of n, then $L_m f \in \Phi(\sigma, n)$. The operators L_m are obviously linear, and they are of a type called "quasi-interpolants," because they involve f only through point-evaluation functionals.

Since σ is a sigmoid function, the following function values converge to 0 when t goes to $+\infty$:

(6)
$$\varepsilon(t) = \max \left\{ \max_{x \geq t} |\sigma(x) - 1|, \max_{x \leq -t} |\sigma(x)| \right\}$$

We are going to prove the following estimate, in which ω is the modulus of continuity of f, and $0 < \delta < 1/2n$:

(7)
$$\|L_m f - f\| \leq \|\sigma\| \omega\left(\frac{1}{n}\right) + \omega(\delta) + \varepsilon(m\delta)\left[\|f\| + n\omega\left(\frac{1}{n}\right)\right]$$

Once (7) has been proved, we can set $\delta = 1/\sqrt{m}$ and then let $m \to \infty$ through the multiples of n. Since $\text{dist}(f, \Phi(\sigma, n)) \leq \|L_m f - f\|$, and the bound in (7) converges to $\|\sigma\| \omega\left(\frac{1}{n}\right)$, this procedure will establish the theorem. Now, to prove (7) we start by writing

(8)
$$(L_m f - f)(x) = f_k - f(x) + f_1[\sigma(mx + m) - 1] + \sum_{i=1}^{k-1} (f_{i+1} - f_i)[\sigma(mx - mx_i) - 1]$$
$$+ \sum_{i=k}^{n-1} (f_{i+1} - f_i)\sigma(mx - mx_i)$$

This equation is verified by elementary algebra. Now fix m and x. Let $0 < \delta < 1/2n$. There are two cases to consider.

In case 1, $|x - x_k| < \delta$ for some k in $\{1, 2, \ldots, n\}$. Write Equation (8) in the form

(9)
$$(L_m f - f)(x) = f_k - f(x) + f_1[\sigma(mx + m) - 1] + \sum_{i=1}^{k-1} (f_{i+1} - f_i)[\sigma(mx - mx_i) - 1]$$
$$+ (f_{k+1} - f_k)\sigma(mx - mx_k) + \sum_{i=k+1}^{n-1} (f_{i+1} - f_i)\sigma(mx - mx_i)$$

For indices i in the range $1 \leq i \leq k - 1$, we have

$$x - x_i \geq x - x_{k-1} \geq x_k - \delta - x_{k-1} = \frac{1}{n} - \delta \geq \frac{1}{n} - \frac{1}{2n} = \frac{1}{2n} > \delta$$

Hence $m(x - x_i) \geq m\delta$ and $|\sigma(mx - mx_i) - 1| \leq \varepsilon(m\delta)$. Similarly, if $k + 1 \leq i \leq n - 1$, then

$$x - x_i \leq x - x_{k+1} \leq x_k + \delta - x_{k+1} = -\frac{1}{n} + \delta \leq -\delta$$

In this case, $m(x - x_i) \leq -m\delta$ and $|\sigma(mx - mx_i)| \leq \varepsilon(m\delta)$. Notice also that $mx + m \geq m$ so that $|\sigma(mx + m) - 1| \leq \varepsilon(m)$. From Equation (9) it now follows that

$$|(L_m f - f)(x)| \leq \omega(\delta) + \|f\|\varepsilon(m) + (k-1)\omega\left(\frac{1}{n}\right)\varepsilon(m\delta)$$
$$+ \omega\left(\frac{1}{n}\right)\|\sigma\| + (n - k - 1)\omega\left(\frac{1}{n}\right)\varepsilon(m\delta)$$
$$\leq \omega\left(\frac{1}{n}\right)\|\sigma\| + \varepsilon(m\delta)\left[\|f\| + n\omega\left(\frac{1}{n}\right)\right] + \omega(\delta)$$

In case 2, it is supposed that $|x - x_i| \geq \delta$ for all i in $\{1, 2, \ldots, n\}$. Select k so that $x_{k-1} \leq x < x_k$. Proceeding as before, we can employ Equation (8) as it stands to obtain

$$|(L_m f - f)(x)| \leq \omega\left(\frac{1}{n}\right) + \|f\|\varepsilon(m\delta) + n\omega\left(\frac{1}{n}\right)\varepsilon(m\delta)$$

$$\leq \omega\left(\frac{1}{n}\right)\|\sigma\| + \varepsilon(m\delta)\left[\|f\| + n\omega\left(\frac{1}{n}\right)\right] + \omega(\delta)$$

Thus, in cases 1 and 2 the same upper bound has been obtained; this establishes Inequality (7), and completes the proof. ∎

For functions in $C[a, b]$, the operator L_m can be defined in a similar way. Let $x_i = a + i(b - a)/n$ and put

$$(L_m f)(x) = f_1 \sigma(mx + m - a) + \sum_{i=1}^{n-1} (f_{i+1} - f_i)\sigma(mx - mx_i)$$

If a and b are integers, and if m is a multiple of n, then $L_m f \in \Phi(\sigma, n)$. In this case, one can prove

(10) $$\text{dist}(f, \Phi(\sigma, n)) \leq \|\sigma\|\omega\left(f; \frac{b - a}{n}\right)$$

The distance function is defined with the supremum norm on $C[a, b]$.

References to ridge functions are [Bar], [BP], [ChuLi], [Cy], [DM1], [DiS], [HS], [Jo], [MM1], [Mh5] [MN2], [Na], [SC], [Su1], and [XLC]. Theorems 2, 3, and 4 are from the paper [SC].

Problems

1. Prove that for the ridge function $g \circ \phi$ to be continuous, it is necessary that $g \in C(\mathbb{R})$. (Assume $\phi \neq 0$.)

2. Give an example of a continuous function on $\mathbb{R}^2$ that is not a linear combination of ridge functions.

3. Prove that the set of functions $\{t \mapsto e^{\lambda t} : \lambda \in \mathbb{R}\}$ is fundamental in $C(\mathbb{R})$.

4. Prove that in an inner-product space the condition $\|x\| = 1 < \|y\|$ implies that $\|x - y/\|y\|\| < \|x - y\|$. *Hint:* Consider the quadratic function $t \mapsto \|x - ty\|^2$.

5. Prove that the bivariate function $x \mapsto \xi_1 \xi_2$ is the sum of two ridge functions. *Hint:* Consider $(\xi_1 \pm \xi_2)^2$. Here, and in Problem 6, $x = (\xi_1, \xi_2)$.

6. Prove that the bivariate function $x \mapsto \max(\xi_1, \xi_2)$ is the sum of two ridge functions.

7. In the proof of Theorem 3, explain in detail why we can assume that $\|\psi_i\| = 1$.

8. Prove that $\{x \mapsto \exp\langle j, x\rangle : j \in \mathbb{Z}_+^s\}$ is fundamental in $C(\mathbb{R}^s)$. Note that $\mathbb{Z}_+ = \{0, 1, 2, \ldots\}$.

9. A function $g \in C(\mathbb{R})$ is called "admissible" if the set $\{x \mapsto g(\alpha t + \beta) : \alpha, \beta \in \mathbb{R}\}$ is fundamental in $C(\mathbb{R})$. Prove or disprove the conjecture that if g is admissible then so is $ag + b$, provided that $a, b \in \mathbb{R}$ and $a \neq 0$.

10. Prove that if G is fundamental in $C(\mathbb{R})$ and if X is a normed linear space, and if $v \in X$, $v \neq 0$ then the set

$$\{g \circ \phi : g \in G, \phi \in X^*, \|\phi\| = 1, \phi(v) > 0\}$$

is fundamental in $C(X)$.

11. Let X be any Hausdorff space. Let G be a subset of $C(X)$ that is closed under multiplication, separates the points of X, and contains the function 1. Prove that span(G) is dense in $C(X)$ in the topology of uniform convergence on compact sets.

12. If span (G) separates the points of X, does it follow necessarily that G separates the points of X? Here $G \subset C(X)$ and X is a compact Hausdorff space.

13. Prove that if μ is a signed Borel measure on $[a, b]$ that assigns zero measure to each closed interval having rational endpoints, then $\mu = 0$.

14. Prove that the set $\{g \circ \phi : g \in C(\mathbb{R}), \phi \in X^*\}$ is identical to the set $\{g \circ \phi : g \in C(\mathbb{R}), \phi \in X^*, \|\phi\| = 1\}$.

15. Let $f = \sum_{i=1}^{n} g_i \circ \phi_i$, where $\phi_i \in X^*$, $\|\phi_i\| = 1$, and each g_i is uniformly continuous on $\mathbb{R}$. Prove that $\omega(f; \delta) \leq \sum_{i=1}^{n} \omega(g_i; \delta)$.

16. Provide a complete proof for the inequality (10).

17. Is Corollary 3 valid when z is restricted to $\mathbb{Z}_+^s$?

18. Suppose that $\omega(f; d) \leq c\delta^2$. What can be said of f?

References

[Bar]   Barron, A. R. "Universal approximation bounds for superpositions of a sigmoidal function." *IEEE Trans. Inform. Theory, 39* (1993), no. 3, 930–945.

[Bla2]   Blatter, J. *Grothendieck Spaces in Approximation Theory.* Memoirs Amer. Math. Soc., vol. 120, Amer. Math. Soc., Providence, RI, 1972.

[BP]   Braess, D., and A. Pinkus. "Interpolation by ridge functions." *J. Approx. Theory 73* (1993), 218–236.

[DC1]   Chen, Debao. "Degree of approximation by superpositions of a sigmoidal function." *Approx. Theory Appl. 9* (1993), 17–28.

[ChuLi] Chui, C. K., and Xin Li. "Approximation by ridge functions and neural networks with one hidden layer." *J. Approx. Theory 70* (1992), 131–141.

[Cy]   Cybenko, G. "Approximation by superpositions of a sigmoidal function." *Math. Control Signals Systems 2* (1989), 203–314.

[DM1]   Dahmen, W., and C. Micchelli. "Some remarks on ridge functions." *Approx. Theory and Its Applications 3* (1987), 139–143.

[DiS]   Diaconis, P., and M. Shahshahani. "On nonlinear functions of linear combinations." *SIAM J. Sci. Stat. Comput 5* (1984), 175–191.

[DS] Dunford, N., and J. T. Schwartz. *Linear Operators, Part I, General Theory.* Interscience, New York, 1958.

[F] Friedman, A. *Foundations of Modern Analysis.* Holt, Rinehart and Winston, New York, 1970. Reprint, Dover, New York, 1982.

[GJ] Gillman, L., and M. Jerison. *Rings of Continuous Functions.* Van Nostrand, New York, 1960. Available from Springer-Verlag, New York.

[HaSo] Hamaker, C., and D. C. Solmon. "The angles between the null spaces of x-rays." *J. Math. Analysis and Applications 62* (1978), 1–23.

[Jo] John, F. *Plane Waves and Spherical Means Applied to Partial Differential Equations.* Interscience, New York, 1955. [QA 377 J62]

[Ke] Kelley, J. L. *General Topology.* Graduate Texts in Mathematics, vol. 27, Springer-Verlag, New York, 1955.

[KN] Kelley, J. L., I. Namioka, et al. *Linear Topological Spaces.* Van Nostrand, Princeton, NJ, 1963.

[LP] Lin, V. Y., and A. Pinkus. "Fundamentality of ridge functions." *J. Approx. Theory 75* (1993), 295–311.

[MN2] Madych, W. R., and S. A. Nelson. "Radial sums of ridge functions: A characterization." *Math. Meth. Appl. Sci. 7* (1985), 90–100.

[Mh5] Mhaskar, H. N. "Approximation of real functions using neural networks." In Proceedings of International Conference on Computational Mathematics, ed. by H. P. Dikshit and C. A. Micchelli. World Scientific Press, Singapore, 1994.

[MM1] Mhaskar, H. N., and C. Micchelli. "Approximation by superpositions of a sigmoidal function." *Advances in Appl. Math. 13* (1992), 350–373.

[Na] Natterer, F. *The Mathematics of Computerized Tomography.* Wiley, New York, 1986.

[Pr] Prolla, J. B. *Weierstrass-Stone: The Theorem.* Peter Lang, Frankfurt, 1993.

[RS] Reed, M., and B. Simon. *Methods of Modern Mathematical Physics, Vol. 1, Functional Analysis.* 2nd ed., Academic Press, New York, 1980.

[Ru0] Rudin, W. *Real and Complex Analysis.* 2nd ed., McGraw-Hill, New York, 1974.

[Ru2] Rudin, W. *Functional Analysis.* McGraw-Hill, New York, 1973.

[Sem] Semadini, Zbigniew. *Banach Spaces of Continuous Functions.* Polish Scientific Publishers, Warsaw, 1971.

[Sim] Simmons, G. F. *Topology and Modern Analysis.* McGraw-Hill, New York, 1963.

[Su1] Sun, Xingping. *Multivariate Interpolation Using Ridge or Related Functions.* Unpublished doctoral dissertation, University of Texas at Austin, August 1990.

[SC1] Sun, Xingping, and E. W. Cheney. "The fundamentality of sets of ridge functions." *Aequat. Math. 44* (1992), 226–235.

[XLC] Xu, Yuan, W. A. Light, and E. W. Cheney. "Constructive methods of approximation by ridge functions and radial functions." *Numerical Algorithms 4* (1993), 205–223.

23
Ridge Function Approximation via Convolutions

In Chapters 20 and 21 we studied methods of approximating functions using convolution with suitable kernels. If $H \in L^1(\mathbb{R}^s)$ and $\int_{\mathbb{R}^s} H(x)\, dx = 1$, then

$$\lim_{\lambda \to \infty} (H_\lambda * f)(x) = \lim_{\lambda \to \infty} \int_{\mathbb{R}^s} H(\lambda x - \lambda y) f(y) \lambda^s \, dy = f(x)$$

when f is a bounded function in $C(\mathbb{R}^s)$. By making further assumptions about H, we found that one can produce certain types of approximations or speed up the convergence of $H_\lambda * f$ as $\lambda \to \infty$.

Now we ask whether ridge function approximations can be produced with suitable kernels. We seek approximations of the form

$$f(x) = \sum_{i=1}^{n} c_i \phi(x u_i + \alpha_i)$$

Notice that ϕ is fixed (does not depend on i). Since a function $x \mapsto \phi(xu)$ cannot be in $L^1(\mathbb{R}^s)$, unless $\phi = 0$, we cannot use such a function as a kernel. We proceed by performing an averaging process over the unit sphere. Assume that $\phi \in C(\mathbb{R})$. Define g on $\mathbb{R}^s$ by setting $g(x)$ equal to the average of $\phi(xu)$ as u runs over $\mathbb{S}^{s-1}$:

$$g(x) = \frac{1}{\omega_{s-1}} \int_{\mathbb{S}^{s-1}} \phi(xu) \, d\mathbb{S}^{s-1}(u) \qquad (x \in \mathbb{R}^s)$$

We require some geometric facts about the spheres $\mathbb{S}^s$, as well as a simpler formula for $g(x)$. These are taken up next.

As usual, we denote by $\mathbb{S}^s$ the unit sphere in $\mathbb{R}^{s+1}$:

$$\mathbb{S}^s = \{x \in \mathbb{R}^{s+1} : \|x\|_2 = 1\}$$

The s-dimensional "area" of $\mathbb{S}^s$ is denoted by ω_s. The volume enclosed by $\mathbb{S}^s$ (that is, the volume of the unit cell in $\mathbb{R}^{s+1}$) is denoted by v_{s+1}.

LEMMA 1. *We have $v_{s+1} = \omega_s/(s+1)$.*

Proof. We can compute the volume inside $\mathcal{S}^s$ by the elementary "method of shells." The shell of radius r will have surface area $r^s\omega_s$ and thickness Δr. Hence

$$v_{s+1} = \int_0^1 r^s\omega_s \, dr = \omega_s/(s+1) \qquad \blacksquare$$

The following lemma was given by Schoenberg in his 1942 paper [S3]. It also appears in Fritz John's book [Jo], page 8. See also Madych's paper [Ma1]. In reading [Jo], one must notice that the author uses ω_s to denote the area of $\mathcal{S}^{s+1}$. We use ω_s as the area of $\mathcal{S}^s$, in harmony with [S3] and [SW], page 9.

LEMMA 2. *For $u \in \mathcal{S}^s$, and for any continuous function F on $[-1, 1]$,*

$$\int_{\mathcal{S}^s} F(xu) \, d\mathcal{S}^s(x) = \omega_{s-1} \int_{-1}^1 F(t)(1 - t^2)^{(s-2)/2} \, dt$$

Proof. Since $\mathcal{S}^s$ is the unit sphere in $\mathbb{R}^{s+1}$, we work in the latter space. Let B_r^{s+1} denote the solid ball in $\mathbb{R}^{s+1}$ having radius r and center 0. The hyperplane

$$P_t = \{x : xu = t\}$$

is of dimension s. Its intersection with B_r^{s+1} is (a translate of) the ball B_ρ^s, where ρ can be determined from Figure 23.1 to be $\rho = \sqrt{r^2 - t^2}$. This ball has volume $\rho^s v_s$, where v_s denotes the volume of the unit ball in $\mathbb{R}^s$. By Lemma 1, $\rho^s v_s = \rho^s \omega_{s-1}/s$. Thus we can use the "method of disks" to compute the integral

$$\int_{B_r^{s+1}} F(xu) \, dx = \int_{-r}^r F(t)(r^2 - t^2)^{s/2}(\omega_{s-1}/s) \, dt$$

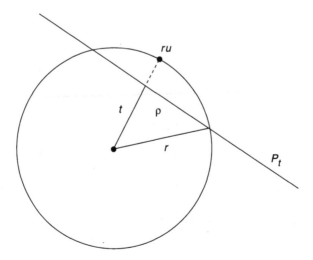

Figure 23.1 *Determination of ρ*

The "method of shells" gives us

$$\int_{B_r^{s+1}} F(xu)\, dx = \int_0^r \int_{\mathcal{S}_t^s} F(xu)\, d\mathcal{S}_t^s(x)\, dt$$

The two integrals on the right in these equations must be equal, and their derivatives must be equal. This leads to

$$\int_{\mathcal{S}_r^s} F(xu)\, d\mathcal{S}_r^s(x) = \int_{-r}^r F(t)(r^2 - t^2)^{(s-2)/2} r\omega_{s-1}\, dt$$

Putting $r = 1$ gives the formula in the lemma. ∎

Returning now to the functions ϕ and g described at the beginning of the chapter, we let $r = \|x\|$ and use Lemma 2 to write $g(x) = g_0(r)$, where

$$g_0(r) = \frac{\omega_{s-2}}{\omega_{s-1}} r^{2-s} \int_{-r}^r \phi(t)(r^2 - t^2)^{(s-3)/2}\, dt$$

$$= \frac{\omega_{s-2}}{\omega_{s-1}} \int_{-1}^1 \phi(rt)(1 - t^2)^{(s-3)/2}\, dt$$

The next two theorems are from [XLC].

THEOREM 1. *Let ϕ and g be as above. Let the dimension s be odd and at least 3. Let $n \in \mathbb{N}$. In order that the function $x \mapsto \|x\|_2^n g(x)$ be in $L^1(\mathbb{R}^s)$, it is sufficient that ϕ have these properties:*

(a)
$$\int_{-\infty}^\infty \phi(t) t^{2j}\, dt = 0 \qquad (0 \le 2j \le s - 3)$$

(b)
$$\int_{-\infty}^\infty \left| \phi(t) t^{n+s-1} \right|\, dt < \infty$$

Proof. Put $\lambda = (s - 3)/2$. This is an integer. Notice that we can assume ϕ to be even, for the odd part of ϕ does not contribute to g. We use the above equation for $g_0(r)$ and the "shell" method in the following calculation.

$$I = \int_{\mathbb{R}^s} \|x\|_2^n |g(x)|\, dx = \int_0^\infty \int_{\mathcal{S}_r^{s-1}} \|x\|_2^n |g(x)|\, d\mathcal{S}_r^{s-1}(x)\, dr$$

$$= \int_0^\infty \int_{\mathcal{S}_r^{s-1}} r^n |g_0(r)|\, d\mathcal{S}_r^{s-1}(x)\, dr$$

$$= \int_0^\infty r^n |g_0(r)| r^{s-1} \omega_{s-1}\, dr$$

$$= \omega_{s-1} \int_0^\infty r^{s+n-1} \frac{\omega_{s-2}}{\omega_{s-1}} r^{2-s} \left| \int_{-r}^r \phi(t)(r^2 - t^2)^\lambda\, dt \right| dr$$

$$= 2\omega_{s-2} \int_0^\infty r^{n+1} \left| \int_0^r \phi(t)(r^2 - t^2)^\lambda\, dt \right| dr$$

The function $t \mapsto (r^2 - t^2)^\lambda$ is a polynomial containing terms t^{2j}, for $0 \leq 2j \leq s - 3$. Thus, by hypothesis (a) in the statement of this theorem, and by the fact that ϕ is even, we have

$$0 = \int_0^\infty \phi(t)(r^2 - t^2)^\lambda \, dt = \left(\int_0^r + \int_r^\infty \right) \phi(t)(r^2 - t^2)^\lambda \, dt$$

Hence

$$I = 2\omega_{s-2} \int_0^\infty r^{n+1} \left| \int_r^\infty \phi(t)(r^2 - t^2)^\lambda \, dt \right| dr$$

The Fubini Theorem is applicable, since the final result is finite. Hence

$$I \leq 2\omega_{s-2} \int_0^\infty |\phi(t)| \int_0^t r^{n+1}(t^2 - r^2)^\lambda \, dr \, dt$$

$$= 2\omega_{s-2} c_{sn} \int_0^\infty |\phi(t)| t^{2\lambda+n+2} \, dt$$

$$= \omega_{s-2} c_{sn} \int_{-\infty}^\infty |\phi(t) t^{n+s-1}| \, dt < \infty$$

Here we have introduced constants c_{sn} such that

$$\int_0^t r^{n+1}(t^2 - r^2)^\lambda \, dr = c_{sn} t^{2\lambda+n+2}$$

This is justified by writing the integral as follows.

$$\int_0^t r^{n+1} \sum_{k=0}^\lambda \binom{\lambda}{k} t^{2k}(-1)^{\lambda-k} r^{2(\lambda-k)} \, dr$$

$$= \sum_{k=0}^\lambda (-1)^{\lambda-k} \binom{\lambda}{k} t^{2k} \int_0^t r^{2\lambda-2k+n+1} \, dr$$

$$= \sum_{k=0}^\lambda (-1)^{\lambda-k} \binom{\lambda}{k} t^{2k} t^{2\lambda-2k+n+2} \Big/ (2\lambda - 2k + n + 2)$$

$$= t^{2\lambda+n+2} \sum_{k=0}^\lambda (-1)^{\lambda-k} \binom{\lambda}{k} \Big/ (2\lambda - 2k + n + 2) \qquad \blacksquare$$

THEOREM 2. *Let ϕ and g be as above. Let s be odd, $s \geq 3$. Assume that ϕ satisfies hypotheses (a) and (b) in Theorem 1. In order that $\int_{\mathbb{R}^s} g \neq 0$ it is necessary and sufficient that $\int_{-\infty}^\infty \phi(t) t^{s-1} \, dt \neq 0$.*

Proof. We can assume that ϕ is even, as in the preceding proof. The calculation in that proof can be repeated without absolute values and with $n = 0$. We obtain, since s is odd,

$$\int_{\mathbb{R}^s} g(x) \, dx = -2\omega_{s-2} \int_0^\infty \phi(t) c_{s0} \, t^{s-1} \, dt$$

$$= -\omega_{s-2} c_{s0} \int_{-\infty}^\infty \phi(t) t^{s-1} \, dt$$

Now $c_{s0} \neq 0$, since in its definition we can let $t = 1$ and obtain

$$c_{s0} = \int_0^1 r(1 - r^2)^\lambda \, dr > 0$$ ∎

LEMMA 3. *A recurrence relation for the volumes v_s is*

$$v_{s+1} = v_s \int_{-1}^1 (1 - t^2)^{s/2} \, dt$$

Proof. We compute the volume v_{s+1} enclosed by S^s by the "method of disks." A hyperplane in $\mathbb{R}^{s+1}$ situated at distance t from the origin intersects S^s in a "hypercircle," which is S_r^{s-1} with $r = \sqrt{1 - t^2}$. See Figure 23.1, and make suitable changes. The disk forming an element of volume has thickness Δt and radius r. Its volume is $r^s v_s \, \Delta t = (1 - t^2)^{s/2} v_s \, \Delta t$. The volume v_{s+1} is thus

$$v_{s+1} = v_s \int_{-1}^1 (1 - t^2)^{s/2} \, dt$$ ∎

LEMMA 4. *Define the Beta function by the equation*

$$B(z, w) = \int_0^1 t^{z-1} (1 - t)^{w-1} \, dt$$

Then

$$B(z, w) = \Gamma(z)\Gamma(w)/\Gamma(z + w)$$

and

$$\int_a^b (x - a)^{z-1}(b - x)^{w-1} \, dx = (b - a)^{z+w-1} B(z, w)$$

Proof. The first formula is proved in [W4], page 375. The second formula is proved by changing the variable in the integral from t to x, where $x = a + (b - a)t$. ∎

In the second formula of Lemma 4, let $a = 1, b = -1, z - 1 = w - 1 = \alpha$. Then

$$\int_{-1}^1 (1 - x^2)^\alpha \, dx = 2^{2\alpha+1} B(\alpha + 1, \alpha + 1)$$

LEMMA 5. *A recurrence relation for the sequence $[\omega_s]$ is*

$$\omega_s = \omega_{s-1} \Gamma\left(\frac{s}{2}\right) \Gamma\left(\frac{1}{2}\right) \Big/ \Gamma\left(\frac{s+1}{2}\right)$$

Proof. By Lemma 1, $\omega_s/\omega_{s-1} = (s + 1)v_{s+1}/sv_s$. By Lemma 3,

$$\frac{\omega_s}{\omega_{s-1}} = \frac{s+1}{s} \int_{-1}^1 (1 - t^2)^{s/2} \, dt = \frac{s+1}{s} \int_{-1}^1 (1 - t)^{s/2}(t + 1)^{s/2} \, dt$$

By Lemma 4,

$$\frac{\omega_s}{\omega_{s-1}} = \frac{s+1}{s} 2^{s+1} B\left(\frac{s}{2}+1, \frac{s}{2}+1\right) = \frac{s+1}{s} 2^{s+1} \Gamma\left(\frac{s}{2}+1\right)^2 \Big/ \Gamma(s+2)$$

Using the identity $\Gamma(z+1) = z\Gamma(z)$, we obtain

$$\frac{\omega_s}{\omega_{s-1}} = \frac{s+1}{s} 2^{s+1} \left[\frac{s}{2}\Gamma\left(\frac{s}{2}\right)\right]^2 \Big/ (s+1)\Gamma(s+1)$$

Now write $\Gamma(s+1) = \Gamma(2([s+1]/2))$ and use the identity

$$\Gamma(2z) = \pi^{-1/2} 2^{2z-1} \Gamma(z)\Gamma\left(z+\frac{1}{2}\right)$$

After simplification we will have

$$\frac{\omega_s}{\omega_{s-1}} = \frac{(s/2)\Gamma(s/2)^2 \pi^{1/2}}{\Gamma((s+1)/2)\Gamma(1+s/2)} = \frac{(s/2)\Gamma(s/2)^2 \pi^{1/2}}{\Gamma((s+1)/2)(s/2)\Gamma(s/2)}$$

Since $\pi^{1/2} = \Gamma(1/2)$, this last expression is equivalent to the one in the lemma. The identities involving the Gamma function are in [AS], pages 255–256.   ∎

LEMMA 6. *We have*

$$\omega_{s-1} = 2\pi^{s/2}/\Gamma(s/2), \quad v_s = 2\pi^{s/2}/s\Gamma(s/2)$$

Proof. When $s = 1$, the first formula asserts that $\omega_0 = 2\sqrt{\pi}/\Gamma(1/2) = 2$. This is correct, because S^0 is the interval $[-1, 1]$. Proceeding by induction, assume that the formula is correct for s. Then by Lemma 5 we have

$$\omega_s = \omega_{s-1}\Gamma(s/2)\Gamma(1/2)/\Gamma((s+1)/2)$$
$$= [2\pi^{s/2}/\Gamma(s/2)]\Gamma(s/2)\Gamma(1/2)/\Gamma((s+1)/2)$$
$$= 2\pi^{(s+1)/2}/\Gamma((s+1)/2)$$

The second formula in the lemma follows from the first by using Lemma 1.   ∎

Problems

1. Refer to Theorem 2 and prove that $\int_{\mathbb{R}^s} g = 1$ if and only if $\int_{-\infty}^{\infty} \phi(t)t^{s-1}\, dt = -[(s-1)\omega_{s-2}]^{-1}$.

2. Express the integral in Lemma 3 in terms of the Beta function.

References

[AS]   Abramowitz, M., and I. A. Stegun. *Handbook of Mathematical Functions.* National Bureau of Standards, Washington, 1964. Reprint, Dover, New York.

[Jo]   John, F. *Plane Waves and Spherical Means Applied to Partial Differential Equations.* Interscience, New York, 1955.

[Ma1]   Madych, W. R. "Summability and approximate reconstruction from Radon transform data." *Contemp. Math. 113* (1990), 189–219.

[S3]   Schoenberg, I. J. "Positive definite functions on spheres." *Duke Math. J. 9* (1942), 96–108.

[SW]   Stein, E. M., and G. Weiss. *Introduction to Fourier Analysis on Euclidean Spaces.* Princeton University Press, Princeton, NJ, 1971.

[W4]   Widder, D. V. *Advanced Calculus.* 2nd ed., Prentice-Hall, Englewood Cliffs, NJ, 1961. Reprint, Dover, New York.

[XLC]   Xu, Yuan, W. A. Light, and E. W. Cheney. "Constructive methods of approximation by ridge functions and radial functions." *Numerical Algorithms 4* (1993), 205–223.

24
Density of Ridge Functions

We can use the theorems in Chapter 23 to answer some questions about the density of ridge functions in $C(\mathbb{R}^s)$. A basic question asks for necessary and sufficient conditions on a function ϕ in $C(\mathbb{R})$ in order that the set

(1) $$\{x \longmapsto \phi(ux + \alpha) : u \in \mathbb{R}^s, \alpha \in \mathbb{R}\}$$

shall be fundamental in $C(\mathbb{R}^s)$. We do not know the answer to this, but can offer sufficient conditions.

Since our theorems depend on "discretizing" convolution transforms, we begin by discussing how Riemann sums are used to approximate integrals. The setting adopted is a compact metric space (T, d) on which a Borel measure μ has been defined. The measure μ can be a signed measure or even complex valued. For a given $\delta > 0$ the open balls $B(\tau, \delta/2)$, for $\tau \in T$, obviously cover T. By compactness, a finite number suffice, say

(2) $$T = \bigcup_{i=1}^{n} B\left(\tau_i, \frac{\delta}{2}\right) = \bigcup_{i=1}^{n} B_i$$

In order to obtain a disjoint cover of T, we define

$$A_1 = B_1, \quad A_2 = B_2 \backslash B_1, \quad A_3 = B_3 \backslash (B_1 \cup B_2), \quad \text{etc.}$$

The sets A_i have these properties:

1. Each A_i is a Borel set

2. $\bigcup_{i=1}^{n} A_i = T$

3. $A_i \cap A_j = \varnothing$ if $i \neq j$

4. $A_i \subset B_i$ and $\text{diam}(A_i) \leq \delta$, where $\text{diam}(A_i) = \sup\{d(t_1, t_2) : t_1, t_2 \in A_i\}$

We say that the family $P = \{A_1, A_2, \ldots, A_n\}$ is a *partition* of T. Corresponding to the given partition, there is a *quadrature* formula based on Riemann sums. We begin by selecting $t_i \in A_i \ (1 \leq i \leq n)$, and approximate any function $f : T \rightarrow \mathbb{R}$ by the function

$$\text{(3)} \qquad \widetilde{f} = \sum_{i=1}^{n} f(t_i) \mathcal{X}_{A_i}$$

Here $\mathcal{X}_A$ denotes the characteristic function of A:

$$\mathcal{X}_A(t) = \begin{cases} 1 & \text{if } t \in A \\ 0 & \text{if } t \notin A \end{cases}$$

The function $\widetilde{f}$ is an example of a *simple* function, which means that it is a measurable function with finite range. The function $\widetilde{f}$ is measurable because each A_i is measurable.

How well does $\widetilde{f}$ approximate f? We can use the modulus of continuity to answer this question. The modulus of continuity of f is defined by

$$\text{(4)} \qquad \omega(f; \delta) = \sup\{|f(t_1) - f(t_2)| : t_1, t_2 \in T, d(t_1, t_2) \leq \delta\}$$

For any t in T there is a unique index k for which $t \in A_k$. Then $\widetilde{f}(t) = f(t_k)$. Hence

$$\text{(5)} \qquad |f(t) - \widetilde{f}(t)| = |f(t) - f(t_k)| \leq \omega(f; \delta)$$

Here we use the fact that $t, t_k \in A_k \subset B(\tau_k, \delta/2)$.

The quadrature formula now arises from approximating $\int f$ by $\int \widetilde{f}$. Thus,

$$\text{(6)} \qquad \int_T f(t) \, d\mu(t) \approx \sum_{i=1}^{n} f(t_i) \mu(A_i)$$

The error involved in using this is

$$\text{(7)} \qquad \left| \int f(t) \, d\mu(t) - \int \widetilde{f}(t) \, d\mu(t) \right| \leq \int |f(t) - \widetilde{f}(t)| \, d|\mu|(t) \leq \omega(f; \delta) |\mu|(T)$$

Thus, we can expect comparable precision for all functions that have the same modulus of continuity.

The first density theorem is as follows. It has been given in [XLC].

THEOREM 1. *Let the dimension s be odd and at least 3. Let ϕ be a continuous function on $\mathbb{R}$ such that*

    a. $\int_{-\infty}^{\infty} |\phi(t) t^{s-1}| \, dt < \infty$

    b. $\int_{-\infty}^{\infty} \phi(t) t^{2j} \, dt = 0 \qquad 0 \leq 2j \leq s - 3$

    c. $\int_{-\infty}^{\infty} \phi(t) t^{s-1} \, dt \neq 0$

Then the set $\{x \mapsto \phi(ux + \alpha) : u \in \mathbb{R}^s, \alpha \in \mathbb{R}\}$ is fundamental in $C(\mathbb{R}^s)$.

Proof. Assume the hypotheses on ϕ, and define g as in Chapter 23:

$$g(x) = \omega_{s-1}^{-1} \int_{\mathbb{S}^{s-1}} \phi(xu) \, d\mathbb{S}^{s-1}(u) \qquad (x \in \mathbb{R}^s)$$

By Theorems 1 and 2 in Chapter 23, $g \in L^1(\mathbb{R}^s)$ and $\int_{\mathbb{R}^s} g(x)\, dx \neq 0$. Because of the nature of the conclusion we wish to draw, there is no loss of generality in assuming $\int_{\mathbb{R}^s} g(x)\, dx = 1$. Recall that our standard topology on $C(\mathbb{R}^s)$, as described in Chapter 22, is the topology of uniform convergence on compact sets. To establish fundamentality, we therefore take any $f \in C(\mathbb{R}^s)$, any compact set $K \subset \mathbb{R}^s$ and any $\varepsilon > 0$. Our task is to find c_i, v_i, and α_i so that

$$(8) \qquad \left| f(x) - \sum_{i=1}^{n} c_i \phi(x v_i + \alpha_i) \right| < \varepsilon \qquad (x \in K)$$

There is no loss of generality in assuming that K is a ball, $K = \{x : \|x\| \leq r\}$. It can also be assumed that $f(x) = 0$ on $\mathbb{R}^s \setminus 2K$. As usual, we set $g_m(x) = m^s g(mx)$. By Theorem 2 in Chapter 20, page 149, there exists an m such that

$$(9) \qquad |(f - g_m * f)(x)| < \varepsilon/3 \qquad (x \in K)$$

The remainder of the proof uses Riemann sums to approximate $g_m * f$. These will produce the desired ridge functions.

For any x we have

$$\begin{aligned}
(g_m * f)(x) &= \int_{2K} g_m(x - y) f(y)\, dy \\
&= \int_{2K} g(mx - my) f(y) m^s\, dy \\
&= \int_{2mK} g(mx - z) f\!\left(\frac{z}{m}\right) dz
\end{aligned}$$

Let P be a partition of $2mK$. Corresponding to this partition there is a Riemann sum for the last integral written above. We assume that P is chosen so that for all $x \in K$

$$(10) \qquad \left| (g_m * f)(x) - \sum_{A \in P} c_A g(mx - z_A) \right| < \varepsilon/3$$

The details for accomplishing this will be provided later.

Since g is also defined by an integral, another Riemann sum is required. Let Q be a partition of $\mathbb{S}^{s-1}$, chosen so that for all $x \in K$

$$(11) \qquad \left| \sum_{A \in P} c_A g(mx - z_A) - \sum_{A \in P} c_A \sum_{B \in Q} e_B \phi((mx - z_A) u_B) \right| < \varepsilon/3$$

Details for this will also be given later. Putting all the inequalities together, we have the ridge function approximation

$$f(x) \approx \sum_{A \in P} \sum_{B \in Q} c_A e_B \phi(mx u_B - z_A u_B)$$

It entails an error of at most ε on K.

Returning to the first quadrature, we define

$$c_A = \int_A f\left(\frac{z}{m}\right) dz \qquad (A \in P)$$

Assume $\text{diam}(A) \leq \delta$ for $A \in P$. If $x \in K$, then

$$\left| (g_m * f)(x) - \sum c_A g(mx - z_A) \right|$$

$$= \left| \int_{2mK} g(mx - z)f(z/m)\, dz - \sum_{A \in P} g(mx - z_A) \int_A f(z/m)\, dz \right|$$

$$\leq \sum_{A \in P} \int_A |g(mx - z) - g(mx - z_A)|\; |f(z/m)|\, dz$$

$$\leq \omega(g, 3mr, \delta) \sum_{A \in P} \int_A |f(z/m)|\, dz$$

$$= m^s \|f\|_1 \omega(g, 3mr, \delta) \leq m^s \|f\|_1 \omega(\phi, 3mr, \delta)$$

In the last step we used Problem 1. The modulus of continuity here is defined (for any $\rho > 0$) by the equation

$$\omega(g, \rho, \delta) = \sup\{|g(x) - g(y)| : \|x\| \leq \rho\,,\ \|y\| \leq \rho\,,\ \|x - y\| \leq \delta\}$$

We select δ so that the final bound above is less than $\varepsilon/3$. This justifies Equation (10).

For the second quadrature, we define

$$e_B = \omega_{s-1}^{-1} \int_B d\mathbb{S}^{s-1}(u) \qquad (B \in Q)$$

Assume that $\text{diam}(B) \leq \theta$ for each $B \in Q$. The quadrature formula (for any function ψ) is

$$(12) \qquad \omega_{s-1}^{-1} \int_{\mathbb{S}^{s-1}} \psi(u)\, d\mathbb{S}^{s-1}(u) \approx \sum_{B \in Q} e_B \psi(u_B)$$

It involves an error of at most $\omega(\psi, 1, \theta)$, as is routinely verified. Therefore, for x in K

$$\left| g(mx - z_A) - \sum_{B \in Q} e_B \phi((mx - z_A)u_B) \right|$$

$$= \left| \omega_{s-1}^{-1} \int_{\mathbb{S}^{s-1}} \phi((mx - z_A)u)\, d\mathbb{S}^{s-1}(u) - \sum_{B \in Q} e_B \phi((mx - z_A)u_B) \right|$$

$$\leq \omega(\phi, 3mr, 3mr\theta)$$

Here we note that

$$|(mx - z_A)u - (mx - z_A)u_B| \leq \|mx - z_A\|\; \|u - u_B\| \leq 3mr\theta$$

and

$$|(mx - z_A)u| \leq \|mx - z_A\| \leq 3mr$$

In the final step of the calculation, we write

$$\left| \sum_{A \in P} c_A g(mx - z_A) - \sum_{A \in P} c_A \sum_{B \in Q} e_B \phi((mx - z_A)u_B) \right| \leq \sum_{A \in P} |c_A| \omega(\phi, 3mr, 3mr\theta)$$

We choose θ so this bound is at most $\varepsilon/3$. This justifies Equation (11). Equation (8) is now proved by combining Equations (9), (10), and (11). ∎

Problems

1. Prove that the functions ϕ and g in Theorem 1 obey the relation $\omega(g, \rho, \delta) \leq \omega(\phi, \rho, \delta)$.

2. Verify the error estimate for the quadrature formula (12).

3. Show that the theory of Riemann sums given in this chapter can be generalized to any compact topological space. (Start with f, and use its continuity to obtain the partition.)

4. Show that $\sum_{A \in P} |c_A| \leq m^s \|f\|_1$.

References

[XLC]   Xu, Yuan, W. A. Light, and E. W. Cheney. "Constructive methods of approximation by ridge functions and radial functions." *Numerical Algorithms 4* (1993), 205–223.

25
Artificial Neural Networks

An artificial neural network is a mathematical model of the human brain. Many different types of neural network models are studied, but we shall describe just one, called a **feed-forward network with one hidden layer.** This network consists of a large number of computing units arranged schematically in three layers as shown in Figure 25.1.

In principle, each unit of the input layer can be connected to each unit of the hidden layer. This connection has associated with it a **weight,** which is a real number. The weight attached to the link from input unit j to unit i on the hidden layer is denoted by w_{ij}. In a typical operation, each unit on the input layer will contain a real number. Let the j-th unit contain the real number ξ_j. Then unit i on the hidden layer will receive from unit j on the input layer the quantity $w_{ij}\xi_j$. The total input that unit i receives from *all* the input units is then $\sum_{j=1}^{s} w_{ij}\xi_j$. Unit i on the hidden layer now processes this input with a continuous sigmoid function σ and outputs the real number $\sigma(\sum_j w_{ij}\xi_j + \alpha_i)$. This output is then transmitted, with a weight θ_i, to the output unit. The total output is then

$$(1) \qquad \sum_{i=1}^{n} \theta_i \sigma\left(\sum_{j=1}^{s} w_{ij}\xi_j + \alpha_i \right) = f(\xi_1, \xi_2, \ldots, \xi_s)$$

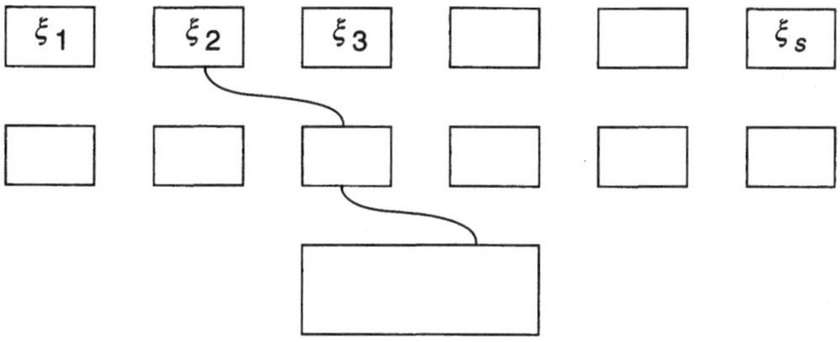

Figure 25.1 *Layers in a Neural Network*

As we know from results in Chapters 22 and 24, any continuous function of *s* variables can be approximated with arbitrary precision on a compact set by a function of the form *f* in Equation (1). Thus, by suitably adjusting the parameters n, α_i, θ_i and w_{ij}, we can reproduce approximately any desired output with the artificial neural network just described. In simulating the human learning process, the weights are varied in an attempt to match the output of Equation (1) with a given intended output. Different algorithms can be employed to arrive at an *optimal* (or at least *satisfactory*) set of weights.

Here we shall analyze one algorithm that applies to such problems. The only modification that we make in the previous description is that we shall require $\theta_i \geq 0$ and $\sum_{i=1}^m \theta_i = 1$. The setting is a normed linear space in which a set G is given. The **convex hull** of G is the set

$$\mathrm{co}(G) = \left\{ \sum_{i=1}^n \theta_i g_i : n \in \mathbb{N},\ \theta_i \geq 0,\ g_i \in G,\ \sum_{i=1}^n \theta_i = 1 \right\}$$

The closure of this set is denoted by $\overline{\mathrm{co}}(G)$.

LEMMA 1. *Let G be a subset of a normed linear space E, and let f be an element of E. In order that $f \in \overline{\mathrm{co}}(G)$ it is necessary and sufficient that there exist a sequence $[g_n]$ in G and a sequence $[f_n]$, with $f_n \in \mathrm{co}\{g_1, \ldots, g_n\}$ for each n, such that $f_n \to f$.*

Proof. The sufficiency is trivial, since $f_n \in \mathrm{co}(G)$ for each n. To prove the necessity, assume that $f \in \overline{\mathrm{co}}(G)$. Then there exist elements $h_n \in \mathrm{co}(G)$ such that $h_n \to f$. Write

$$h_n = \sum_{i=1}^{k(n)} \theta_{ni} \bar{g}_{ni} \qquad \left(\theta_{ni} \geq 0, \quad \sum_{i=1}^{k(n)} \theta_{ni} = 1, \quad \bar{g}_{ni} \in G \right)$$

Write the countable set $\{\bar{g}_{ni}\}$ as a sequence $[g_n]$. For each n let f_n be the best approximation to f of the form $\sum_{i=1}^n \alpha_{ni} g_i$, with $\alpha_{ni} \geq 0$ and $\sum_{i=1}^n \alpha_{ni} = 1$. To see that $f_n \to f$, let $\varepsilon > 0$. Select m so that $\|h_m - f\| < \varepsilon$. Select n so that $\{\bar{g}_{m1}, \bar{g}_{m2}, \ldots, \bar{g}_{m,k(m)}\} \subset \{g_1, g_2, \ldots, g_n\}$. Then, by the definition of f_n, we have

$$\varepsilon > \|h_m - f\| \geq \|f_n - f\| \geq \|f_{n+1} - f\| \geq \cdots \qquad \blacksquare$$

Remark. We do not know whether Lemma 1 remains true if we add the condition that f_n have the form $\lambda f_{n-1} + (1 - \lambda) g_n$. As we shall see later, this condition may be added if the space is an inner product space.

It is natural to attempt an algorithmic proof of Lemma 1. One such algorithm, called the **Greedy Algorithm**, goes as follows. The first element, f_1, is chosen as a point of G as close as possible to f. Since such a point may not exist, we assign a positive tolerance ε and select $f_1 \in G$ so that

$$\|f - f_1\| < \inf_{g \in G} \|f - g\| + \varepsilon$$

If f_{n-1} has been chosen in $\mathrm{co}(G)$, we let $f_n = \bar{\lambda} f_{n-1} + (1 - \bar{\lambda})\bar{g}$ where $(\bar{\lambda}, \bar{g})$ is chosen in $[0, 1] \times G$ to make $\|f - f_n\|$ nearly minimal:

$$\|f - f_n\| < \inf_{0 \leq \lambda \leq 1} \inf_{g \in G} \|f - \lambda f_{n-1} - (1 - \lambda)g\| + \varepsilon/2^n$$

We do not know whether this algorithm produces a sequence $[f_n]$ converging to f in any normed linear space. However, by a theorem of Barron [Bar], the algorithm succeeds in an inner-product space. A generalization of the algorithm and theorem due to Lee, Bartlett, and Williamson [LBW] will be presented here. This generalization solves the problem of constructing a point in a closed convex subset of Hilbert space closest to a given point outside that set. Some lemmas are needed first.

LEMMA 2. *Let f, g, h, k be points in an inner-product space, and let λ and μ be real numbers whose sum is 1. Then*

$$\|\lambda h + \mu g - f\|^2 - \|f - k\|^2 = \lambda^2\|h - k\|^2 + \mu^2\|g - k\|^2$$
$$+ 2\lambda\mu\langle h - k, g - k\rangle + 2\langle\lambda h + \mu g - k, k - f\rangle$$

Proof.

$$\|\lambda h + \mu g + f\|^2 - \|f - k\|^2$$
$$= \|(\lambda h + \mu g - k) + (k - f)\|^2 - \|f - k\|^2$$
$$= \|\lambda h + \mu g - k\|^2 + 2\langle\lambda h + \mu g - k, k - f\rangle$$
$$= \|\lambda(h - k) + \mu(g - k)\|^2 + 2\langle\lambda h + \mu g - k, k - f\rangle$$
$$= \lambda^2\|h - k\|^2 + \mu^2\|g - k\|^2 + 2\lambda\mu\langle h - k, g - k\rangle + 2\langle\lambda h + \mu g - k, k - f\rangle \quad\blacksquare$$

LEMMA 3. *Let G be a bounded set in an inner-product space, E, and define $b = \sup\{\|g\| : g \in G\}$. Let f, $h \in E$ and $k \in \operatorname{co}(G)$. If $\lambda + \mu = 1$, then*

$$\inf_{g \in G} \|\lambda h + \mu g - f\|^2 - \|f - k\|^2 \le \lambda^2\|h - k\|^2 + \mu^2 b^2 + 2\lambda\langle h - k, k - f\rangle$$

Proof. Since k is in $\operatorname{co}(G)$, it can be expressed as $k = \sum_1^n \theta_i g_i$, where $\theta_i > 0$, $g_i \in G$, and $\sum_1^n \theta_i = 1$. In the following calculation we use Lemma 2 and the formula for k in crucial places.

$$\inf_{g \in G} \|\lambda h + \mu g - f\|^2 - \|f - k\|^2$$

$$\le \min_i \|\lambda h + \mu g_i - f\|^2 - \|f - k\|^2$$

$$\le \sum_{i=1}^n \theta_i\big[\|\lambda h + \mu g_i - f\|^2 - \|f - k\|^2\big]$$

$$= \sum_{i=1}^n \theta_i\big[\lambda^2\|h - k\|^2 + \mu^2\|g_i - k\|^2 + 2\lambda\mu\langle h - k, g_i - k\rangle$$
$$+ 2\langle\lambda h + \mu g_i - k, k - f\rangle\big]$$

$$= \lambda^2\|h - k\|^2 + \mu^2 \sum_{i=1}^n \theta_i\|g_i - k\|^2 + 2\langle\lambda h + \mu k - k, k - f\rangle$$

$$= \lambda^2\|h - k\|^2 + \mu^2 \sum_{i=1}^n \theta_i\big[\|g_i\|^2 + \|k\|^2 - 2\langle g_i, k\rangle\big] + 2\langle\lambda h - \lambda k, k - f\rangle$$

$$= \lambda^2\|h - k\|^2 + \mu^2\Big[\sum_{i=1}^n \theta_i\|g_i\|^2 + \|k\|^2 - 2\langle k, k\rangle\Big] + 2\lambda\langle h - k, k - f\rangle$$

$$\le \lambda^2\|h - k\|^2 + \mu^2[b^2 - \|k\|^2] + 2\lambda\langle h - k, k - f\rangle$$

$$\le \lambda^2\|h - k\|^2 + \mu^2 b^2 + 2\lambda\langle h - k, k - f\rangle \quad\blacksquare$$

LEMMA 4. *Let G, b, h, and f be as in Lemma 3. Let* $\lambda + \mu = 1$ *and* $0 \le \lambda \le 1$. *Then*

$$\inf_{g \in G} \|\lambda h + \mu g - f\|^2 \le \mu \inf_{k \in \mathrm{co}(G)} \|f - k\|^2 + \lambda \|h - f\|^2 + \mu^2 b^2$$

Proof. Note first that

$$\|h - f\|^2 - \|k - f\|^2 = \|(h - k) + (k - f)\|^2 - \|k - f\|^2$$
$$= \|h - k\|^2 + 2\langle h - k, k - f \rangle$$

Using this equation and Lemma 3, we have

$$\inf_{g \in G} \|\lambda h + \mu g - f\|^2 \le \|k - f\|^2 + \lambda^2 \|h - k\|^2 + \mu^2 b^2 + 2\lambda \langle h - k, k - f \rangle$$

$$\le \|k - f\|^2 + \lambda \|h - k\|^2 + 2\lambda \langle h - k, k - f \rangle + \mu^2 b^2$$

$$= \|k - f\|^2 + \lambda \left[\|h - f\|^2 - \|k - f\|^2 \right] + \mu^2 b^2$$

$$= \mu \|k - f\|^2 + \lambda \|h - f\|^2 + \mu^2 b^2$$

The assertion in the lemma is obtained by taking an infimum in k.   ∎

A suitable form of the Greedy Algorithm can be designed to find the point in a closed convex set closest to a given point. The setting is a Hilbert space, and we suppose as given a bounded set G. A point f outside $\mathrm{co}(G)$ is prescribed, and we seek the unique point f^* in $\overline{\mathrm{co}}(G)$ closest to f.

In describing the algorithm and proving its rate of convergence, we follow Lee, Bartlett, and Williamson [LBW]. Their work was based on the pioneering work of Barron [Bar2] and Jones [Jon].

Since G is bounded, we can define b to be the least upper bound of $\|g\|$ as g ranges over G. Define $\mu_n = 2/(n + 1)$, $\lambda_n = 1 - \mu_n$, and $f_0 = 0$. Then for $n = 1, 2, 3, \ldots$ we select $g_n \in G$ so that

$$\|\lambda_n f_{n-1} + \mu_n g_n - f\|^2 \le \inf_{g \in G} \|\lambda_n f_{n-1} + \mu_n g - f\|^2 + \mu_n^2$$

This n-th step is completed by defining $f_n = \lambda_n f_{n-1} + \mu_n g_n$.

THEOREM. *The sequence* $[f_n]$ *produced by the Greedy Algorithm lies in the convex hull of G and converges to the unique point of* $\overline{\mathrm{co}}(G)$ *closest to* f. *Moreover, for each n,*

(2) $$\|f - f_n\|^2 - d^2 \le \frac{4(b^2 + 1)}{n}   \qquad d = \mathrm{dist}(f, \overline{\mathrm{co}}(G))$$

Proof. In the inequality of Lemma 4, let $\lambda = 0$ and $\mu = 1$ to conclude that

$$\inf_{g \in G} \|g - f\|^2 \le d^2 + b^2$$

Referring to the definition of f_1 in the algorithm, we have

$$\|f_1 - f\| \leq \inf_{g \in G} \|g - f\| + 1 \leq d^2 + b^2 + 1$$

Thus the inequality in (2) is true for $n = 1$.

Now assume that f_{n-1} is in $\mathrm{co}(G)$ and satisfies the inequality (2). Thus $\|f_{n-1} - f\|^2 \leq 4(b^2 + 1)/(n - 1)$. The new element f_n will be of the form $\lambda_n f_{n-1} + \mu_n g_n$, and will therefore be in $\mathrm{co}(G)$. It remains to be shown that the inequality (2) is valid for f_n. By the definition of f_n in the algorithm,

$$\|f_n - f\|^2 \leq \inf_{g \in G} \|\lambda_n f_{n-1} + \mu_n g - f\|^2 + \mu_n^2$$

By Lemma 4, using $h = f_{n-1}$,

$$\|f_n - f\|^2 \leq \mu_n d^2 + \lambda_n \|f_{n-1} - f\|^2 + \mu_n^2 b^2 + \mu_n^2$$

From this and the fact that $\mu_n d^2 - d^2 = -\lambda_n d^2$ we have

$$\|f_n - f\|^2 - d^2 \leq \lambda_n \left[\|f_{n-1} - f\|^2 - d^2 \right] + \mu_n^2 (b^2 + 1)$$

$$\leq \lambda_n \left[\frac{4(b^2 + 1)}{n - 1} \right] + \mu_n^2 (b^2 + 1)$$

$$= 4(b^2 + 1) \left[\frac{1}{n + 1} + \frac{1}{(n + 1)^2} \right]$$

$$< 4(b^2 + 1) \frac{1}{n} \qquad \blacksquare$$

COROLLARY. *Let G be a bounded set in an inner-product space. There is a constant c depending only on G such that for each f in the closed convex hull of G, and for each n,*

(2)
$$\inf \left\| f - \sum_{i=1}^{n} \lambda_i g_i \right\| \leq c n^{-1/2}$$

Here the infimum is over all $g_1, g_2, \ldots, g_n \in G$ and over all nonnegative $\lambda_1, \lambda_2, \ldots, \lambda_n$ whose sum is 1.

We do not know whether a better order of approximation is possible, such as cn^{-1}. This corollary is ascribed to B. Maurey by Barron, who refers to the paper by Pisier [Pi]. If G is m-dimensional, then the infimum in Equation (2) is 0 for $n > m$, by a famous theorem of Carathéodory. See, for example, [C1].

A variant of the Greedy Algorithm goes as follows. (Its verification is left to the problems.) Let $G, f, \gamma, c, \varepsilon_n$, and f_1 be as before. At the nth step, with f_{n-1} in hand, define $\alpha = \|f - f_{n-1}\|$ and $v = f + (\gamma/\alpha)(f - f_{n-1})$. Select g_n in G so that

$$\|v - g_n\|^2 \leq \inf_{g \in G} \|v - g\|^2 + \varepsilon_n$$

Then let

$$f_n = \frac{\gamma}{\gamma + \alpha} f_{n-1} + \frac{\alpha}{\gamma + \alpha} g_n$$

An open problem is whether the Greedy Algorithm works in other normed linear spaces.

Problems

1. In $\mathbb{R}^2$, let $G = \{(s, t) : |s|t \geq 1\}$. Prove that G is closed but $\text{co}(G)$ is not.

2. Let $\{u_n\}$ be an orthonormal base in a Hilbert space. Prove that $\overline{\text{co}}\{u_n : n \in \mathbb{N}\}$ is the set of all points $\sum_{n=1}^{\infty} \alpha_n u_n$ in which $\alpha_n \geq 0$ and $\sum_{n=1}^{\infty} \alpha_n = 1$.

3. Carathéodory's theorem asserts that if G is an m-dimensional set, then each h in $\text{co}(G)$ can be expressed as a convex combination of $m + 1$ points in G. Prove the assertion in the text that if G is m-dimensional and $f \in \overline{\text{co}}(G)$ then $\inf \| f - \sum_{i=0}^{m} \lambda_i g_i \| = 0$, where the infimum is over all $g_0, g_1, \ldots, g_m$ in G and nonnegative λ_i summing to 1.

4. Let G be a bounded subset of a normed linear space, Φ a continuous linear functional of norm 1, $f \in \overline{\text{co}}(G)$, $h \in \text{co}(G)$, and $0 \leq \lambda \leq 1$. Prove that

$$\inf_{g \in G} \Phi[\lambda h + (1 - \lambda)g - f] \leq \lambda \| h - f \|$$

5. In an inner-product space, if $f = \sum_{i=1}^{n} a_i g_i$ and $\sum_{i=1}^{n} a_i = 1$, then

$$\| f \|^2 = \sum_{i=1}^{n} a_i(\| g_i \|^2 - \| g_i - f \|^2)$$

6. Fix $v \in E$ and $f \in \overline{\text{co}}(G)$. Prove that $\| v - f \| \leq \sup_{g \in G} \| v - g \|$.

7. Prove that in the greedy algorithm, each f_n is a convex linear combination of at most n points in G.

8. Using the definition of $\overline{\text{co}}(G)$ given in the text, prove that $\overline{\text{co}}(G)$ is the intersection of all the closed convex sets containing G.

9. Let G be a bounded set in an inner-product space E. Suppose that 0 is in the closed convex hull of G. Prove that for any h in E and for any real λ,

$$\inf_{g \in G} \| \lambda h + (1 - \lambda)g \|^2 \leq \lambda^2 \| h \|^2 + (1 - \lambda)^2 b$$

where $b = \sup\{\| g \| : g \in G\}$.

10. Let G be a bounded sequence $\{g_1, g_2, \ldots\}$ in a normed linear space, and suppose that $f \in \overline{\text{co}}(G)$. Let $G_n = \text{co}\{g_1, g_2, \ldots, g_n\}$, and let f_n be the point of G_n closest to f. Prove that $f_n \to f$. Prove that no order of convergence can be established unless further assumptions are made. Explain why this does not contradict the theorem, where an order of convergence $\mathcal{O}(1/\sqrt{n})$ is proved.

11. Prove that for any set H in a normed linear space, $\inf_{h \in H} \| f - h \|$ is a continuous function of f.

12. Make suitable changes in the algorithm and the theorem to cope with a set that is not bounded.

References

[Bar]   Barron, A. R. "Universal approximation bounds for superpositions of a sigmoidal function." *IEEE Trans. Inform. Theory 39* (1993), no. 3, 930–945.

[Bar2]   Barron, A. R. "Approximation bounds for superpositions of a sigmoidal function." *Proc. IEEE Internat. Symp. Inform. Theory,* Budapest, June 1991.

[C1]   Cheney, E. W. *Introduction to Approximation Theory.* McGraw-Hill, New York, 1966. 2nd ed., American Mathematical Society, Providence.

[CCL]   Chen, T., H. Chen, and R. Liu. "A constructive proof and extension of Cybenko's approximation theorem." In *Computing Science and Statistics: Proceedings of the 22nd Symposium on the Interface.* Springer-Verlag, New York, 1991, pp. 163–168.

[ChuLi]   Chui, C. K., and Xin Li. "Approximation by ridge functions and neural networks with one hidden layer." *J. Approx. Theory 70* (1992), 131–141.

[CLM1]   Chui, C. K., Xin Li, and H. N. Mhaskar. "Neural networks for localized approximation." *Math. Comp. 63* (1994), 607–623.

[Din]   Dingankar, A. *Approximation of Functionals on Hilbert Spaces.* Unpublished doctoral dissertation, Dept. of Elect. and Comput. Eng., University of Texas at Austin, 1993.

[Hol]   Holmes, R. B. *A Course on Optimization and Best Approximation.* Lecture Notes in Math., vol. 257. Springer-Verlag, Berlin, 1972.

[Jon]   Jones, L. K. "A simple lemma on greedy approximation in Hilbert space and convergence rates for projection pursuit regression and neural network training." *Annals of Stat. 20* (1992), 608–613.

[Jon2]   Jones, L. K. "Constructive approximations for neural networks by sigmoidal functions." *Proc. of the IEEE 78* (1990), 1586–1589.

[LBW]   Lee, Wee Sun, P. L. Bartlett, and R. C. Williamson. "Efficient agnostic learning of neural networks with bounded fan-in." *IEEE Trans. Inform. Theory 42* (1996), 2118–2132.

[L2]   Light, W. A. "Ridge functions, sigmoidal functions, and neural networks." In *Approximation Theory VII,* ed. by E. W. Cheney, C. K. Chui, and L. L. Schumaker. Academic Press, New York, 1992, pp. 163–206.

[Mh1]   Mhaskar, H. N. "Approximation properties of a multilayered feedforward artificial neural network." *Advances in Comp. Math. 1* (1993), 61–80.

[Mh5]   Mhaskar, H. N. "Approximation of real functions using neural networks." In *Proceedings of International Conference on Computational Mathematics,* ed. by H. P. Dikshit and C. A. Micchelli. World Scientific Press, Singapore, 1994.

[MM1]   Mhaskar, H. N., and C. Micchelli. "Approximation by superpositions of a sigmoidal function." *Advances in Appl. Math. 13* (1992), 350–373.

[MM3]   Mhaskar, H. N., and C. A. Micchelli. "Dimension independent bounds on the degree of approximation by neural networks." *International Business Machines 38* (1994), 277–284.

[ParS]   Park, Joo-Young, and I. W. Sandberg. "Universal approximation using radial basis function networks." *Neural Comp. 3* (1991), 246–257.

[Pi] Pisier, G. "Rémarques sur un resultat non publié de B. Maury." *Sem. d'Analyse Fonc-tionelle 1980–1981,* vols. 1–12. Ecole Poly., Centre de Math., Palaiseau.

[Sand1] Sandberg, I. W. "General structures for classification." *IEEE Trans. Circuits and Systems 41* (1994), 372–376.

[Sand2] Sandberg, I. W. "Approximations for nonlinear functionals." *IEEE Trans. Circuits and Systems 39* (1992), 65–67.

[SD] Sandberg, I. W., and A. Dingankar. "On approximation of linear functionals on L_p spaces. Preprint, 1994.

[SC1] Sun, Xingping, and E. W. Cheney. "The fundamentality of sets of ridge functions." *Aequat. Math. 44* (1992), 226–235.

26
Chebyshev Centers

This topic arises tangentially in Chapters 25 and 27. Here we shall provide a few results and give references.

Given a bounded set K in a normed linear space X, we seek the smallest closed ball $B(x_0, r)$ that contains K. If such a smallest ball exists, we say that x_0 is the **Chebyshev center** of K and that r is the **Chebyshev radius** of K. For simplicity we drop the adjective "Chebyshev" and refer to the **center** and **radius** of K.

The formal definition of the radius of K is

(1) $$r = r(K) = \inf_{x \in X} \sup_{v \in K} \|x - v\|$$

If it is necessary to emphasize the ambient space X, we will use the notation $r_X(K)$. Since K is bounded (by assumption) we have (by considering $x = 0$)

$$r \le \sup_{v \in K} \|v\| < \infty$$

The formal definition of a center is this: if there exists an $x_0 \in X$ such that

$$r = \sup_{v \in K} \|x_0 - v\|$$

then we call x_0 a *center* of K. It is an easy exercise to prove that a bounded set in a finite-dimensional normed space has at least one center (Problem 1.)

Example 1. In the Euclidean space $\mathbb{R}^2$, the center of an acute triangle is the center of its circumscribing triangle.

Example 2. In the space $\left(\mathbb{R}^2, \|\cdot\|_\infty\right)$, the centers of the set $\{(\lambda, 0) : |\lambda| \le 1\}$ fill the set $\{(0, \mu) : |\mu| \le 1\}$.

Recall the notation $\text{co}(K)$ for the **convex hull** of a set K, and $\overline{\text{co}}(K)$ for the closure of $\text{co}(K)$.

THEOREM 1. *For a bounded set K in a normed space, we have*

(2) $$r(K) = r\big(\overline{\text{co}}(K)\big)$$

Furthermore, the centers of these sets are the same.

Proof. If $K \subset B(x, r)$ then $\text{co}(K) \subset B(x, r)$ because $B(x, r)$ is convex. Likewise, $\overline{\text{co}}(K) \subset B(x, r)$ because $B(x, r)$ is closed. By taking infima, we obtain (2). By taking r to be the radius of K we see that x is a center of K and hence is a center of $\overline{\text{co}}(K)$. On the other hand, if x is a center of $\overline{\text{co}}(K)$ then

$$\sup_{v \in K} \|x - v\| \leq \sup_{v \in \overline{\text{co}}(K)} \|x - v\| = r\big(\overline{\text{co}}(K)\big) = r(K)$$

This shows that x is a center of K.   ∎

THEOREM 2. *Let $P : Z \twoheadrightarrow X$ be a projection of norm 1 of a normed space Z onto a subspace X. If K is a bounded set in Z, then*

$$r_Z(P(K)) \leq r_X(P(K)) \leq r_Z(K)$$

If z is a center of K and $K \subset X$, then Pz is also a center of K.

Proof. We have

$$r_Z(P(K)) = \inf_{z \in Z} \sup_{y \in P(K)} \|z - y\| \leq \inf_{x \in X} \sup_{y \in P(K)} \|x - y\|$$

$$= r_X(P(K)) = \inf_{z \in Z} \sup_{v \in K} \|Pz - Pv\| \leq \inf_{z \in Z} \sup_{v \in K} \|z - v\|$$

$$= r_Z(K)$$

If z is a center of K and $K \subset X$, then Pz is a center of K, because

$$\sup_{v \in K} \|Pz - v\| = \sup_{v \in K} \|P(z - v)\| \leq \sup_{v \in K} \|z - v\| = r_Z(K)$$ ∎

COROLLARY. *Let P be a projection of norm 1 from a normed linear space Z onto a subspace X. If each bounded set in Z has a center then the same is true of X.*

Proof. Let K be a bounded set in X. Since it is also a bounded set in Z, it has a center, z, in Z. By Theorem 2, Pz is a center of K in X.   ∎

THEOREM 3. *In any conjugate Banach space, each bounded set has a center.*

Proof. Let $X = Y^*$ and let K be a bounded set in X. By the definition of $r(K)$, there is a sequence $[x_n]$ in X such that

$$\sup_{v \in K} \|x_n - v\| < r(K) + \frac{1}{n}$$

The sequence is bounded. Since closed balls in X are compact in the weak*-topology, the sequence $[x_n]$ has a weak*-cluster point, x. Since the norm is lower semicontinuous in the weak*-topology, $\|x - v\| \leq r(K)$ for all $v \in K$. Hence x is a center of K. ∎

COROLLARY. *If X is a Banach space such that there is a projection of norm 1 from X^{**} onto (the canonical image of) X, then each bounded set in X has a center.*

Proof. By Theorem 3, each bounded set in X^{**} has a center. By the preceding corollary, the same is true of X. ∎

COROLLARY. *In a Hilbert space or a uniformly convex space or a reflexive space, each bounded set has a center.*

THEOREM 4. *Let T be a compact Hausdorff space. Then each bounded set in $C(T)$ has a center.*

Proof. Let K be a bounded set in $C(T)$. Define

$$v(t) = \inf_{\mathcal{N}} \sup_{s \in \mathcal{N}} \sup_{z \in K} z(s)$$

and

$$u(t) = \sup_{\mathcal{N}} \inf_{s \in \mathcal{N}} \inf_{z \in K} z(s)$$

In these equations, $\mathcal{N}$ runs over all neighborhoods of t. Let us prove that v is upper semicontinuous. Thus for any λ we must show that $\{t : v(t) < \lambda\}$ is open. Let $v(t_0) < \lambda$. By the definition of $v(t_0)$, there is an open neighborhood $\mathcal{N}_0$ of t_0 such that

$$\sup_{\sigma \in \mathcal{N}_0} \sup_{z \in K} z(\sigma) < \lambda$$

If $s \in \mathcal{N}_0$ then $\mathcal{N}_0$ is a neighborhood of s. Consequently

$$v(s) = \inf_{\mathcal{N}} \sup_{\sigma \in \mathcal{N}} \sup_{z \in K} z(\sigma) \leq \sup_{\sigma \in \mathcal{N}_0} \sup_{z \in K} z(\sigma) < \lambda$$

Thus the neighborhood $\mathcal{N}_0$ lies in the set $\{t : v(t) < \lambda\}$, and the latter is open. A similar proof shows that u is lower semicontinuous. Put $r = \|v - u\|/2$. Since $v - u \leq \|v - u\| = 2r$ we have $v - r \leq u + r$. Now recall the **Hahn-Tong Theorem:** If g and h are real functions on a normal space such that $g \leq h$ and such that g is upper semicontinuous, while h is lower semicontinuous, then there is a continuous f satisfying $g \leq f \leq h$. [Sem], page 100. Applying this in the present situation, we find $x \in C(T)$ for which $v - r \leq x \leq u + r$. We shall prove that r is the radius of K and that x is a center of K, via the following two assertions.

ASSERTION. *For each $y \in C(T)$*

$$\sup_{w \in K} \|y - w\| \geq r$$

To prove this, let $\varepsilon > 0$. Let t be a point where $(v - u)(t) = \|v - u\|$. Let $\mathcal{N}$ be a neighborhood of t such that $|y(s) - y(t)| < \varepsilon$ when $s \in \mathcal{N}$. Now either $v(t) - y(t) \geq r$ or

$y(t) - u(t) \geq r$, for if both of these inequalities are reversed, their sum would give us $v(t) - u(t) < 2r = \|v - u\|$. Suppose, for definiteness, that $v(t) \geq y(t) + r$. By the definition of $v(t)$, we then have

$$\sup_{s \in \mathcal{N}} \sup_{z \in K} z(s) \geq y(t) + r$$

Select $s \in \mathcal{N}$ so that

$$\sup_{z \in K} z(s) \geq y(t) + r - \varepsilon$$

Select $z \in K$ so that $z(s) \geq y(t) + r - 2\varepsilon$. By the choice of $\mathcal{N}$, we then have

$$\sup_{w \in K} \|y - w\| \geq \|y - z\| \geq z(s) - y(s) \geq z(s) - y(t) - \varepsilon \geq r - 3\varepsilon$$

Since ε was arbitrary, the assertion is proved.

ASSERTION. *We have* $\|x - w\| \leq r$ *for all* $w \in K$.

To prove this, it suffices to prove that $-r \leq x - w \leq r$. If $\mathcal{N}$ is a neighborhood of t, then

$$w(t) \leq \sup_{z \in K} z(t) \leq \sup_{s \in \mathcal{N}} \sup_{z \in K} z(s)$$

Hence, by taking an infimum in $\mathcal{N}$, $w(t) \leq v(t)$. Similarly, $w(t) \geq u(t)$, so $-v \leq -w \leq -u$. But we also have $v - r \leq x \leq u + r$. On adding these two inequalities, we obtain $-r \leq x - w \leq r$. ∎

COROLLARY. *For any topological space T, the space* $C(T)$ *of bounded continuous real-valued functions on T has the property that bounded sets have Chebyshev centers.*

Proof. The space $C(T)$ is an abstract M-space with unit. See [KN] or [Sem]. By Kakutani's Theorem, there exists a compact Hausdorff space S such that $C(T)$ and $C(S)$ are isometrically isomorphic. If $L : C(S) \to C(T)$ is the isometry, and if x is a center of K, then Lx is a center of $L(K)$. ∎

Problems

1. Prove that bounded sets in finite-dimensional normed spaces have centers.

2. Prove the assertion in Example 1.

3. Find the center of an arbitrary triangle in $\mathbb{R}^2$ with the ∞-norm.

4. Prove that u, in the proof of Theorem 4, is lower semicontinuous.

5. The Chebyshev radius of a circle or sphere in a Euclidean space is its ordinary radius, and the Chebyshev center is the ordinary center. The Chebyshev radius of an acute triangle is the radius of its circumscribing circle. These statements may not be true for other norms, however.

References

[AMSa] Amir, D., J. Mach, and K. Saatkamp. "Existence of Chebyshev centers . . . ." *Trans. Amer. Math. Soc. 271* (1982), 513–524.

[AZ]   Amir, D., and Z. Ziegler. "Construction of elements of the relative Chebyshev center." In *Approximation Theory and Applications,* ed. by Z. Ziegler. Academic Press, New York, 1981, pp. 1–12.

[FC2]   Franchetti, C., and E. W. Cheney. "Simultaneous approximation and restricted Chebyshev centers in function spaces." In *Approximation Theory and Applications,* ed. by Z. Ziegler. Academic Press, New York, 1981, pp. 65–88.

[FC3]   Franchetti, C., and E. W. Cheney. "The embedding of proximinal sets." *J. Approx. Theory 48* (1986), 213–225.

[Hol]   Holmes, R. B. *A Course on Optimization and Best Approximation.* Lecture Notes in Math., vol. 257. Springer-Verlag, Berlin, 1972.

[KN]   Kelley, J. L., I. Namioka, et al. *Linear Topological Spaces.* Van Nostrand, Princeton, NJ, 1963.

[Sem]   Semadini, Z. *Banach Spaces of Continuous Functions.* Polish Scientific Publishers, Warsaw, 1971.

27
Optimal Reconstruction of Functions

This topic concerns the practical problem of reconstructing a function from a partial information about it. We already alluded to this general question in Chapter 19, where the problem of reconstructing a function from a knowledge of its line integrals was central. The first systematic discussion of the general problem was given by Golomb and Weinberger in [GolW]. This remarkable paper has nine chapters and innumerable examples from diverse fields. Here we take a very general view, as espoused in some recent work of Micchelli and Rivlin [MR1 and MR2]. The material is taken from those two papers and from [Pin2].

Let X, Y, and Z be three normed linear spaces. Let K be a subset of X. Let U be a linear operator from X into Z, and let $\mathfrak{I}$ be a linear operator from X to Y (or from a subspace of X containing K). The operator $\mathfrak{I}$ is the **information operator.** For $x \in K$, we wish to estimate Ux from the knowledge of $\mathfrak{I}x$. For carrying out this task, we consider all maps T from $\mathfrak{I}(K)$ into Z. Such a map is an **algorithm.** With it there is associated an **error,** defined by

$$E(T) = \sup_{x \in K} \|T(\mathfrak{I}x) - Ux\|$$

The **intrinsic error** in the problem is defined to be

$$E^* = \inf_T E(T)$$

where the infimum is over all maps $T : \mathfrak{I}(K) \to Z$. (These maps may be nonlinear.) If an algorithm T has the property $E(T) = E^*$, then it is an **optimal algorithm.** Notice that the problem is completely defined by K, $\mathfrak{I}$, and U.

LEMMA 1. *If K is symmetric (that is, $-x \in K$ when $x \in K$), then*

$$E^* \geq \sup\{\|Ux\| : x \in K, \mathfrak{I}x = 0\}$$

Proof. Let $x \in K$ and $\Im x = 0$. Then $-x \in K$ and $\Im(-x) = 0$. Let T be any algorithm. From the definition of E,

$$E(T) \geq \max\{\|T(\Im x) - Ux\|, \|T(\Im(-x)) - U(-x)\|\}$$
$$= \max\{\|T(0) - Ux\|, \|T(0) + Ux\|\}$$
$$\geq \frac{1}{2}\{\|Ux - T(0)\| + \|Ux + T(0)\|\}$$
$$\geq \frac{1}{2}\|Ux - T(0) + Ux + T(0)\|$$
$$= \|Ux\|$$

By taking an infimum with respect to T and a supremum with respect to x, we obtain the desired inequality. ∎

Example 1. We let $X = C[a, b]$ (the space of continuous functions on $[a, b]$ with supremum norm). Let $Y = \mathbb{R}^n$ and $Z = \mathbb{R}$. The set K is

$$K = \{x \in X : |x(t) - x(s)| \leq |s - t| \text{ for all } s \text{ and } t\}$$

We select $t_0, t_1, \ldots, t_n$ in $[a, b]$ such that $t_1 < t_2 < \cdots < t_n$ and define

$$\Im x = (x(t_1), x(t_2), \ldots, x(t_n))$$
$$Ux = x(t_0)$$

The problem is to find an algorithm by which we can guess $x(t_0)$ from the knowledge of $x(t_i)$, $(1 \leq i \leq n)$, and from the fact that $x \in K$. Using Lemma 1, we have

$$E^* \geq \sup\{|x(t_0)| : x \in K, x(t_i) = 0, 1 \leq i \leq n\}$$

Consider an $x \in K$ that vanishes at $t_1, t_2, \ldots, t_n$. We have, for $1 \leq i \leq n$,

$$|x(t)| = |x(t) - x(t_i)| \leq |t - t_i|$$

Hence

$$|x(t)| \leq \min_{1 \leq i \leq n} |t - t_i| =: w(t)$$

Notice that $w \in K$ and $w(t_i) = 0$ for $1 \leq i \leq n$. Hence $E^* \geq w(t_0)$. An optimal algorithm is obtained as follows. Select $k \in \{1, 2, \ldots, n\}$ so that $w(t_0) = |t_0 - t_k|$. Define $T(\Im x) = x(t_k)$. Then the optimality follows from

$$E(T) = \sup_{x \in K} \|T(\Im x) - Ux\|$$
$$= \sup_{x \in K} |x(t_k) - x(t_0)|$$
$$\leq |t_k - t_0| = w(t_0) \leq E^*$$ ∎

Example 2. We retain the setting of Example 1 except that $Z = C[a, b]$ and $Ux = x$. Assume that $t_1 < t_2 < \cdots < t_n$. Using notation and results from Example 1, we have (since $w \in K$ and $w(t_i) = 0$),

$$E^* \geq \sup\{\|x\| : x \in K, x(t_i) = 0, 1 \leq i \leq n\} \geq \|w\|$$

An examination of the graph of w shows that

$$\|w\| = \max\left\{t_1 - a, \ \frac{t_2 - t_1}{2}, \ \frac{t_3 - t_2}{2}, \ \ldots, \ \frac{t_n - t_{n-1}}{2}, \ b - t_n\right\}$$

Consider now the first-degree interpolating spline, y, whose graph connects the points $(a, x(t_1)), (t_1, x(t_1)), \ldots, (t_n, x(t_n)), (b, x(t_n))$. Note the two horizontal segments at the ends. We define an algorithm T by $T(\Im x) = y$. Notice that the construction of y requires only $\Im x$. Let t be any point in $[a, b]$. If $a \le t \le t_1$, then

$$|x(t) - y(t)| = |x(t) - x(t_1)| \le |t - t_1| \le t_1 - a = w(a) \le \|w\|$$

If $t_i \le t \le t_{i+1}$ for some i, then

$$t = \theta t_i + (1 - \theta)t_{i+1} \qquad \theta = (t_{i+1} - t)/(t_{i+1} - t_i)$$

On this interval, $y(t) = \theta x(t_i) + (1 - \theta)x(t_{i+1})$. Hence

$$x(t) - y(t) = \theta[x(t) - x(t_i)] + (1 - \theta)[x(t) - x(t_{i+1})]$$

Therefore,

$$
\begin{aligned}
|x(t) - y(t)| &\le \theta|x(t) - x(t_i)| + (1 - \theta)|x(t) - x(t_{i+1})| \\
&\le \theta|t - t_i| + (1 - \theta)|t - t_{i+1}| \\
&= \theta(t - t_i) + (1 - \theta)(t_{i+1} - t)
\end{aligned}
$$

It follows that

$$\frac{|x(t) - y(t)|}{t_{i+1} - t_i} \le \theta(1 - \theta) + (1 - \theta)\theta = 2\theta(1 - \theta) \le \frac{1}{2}$$

Hence,

$$|x(t) - y(t)| \le \frac{1}{2}(t_{i+1} - t_i) = w\left(\frac{t_i + t_{i+1}}{2}\right) \le \|w\|$$

Finally, if t is in $[t_n, b]$, we have

$$|x(t) - y(t)| = |x(t) - x(t_n)| \le |t - t_n| \le b - t_n \le \|w\|$$

Thus

$$\|T(\Im x) - Ux\| = \|y - x\| \le \|w\| \le E^*$$

This proves that T is optimal.   ∎

Example 3. We retain the setting of the preceding example, except that now $Ux = \int_a^b x(t)\, dt$. By Lemma 1,

$$
\begin{aligned}
E^* &\ge \sup\{\|Ux\| : x \in K, \Im x = 0\} \\
&= \sup\left\{\left|\int_a^b x(t)\, dt\right| : x \in K, \Im x = 0\right\} \\
&\ge \int_a^b w(t)\, dt = \|w\|_1
\end{aligned}
$$

Define an algorithm T by setting $T(\mathfrak{I}x) = \int_a^b y(t)\, dt$, where y is the piecewise constant interpolant described by $y(t) = x(t_i)$ if $|t - t_i| = w(t)$. (In case i is not uniquely determined by t, choose one value.)

$$\|T(\mathfrak{I}x) - Ux\| = \left| \int_a^b y(t)\, dt - \int_a^b x(t)\, dt \right|$$

$$\leq \int_a^b |y(t) - x(t)|\, dt$$

Now for each t there is an i such that $y(t) = x(t_i)$ and $|t - t_i| = w(t)$. Hence

$$|x(t) - y(t)| = |x(t) - x(t_i)| \leq |t - t_i| = w(t)$$

The above calculation then shows that T is optimal, since

$$\|T(\mathfrak{I}x) - Ux\| \leq \int_a^b w(t)\, dt = \|w\|_1 \leq E^* \qquad \blacksquare$$

LEMMA 2. *If K is symmetric and convex, then*

$$E^* \leq 2 \sup\{\|Ux\| : x \in K, \mathfrak{I}x = 0\}$$

Proof. By the Axiom of Choice, there is mapping $V : \mathfrak{I}(K) \to K$ such that $\mathfrak{I}Vy = y$ for all $y \in \mathfrak{I}(K)$. To draw this conclusion formally, observe that $\mathfrak{I}^{-1}(y)$ is nonempty when $y \in \mathfrak{I}(K)$. Hence there is a choice function, c, whose salient property is $c(\mathfrak{I}^{-1}(y)) \in \mathfrak{I}^{-1}(y)$ for all $y \in \mathfrak{I}(K)$. Then V is given by $V(y) = c(\mathfrak{I}^{-1}(y))$. Define the algorithm $T = UV$. Then

$$E^* \leq E(T) = \sup_{x \in K} \|T(\mathfrak{I}x) - Ux\| = \sup_{x \in K} \|UV\mathfrak{I}x - Ux\|$$

$$= \sup_{x \in K} \|U(V\mathfrak{I}x - x)\|$$

Notice that $\mathfrak{I}(V\mathfrak{I}x - x) = \mathfrak{I}x - \mathfrak{I}x = 0$. Also, for $x \in K$, $V\mathfrak{I}x - x \in K - K = K + K = 2K$ by the symmetry and convexity of K. Hence $\frac{1}{2}(V\mathfrak{I}x - x)$ is in K and in the kernel of $\mathfrak{I}$. Hence

$$E^* \leq 2 \sup \left\| U\left[\frac{1}{2}(V\mathfrak{I}x - x) \right] \right\|$$

$$\leq 2 \sup\{\|Uz\| : z \in K, \mathfrak{I}z = 0\} \qquad \blacksquare$$

LEMMA 3. *Let K be symmetric, and suppose that there is a map $F : \mathfrak{I}(K) \to X$ such that $x - F\mathfrak{I}x \in K$ and $\mathfrak{I}(x - F\mathfrak{I}x) = 0$ for all $x \in K$. Then UF is an optimal algorithm.*

Proof. Use Lemma 1 to obtain

$$E(UF) = \sup_{x \in K} \|UF\mathfrak{I}x - Ux\|$$

$$= \sup_{x \in K} \|U(F\mathfrak{I}x - x)\|$$

$$\leq \sup\{\|Uz\| : z \in K, \mathfrak{I}z = 0\}$$

$$\leq E^* \qquad \blacksquare$$

Example 4. Let

$$X = Z = \{x \in C^{m-1}[a, b] : x^{(m)} \in L^2[a, b]\}$$

and $Y = \mathbb{R}^n$. Define

$$K = \{x \in X : \|x^{(m)}\|_2 \leq 1\}$$

Select nodes t_i such that $a \leq t_1 < t_2 < \cdots < t_n \leq b$, and let the information operator be defined by $\mathfrak{I}x = (x(t_1), x(t_2), \ldots, x(t_n))$. We try to reconstruct x from $\mathfrak{I}x$. Consider the natural spline interpolating operator S that produces a spline of degree $2m - 1$. It turns out that S is the optimal algorithm. The interpolation property of S is expressed by $\mathfrak{I}S\mathfrak{I} = \mathfrak{I}$. In spline theory (see, for example, [KinC], page 386), it is proved that (with $s = S\mathfrak{I}x$)

$$\|x^{(m)}\|_2^2 = \|s^{(m)}\|_2^2 + \|x^{(m)} - s^{(m)}\|_2^2 \geq \|x^{(m)} - s^{(m)}\|_2^2$$

Thus S has the properties required of F in Lemma 3:

$$\mathfrak{I}(x - S\mathfrak{I}x) = 0 \qquad x - S\mathfrak{I}x \in K \qquad (x \in K)$$

Hence, by Lemma 3, S is optimal.   ∎

Example 5. Let $X = C[a, b]$, $Y = \mathbb{R}^n$, $Z = \mathbb{R}$, and

$$\begin{aligned} K &= \{x \in X : x \in C^n[a, b], \|x^{(n)}\|_\infty \leq 1\} \\ \mathfrak{I}x &= (x(t_1), x(t_2), \ldots, x(t_n)) \\ Ux &= x(t_0) \end{aligned}$$

Here $t_0, t_1, \ldots, t_n$ are distinct points in $[a, b]$, not necessarily in order. Define $w(t) = \Pi_{i=1}^n (t - t_i)/n!$. Notice that $w(t) = (t^n + \text{lower-order terms})/n!$. From this it follows that $w^{(n)}(t) = 1$. Also, we have $\mathfrak{I}w = 0$. From Lemma 1,

$$E^* \geq \sup\{\|Ux\| : x \in K, \mathfrak{I}x = 0\} \geq \|Uw\| = |w(t_0)|$$

If x is a given element of K, let p be the polynomial of degree at most $n - 1$ that interpolates x at $t_1, t_2, \ldots, t_n$. The classical theory, [KC], page 284, asserts the existence of a point ξ for which

$$x(t_0) - p(t_0) = x^{(n)}(\xi)w(t_0)$$

The construction of p requires only $\mathfrak{I}x$. Hence there is a map T such that $T\mathfrak{I}x = p(t_0)$. This T is optimal, since

$$\begin{aligned} E(T) &= \sup_{x \in K} \|Ux - T\mathfrak{I}x\| = \sup_{x \in K} |x(t_0) - p(t_0)| \\ &\leq \|x^{(n)}\|_\infty |w(t_0)| \leq |w(t_0)| \leq E^* \end{aligned}$$ ∎

Example 6. This is the same as Example 5 except that $Z = C[a, b]$ and $Ux = x$. As before, we find that

$$E^* \geq \|Uw\|_\infty = \|w\|_\infty$$

Now let T be defined so that $T\mathfrak{I}x = p$. Then

$$E(T) = \sup_{x \in K} \|x - p\|_\infty \leq \sup_{x \in K} \|x^{(n)}\|_\infty \|w\|_\infty = \|w\|_\infty \leq E^*$$

Thus Lagrange interpolation is the optimal algorithm. ■

Example 7. Let T be a metric space of finite diameter. Thus there is an M such that $d(s, t) \le M$ for s and t in T. Let $X = \ell^\infty(T)$, the space of all bounded functions from T into $\mathbb{R}$. If $x \in X$, we define $\|x\| = \sup_t |x(t)|$. Imitating Example 2, we set $Y = \mathbb{R}^n$, $Z = X$, and

$$K = \{x \in X : |x(t) - x(s)| \le d(s, t)\}$$

Let $t_1, t_2, \ldots, t_n$ be n distinct points in T, and define $\Im x = (x(t_1), x(t_2), \ldots, x(t_n))$. We wish to find an optimal algorithm for recovering x from $\Im x$. Let $w(t) = \min_i d(t, t_i)$ and

$$T_i = \{t \in T : d(t, t_i) = w(t)\} \qquad (1 \le i \le n)$$

As in Example 1, one finds that $w \in K$ and $\Im w = 0$. Hence by Lemma 1,

$$E^* \ge \sup\{\|Ux\| : x \in K, \Im x = 0\} \ge \|Uw\| = \|w\|$$

Given $x \in K$, define a simple function $y = T\Im x$ by the rule $y(t) = x(t_i)$, where i is the first index for which $t \in T_i$. Then, assuming this state of affairs, we have

$$|x(t) - y(t)| = |x(t) - x(t_i)| \le d(t, t_i) = w(t)$$

Consequently, T is optimal, because

$$E(T) = \sup_{x \in K} \|Ux - T\Im x\| = \sup_{x \in K} \|x - y\| \le \|w\|$$ ■

Example 8. Let $t_0, t_1, \ldots, t_s$ be $s + 1$ points in the sphere $\mathcal{S}^{s-1}$, not lying on a hyperplane. Then the set $Q = \text{co}\{t_0, t_1, \ldots, t_s\}$ is a simplex. Assume that $0 \in Q$. Let $w(t) = \min \|t - t_i\|_2$, where $t \in \mathbb{R}^s$. Let $X = \ell^\infty(Q)$ (as defined in Example 7). Set

$$K = \{x \in X : |x(t) - x(s)| \le \|t - s\|_2 \text{ for all } t, s \text{ in } Q\}$$
$$\Im x = (x(t_0), x(t_1), \ldots, x(t_s))$$
$$Ux = x$$

What are the optimal algorithms? By Lemma 1,

$$E^* \ge \sup\{\|Ux\| : x \in K, \Im x = 0\} \ge \|w\| \ge w(0) = 1$$

Given $x \in K$, let y be the polynomial of degree 1 that interpolates x at the vertices of the simplex Q. Put $T\Im x = y$. If $t \in Q$, then $t = \sum_{i=0}^s \theta_i t_i$, with $\theta_i \ge 0$ and $\sum \theta_i = 1$. Also, $y(t) = \sum \theta_i x(t_i)$. Thus we can estimate the error by writing

$$|y(t) - x(t)| = \left| \sum \theta_i [x(t_i) - x(t)] \right| \le \sum \theta_i |x(t_i) - x(t)|$$
$$\le \sum \theta_i \|t_i - t\|_2 = \sum \theta_i^{1/2} \theta_i^{1/2} \|t_i - t\|_2$$
$$\le \left(\sum \theta_i \right)^{1/2} \left(\sum \theta_i \|t_i - t\|_2^2 \right)^{1/2} = \left(\sum \theta_i \left[\|t_i\|_2^2 - 2tt_i + \|t\|_2^2 \right] \right)^{1/2}$$
$$= \left(\sum \theta_i \left[1 - 2tt_i + \|t\|_2^2 \right] \right)^{1/2}$$
$$= (1 - 2t^2 + \|t\|_2^2)^{1/2} = (1 - \|t\|_2^2)^{1/2} \le 1$$

From this it follows that T is optimal, since

$$E(T) = \sup_{x \in K} \|T\mathfrak{I}x - x\| \leq 1 \leq E^* \qquad \blacksquare$$

In Chapter 26, we mentioned the Chebyshev radius and center of a set. These concepts are useful here too. If A is a bounded set in a normed linear space, then the number

$$r = r(A) = \inf_{x \in X} \sup_{a \in A} \|x - a\|$$

is called the **Chebyshev radius** of A. Any point x for which

$$\sup_{a \in A} \|x - a\| = r$$

is called a **Chebyshev center** of A.

THEOREM. *Let an optimal recovery problem be defined by the triple $(K, \mathfrak{I}, U)$ as before. Let r denote the Chebyshev radius function. Then*

$$E^* \geq \sup_{y \in \mathfrak{I}(K)} r\big(U(\mathfrak{I}^{-1}(y))\big)$$

If, for each $y \in \mathfrak{I}(K)$, Ty is a Chebyshev center of $U(\mathfrak{I}^{-1}(y))$, then T is an optimal algorithm.

Proof. For *any* algorithm $T : \mathfrak{I}(K) \to Z$, we have

$$E(T) = \sup_{x \in K} \|T\mathfrak{I}x - Ux\|$$

$$= \sup_{y \in \mathfrak{I}(K)} \sup_{x \in \mathfrak{I}^{-1}(y)} \|T\mathfrak{I}x - Ux\|$$

$$= \sup_{y \in \mathfrak{I}(K)} \sup_{x \in \mathfrak{I}^{-1}(y)} \|Ty - Ux\|$$

$$= \sup_{y \in \mathfrak{I}(K)} \sup_{z \in U(\mathfrak{I}^{-1}(y))} \|Ty - z\|$$

$$\geq \sup_{y \in \mathfrak{I}(K)} r\big(U(\mathfrak{I}^{-1}(y))\big)$$

This final inequality becomes an equality if Ty is a Chebyshev center of $U(\mathfrak{I}^{-1}(y))$. $\blacksquare$

If K is a bounded set, then so is $\mathfrak{I}^{-1}(y)$, since $\mathfrak{I}$ is regarded as having K as its domain. If U is a bounded linear transformation, then $U\mathfrak{I}^{-1}(y)$ is a bounded subset of Z. If Z is a Hilbert space (or, more generally, a reflexive Banach space), then $U\mathfrak{I}^{-1}(y)$ will have a Chebyshev center (see Chapter 26). Under these conditions, the preceding theorem guarantees the existence of an optimal recovery algorithm. For methods of constructing Chebyshev centers, see the paper of Amir and Ziegler [AZ].

Problems

1. In Example 1, T will predict the correct value of $x(t_0)$ for all x on the intersection of K with a hyperplane. Prove this. (Recall that a hyperplane has codimension 1 and hence occupies the entire space except for one dimension.)

2. Extend Example 8 to include the case when $0 \notin Q$.

3. Prove that in any normed linear space, x_0 is the Chebyshev center of the closed ball

$$\{x : \|x - x_0\| \le r\}$$

Prove also the uniqueness of the Chebyshev center.

4. Prove that if A is a bounded set in a normed linear space, then the function

$$f(x) = \sup_{a \in A} \|x - a\|$$

is continuous. Prove, in fact, Lipschitz continuity with constant 1.

5. (Continuation) Use Problem 4 to prove that in any finite-dimensional normed space, a bounded set has at least one Chebyshev center.

6. Given a set A in a normed space, $c(A)$ denotes the set of all Chebyshev centers of A. Find a set (not empty nor consisting of a single point) such that $c(c(A)) = A$.

7. Prove that the Chebyshev center of an acute triangle in the Euclidean plane is the center of the circle that passes through its three vertices.

References

[AZ] Amir, D., and Z. Ziegler. "Construction of elements of the relative Chebyshev center." In *Approximation Theory and Applications,* ed. by Z. Ziegler. Academic Press, New York, 1981, pp. 1–12.

[GolW] Golomb, M., and H. F. Weinberger. "Optimal approximation and error bounds." In *On Numerical Approximation,* ed. by R. E. Langer. University of Wisconsin Press, Madison, WI, 1959, pp. 117–190.

[Hol] Holmes, R. B. *A Course on Optimization and Best Approximation.* Lecture Notes in Math., vol. 257. Springer-Verlag, Berlin, 1972.

[KinC] Kincaid, D., and W. Cheney. *Numerical Analysis.* 2nd ed., Brooks/Cole, Pacific Grove, CA, 1996.

[MR1] Micchelli, C. A., and T. J. Rivlin. "Lectures on optimal recovery." In *Numerical Analysis Lancaster 1984,* ed. by P. R. Turner. Lecture Notes in Math., vol. 1129. Springer-Verlag, Berlin, 1985, pp. 21–93.

[MR2] Micchelli, C. A., and T. J. Rivlin. "A survey of optimal recovery." In *Optimal Estimation in Approximation Theory,* ed. by C. A. Micchelli and T. J. Rivlin. Plenum Press, New York, 1977, pp. 1–54.

[Pin2] Pinkus, A. "N-Widths and optimal recovery." In *Approximation Theory,* ed. by C. de Boor. Amer. Math. Soc. Proc. Symp. Appl. Math., vol. 36, Providence, RI, 1986, pp. 51–66.

[TW] Traub, J. F., and H. Wozniakowski. *A General Theory of Optimal Algorithms.* Academic Press, New York, 1980.

28
Algorithmic Orthogonal Projections

Any closed subspace V in a Hilbert space X has associated with it an **orthogonal projection,** $P : X \longrightarrow V$. It is a linear map completely characterized by the property

$$(1) \qquad\qquad x - Px \perp V \qquad\qquad (x \in X)$$

Assuming for the moment that such a map exists, we can show that Px is the unique point in V closest to x. To do so, let v be any point in V. By Property (1) and the Pythagorean Law, we have

$$\|x - v\|^2 = \|(x - Px) + (Px - v)\|^2 = \|x - Px\|^2 + \|Px - v\|^2 \geq \|x - Px\|^2$$

The existence of Px comes from the basic theorem that for any closed convex set in a Hilbert space, each point of the space has a unique closest point in the convex set. To prove that $x - Px \perp V$, let $v \in V$, $v_0 = Px$, and $\lambda > 0$. Then, because v_0 is the closest point to x, we have

$$0 \leq \|x - v_0 + \lambda v\|^2 - \|x - v_0\|^2 = 2\langle x - v_0, \lambda v \rangle + \lambda^2 \|v\|^2$$

Since $\lambda > 0$, this leads to

$$0 \leq 2\langle x - v_0, v \rangle + \lambda \|v\|^2$$

Letting $\lambda \downarrow 0$, we obtain $\langle x - v_0, v \rangle \geq 0$. Applying this result to $-v$, we find that $\langle x - v_0, v \rangle \leq 0$. Hence $x - v_0 \perp v$, and so $x - v_0 \perp V$. To prove that P is linear, we note that $(\alpha x + \beta y) - (\alpha Px + \beta Py) \perp V$, because for $v \in V$

$$\langle \alpha x + \beta y - \alpha Px - \beta Py, v \rangle = \alpha \langle x - Px, v \rangle + \beta \langle y - Py, v \rangle = 0$$

Therefore $\alpha Px + \beta Py = P(\alpha x + \beta y)$, by the unicity of the closest point.

The easiest theoretical construction of P is to use any orthonormal base $\{u_\alpha\}$ for V. Then

(2)
$$Px = \sum \langle x, u_\alpha \rangle u_\alpha$$

In principle, the orthonormal base can be finite, denumerable, or nondenumerable. See [Hal] for these matters. It is easy to verify that the map P defined in Equation (2) has Property (1): For each u_β, we have

$$\langle x - Px, u_\beta \rangle = \langle x, u_\beta \rangle - \left\langle \sum_\alpha \langle x, u_\alpha \rangle u_\alpha, u_\beta \right\rangle = \langle x, u_\beta \rangle - \sum_\alpha \langle x, u_\alpha \rangle \langle u_\alpha, u_\beta \rangle$$

$$= \langle x, u_\beta \rangle - \langle x, u_\beta \rangle = 0$$

The limitation of Equation (2) for practical computation lies in the immense labor of constructing the orthonormal base and in computing all the inner products in the series. An alternative procedure for computing Px has recently been given by Davis, Mallat and Zhang [DMZ], and we shall present their algorithm here.

We assume that the subspace V is made accessible to us by a set of generators, all of norm 1. Thus, $V = \overline{\text{span}}(S)$, where

$$S \subset \{x \in X : \|x\| = 1\}$$

(The notation $\overline{\text{span}}$ signifies the closure of the linear span of a set.) It is explicitly permitted that S be *redundant*, that is, it may contain many more vectors than are needed to generate V. In other words, S may be linearly dependent. Now let x be any element of X. We wish to construct Px, the point in V closest to x.

ALGORITHM. *Given x, define $x_1 = x$. Select a number $\alpha \in (0, 1)$. Proceed inductively. If x_n is known, select $s_n \in S$ so that*

(3)
$$|\langle x_n, s_n \rangle| \geq \alpha \sup_{s \in S} |\langle x_n, s \rangle|$$

Define $v_n = \langle x_n, s_n \rangle s_n$ and $x_{n+1} = x_n - v_n$.

What is the motivation behind this algorithm? If we look at Equation (2), we see that we must take a sum of components $\langle x, s \rangle s$ of the vector x. It is natural to start with the *largest* of these, and proceed to take smaller ones. Since

$$\|\langle x, s \rangle s\| = |\langle x, s \rangle| \, \|s\| = |\langle x, s \rangle|$$

we seek an $s \in S$ for which $|\langle x, s \rangle|$ is large. Ideally, we would select $s \in S$ so that $|\langle x, s \rangle|$ is a maximum, but such an s need not exist. Hence the presence of α in the algorithm. One hopes that

(4)
$$Px = \sum_{n=1}^{\infty} v_n = \sum_{n=1}^{\infty} \langle x_n, s_n \rangle s_n$$

and this turns out to be true. Notice the similarity between Equations (2) and (4).

LEMMA 1. *The vectors in the algorithm have these properties:*

$$(5) \qquad\qquad x_{n+1} \perp s_n, \quad x_{n+1} \perp v_n$$

$$(6) \qquad\qquad \|v_n\|^2 = |\langle x_n, s_n \rangle|^2 = \|x_n\|^2 - \|x_{n+1}\|^2$$

$$(7) \qquad\qquad x_n = \sum_{i=n}^{m-1} v_i + x_m \qquad (m \geq n)$$

$$(8) \qquad\qquad \|x\|^2 = \sum_{i=1}^{n} \|v_i\|^2 + \|x_{n+1}\|^2$$

Proof. To prove (5), simply write

$$\langle x_{n+1}, s_n \rangle = \langle x_n - v_n, s_n \rangle = \langle x_n, s_n \rangle - \langle v_n, s_n \rangle = \langle x_n, s_n \rangle - \langle x_n, s_n \rangle \langle s_n, s_n \rangle = 0$$

To prove (6), use (5) and the Pythagorean Law:

$$\|x_n\|^2 = \|x_{n+1} + v_n\|^2 = \|x_{n+1}\|^2 + \|v_n\|^2$$

To prove (7), write

$$x_n = (x_n - x_{n+1}) + (x_{n+1} - x_{n+2}) + \cdots + (x_{m-1} - x_m) + x_m$$
$$= v_n + v_{n+1} + \cdots + v_{m-1} + x_m$$

To prove (8), use (6) as follows:

$$\|x\|^2 = (\|x_1\|^2 - \|x_2\|^2) + (\|x_2\|^2 - \|x_3\|^2) + \cdots + (\|x_n\|^2 - \|x_{n+1}\|^2) + \|x_{n+1}\|^2$$
$$= \|v_1\|^2 + \|v_2\|^2 + \cdots + \|v_n\|^2 + \|x_{n+1}\|^2 \qquad\blacksquare$$

LEMMA 2. *The following inequality holds:* $|\langle v_i, x_m \rangle| \leq \alpha^{-1} \|v_i\| \, \|v_m\|$

Proof. The inequality follows from this calculation:

$$\|v_i\| \, \|v_m\| = |\langle x_i, s_i \rangle| \, |\langle x_m, s_m \rangle| \geq |\langle x_i, s_i \rangle| \, \alpha \, |\langle x_m, s_i \rangle| = \alpha \left| \langle \langle x_i, s_i \rangle s_i, x_m \rangle \right|$$
$$= \alpha |\langle v_i, x_m \rangle| \qquad\blacksquare$$

LEMMA 3. *If* $w \in \ell^2$ *and* $w \geq 0$, *then for each* n

$$\inf_{k>n} w_k \sum_{j=1}^{k} w_j = 0$$

Proof. Fix n and let $\varepsilon > 0$. Put $z_k = \sum_{i=1}^{k} w_i$. Select $m > n$ so that $\sum_{i>m} w_i^2 < \varepsilon/2$. Select $p > m$ so that $w_p z_m < \varepsilon/2$. Select $j \in \{m+1, m+2, \ldots, p\}$ so that $w_j = \min\{w_{m+1}, w_{m+2}, \ldots, w_p\}$. Then

$$\inf_{k>n} w_k z_k \leq w_j z_j \leq w_j z_p = w_j \left(z_m + \sum_{i=m+1}^{p} w_i \right) = w_j z_m + \sum_{i=m+1}^{p} w_j w_i$$

$$\leq w_p z_m + \sum_{i=m+1}^{p} w_i^2 \leq \varepsilon/2 + \varepsilon/2 = \varepsilon$$

Since ε was arbitrary, $\inf_{k>n} w_k z_k = 0.$ ■

LEMMA 4. *If $q > k$, then the points x_k and x_q produced in the algorithm satisfy*

$$\|x_k - x_q\|^2 \leq \|x_k\|^2 - \|x_q\|^2 + \frac{2}{\alpha}\|v_q\| \sum_{i=k}^{q-1} \|v_i\|$$

Proof. Since

$$\|x_k\|^2 = \|(x_k - x_q) + x_q\|^2$$
$$= \|x_k - x_q\|^2 + 2\langle x_k - x_q, x_q\rangle + \|x_q\|^2$$

we have, with the help of Equation (7) and Lemma 2,

$$\|x_k - x_q\|^2 - \|x_k\|^2 + \|x_q\|^2 = -2\langle x_k - x_q, x_q\rangle$$

$$= -2\sum_{i=k}^{q-1} \langle v_i, x_q\rangle \leq 2 \sum_{i=k}^{q-1} |\langle v_i, x_q\rangle|$$

$$\leq 2\alpha^{-1}\|v_q\| \sum_{i=k}^{q-1} \|v_i\|$$ ■

LEMMA 5. *The sequence $[x_n]$ produced by the algorithm is a Cauchy sequence.*

Proof. Let $\varepsilon > 0$. By Equation (6) it is apparent that the sequence $[\|x_n\|]$ is monotonically decreasing to a limit, $L \geq 0$. Select N so that $\|x_N\|^2 < L^2 + \varepsilon^2$. Let $m > n > N$. By Equation (8), $\sum_{i=1}^{\infty} \|v_i\|^2 < \infty$. Hence Lemma 3 can be applied to obtain an index $q > m$ such that $\|v_q\|\sum_{i=1}^{q} \|v_i\| < \varepsilon^2$. By Lemma 4, for each k in the range $N < k < q$ we will have

$$\|x_k - x_q\|^2 \leq \|x_k\|^2 - \|x_q\|^2 + \frac{2}{\alpha}\|v_q\| \sum_{i=k}^{q-1} \|v_i\|$$

$$< (L^2 + \varepsilon^2) - L^2 + \frac{2}{\alpha}\varepsilon^2 = \varepsilon^2\left(1 + \frac{2}{\alpha}\right)$$

We can use this inequality with $k = n$ and $k = m$ to obtain

$$\|x_m - x_n\| \leq \|x_m - x_q\| + \|x_q - x_n\| < 2\varepsilon\sqrt{1 + \frac{2}{\alpha}}$$ ■

THEOREM. *The series $\sum_{n=1}^{\infty} v_n$ produced by the algorithm converges to the orthogonal projection of x onto $V = \overline{\text{span}}(S)$.*

Proof. By Lemma 5, the sequence $[x_n]$ has the Cauchy property, and we are at liberty to define $y = \lim x_n$. By putting $n = 1$ and letting $m \to \infty$ in Equation (7), we obtain

$$x = \sum_{i=1}^{\infty} v_i + y$$

Thus the series in question converges. Since $v_i \in V$ and V is closed, $\sum_{i=1}^{\infty} v_i \in V$. It remains to prove that $y \perp V$. By Equation (6), $\langle x_n, s_n \rangle \to 0$. For any $s \in S$, we have (by the definition of s_n)

$$|\langle x_n, s_n \rangle| \geq \alpha |\langle x_n, s \rangle|$$

Let $n \to 0$ in this inequality to get $\langle y, s \rangle = 0$ for all $s \in S$. Hence $y \perp V$.   ∎

Problems

1. Describe what happens in the algorithm if S is the unit sphere in the Hilbert space.

2. Prove that the algorithm is "homogeneous." This means that if it is applied to x and yields $y = \sum_{n=1}^{\infty} v_n$, then with suitable choices of s_n the algorithm operating on λx will produce λy.

3. Describe what happens in the algorithm if S is an orthonormal base for V.

4. Describe what happens in the algorithm when $V = X$.

5. Describe what happens in the algorithm if x is in V.

6. Apply the algorithm in this setting:

$$X = \mathbb{R}^2, \quad S = \{s_1, s_2, \ldots\}, \quad s_{2n} = (1, 0), \quad s_{2n+1} = (3/5, 4/5), \quad x = (1, 1)$$

Prove, in particular, that for $n \geq 1$, x_{2n+1} has 0 as its first component. Prove that $\|x_{n+1}\| = (3/5)\|x_n\|$ for $n \geq 2$.

References

[Hal]   Halmos, P. *Introduction to Hilbert Space and the Theory of Spectral Multiplicity.* Chelsea, New York, 1954. Amer. Math. Soc., Providence, RI, 1998.

[DMZ]   Davis, G., S. Mallat, and Zhifeng Zhang. "Adaptive time-frequency approximations with matching pursuits." In *Wavelets: Theory, Algorithms, and Applications,* ed. by C. L. Chui, L. Montefusco, and L. Puccio. Academic Press, New York, 1994, pp. 271–293.

29

Cardinal B-Splines and the Sinc Function

The **cardinal B-splines** mentioned in the title are special spline functions having the integers as knots. They are defined as follows:

$$B_1(x) = \begin{cases} 1 & \text{if } -\frac{1}{2} \leq x < \frac{1}{2} \\ 0 & \text{otherwise} \end{cases}$$

$$B_2 = B_1 * B_1$$

$$B_3 = B_1 * B_1 * B_1, \text{ and so on}$$

It is clear that B_1 is a unit step function, and is piecewise constant. It turns out that B_2 is piecewise linear, B_3 is piecewise quadratic, and so forth. We calculate B_2 as follows:

$$B_2(x) = \int_{-\infty}^{\infty} B_1(y) B_1(x-y)\, dy$$

$$= \int_{-1/2}^{1/2} B_1(x-y)\, dy$$

$$= \begin{cases} 1+x & -1 \leq x \leq 0 \\ 1-x & 0 < x \leq 1 \\ 0 & \text{elsewhere} \end{cases}$$

Figure 29.1 may help in visualizing the last integral.

If the **support** of a function f is defined as the closure of $\{x : f(x) \neq 0\}$, then we can assert that the support of B_n is $[-n/2, n/2]$, for $n = 1, 2, 3, \ldots$. This can be proved by induction.

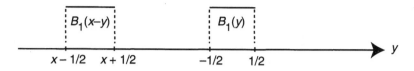

Figure 29.1   *Translates of the Function B_1*

An important function, called the **sinc function**, is defined by the equation

(1) $$\text{sinc}(x) = \frac{\sin(\pi x)}{\pi x}$$

Of course, for continuity we define $\text{sinc}(0) = 1$. This function is an entire function (has no singularities). Its Maclaurin series is easily obtained from

$$\frac{\sin(x)}{x} = \frac{1}{x}\left[x - \frac{x^3}{3!} + \frac{x^5}{5!} - \cdots\right] = 1 - \frac{x^2}{3!} + \frac{x^4}{5!} - \frac{x^6}{7!} + \cdots$$

Notice that the sinc function is even. Figure 29.2 shows the graph of sinc.

What is the connection between the cardinal B-splines and the sinc function? We find out by computing the Fourier transform of B_1:

$$\hat{B}_1(x) = \int_{-\infty}^{\infty} B_1(x)e^{-2\pi ixy}\,dy = \int_{-1/2}^{1/2} e^{-2\pi ixy}\,dy$$

$$= \frac{e^{-2\pi ixy}}{-2\pi ix}\bigg|_{y=-1/2}^{y=1/2} = \frac{e^{-i\pi x} - e^{i\pi x}}{-2\pi ix}$$

$$= \frac{\sin(\pi x)}{\pi x} = \text{sinc}(x)$$

Now recall the Plancherel Theorem, which asserts that the Fourier transform is an isometry on $L^2(\mathbb{R})$. This implies that

$$\|\text{sinc}\|_2 = \|\hat{B}_1\|_2 = \|B_1\|_2 = 1$$

Thus the sinc function belongs to $L^2(\mathbb{R})$. It does not belong to $L^1(\mathbb{R})$, however (see Problem 3 of this chapter).

Recall the computational rule $(f * g)^{\wedge} = \hat{f}\,\hat{g}$. It leads to the formula

$$\hat{B}_n = (B_1 * \cdots * B_1)^{\wedge} = \hat{B}_1 \cdots \hat{B}_1 = (\hat{B}_1)^n = \text{sinc}^n$$

All the powers $\text{sinc}^2, \text{sinc}^3,\dots$ belong to $L^1(\mathbb{R})$. Indeed, we have

$$\|\text{sinc}^2\|_1 = \int \text{sinc}^2 = \|\text{sinc}\|_2^2 = 1$$

Since $|\text{sinc}(x)| \le 1$ for all x, we have $|\text{sinc}|^n \le |\text{sinc}|^2$ if $n \ge 2$. Hence $\|\text{sinc}^n\|_1 \le 1$ for $n \ge 2$.

It is straightforward to extend the foregoing discussion to a multidimensional setting. If x is a point in $\mathbb{R}^s$ written as $x = (\xi_1, \xi_2, \dots, \xi_s)$, then we define

$$\text{sinc}\,x = \prod_{j=1}^{s} \frac{\sin(\pi\xi_j)}{\pi\xi_j}$$

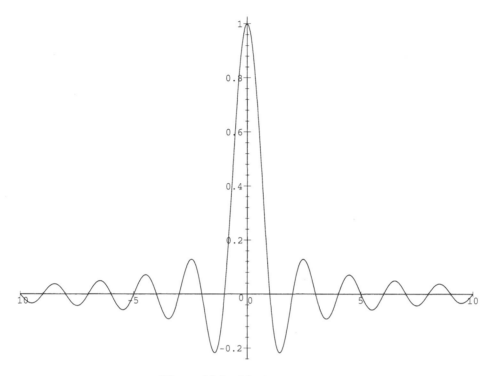

Figure 29.2 *The Sinc Function*

Let a unit cube in $\mathbb{R}^s$ be defined by

$$Q = \left\{ x \in \mathbb{R}^s : \|x\|_\infty \leq 1/2 \right\}$$

The characteristic function of Q is

$$\mathcal{X}_Q(x) = \begin{cases} 1 & \text{if } x \in Q \\ 0 & \text{if } x \notin Q \end{cases}$$

A connection with the B-splines of degree 0 (order 1) is

$$\mathcal{X}_Q(x) = \prod_{j=1}^{s} B_1(\xi_j)$$

An easy calculation (Problem 5) reveals that for $x \in \mathbb{R}^s$,

$$\hat{\mathcal{X}}_Q(x) = \int_{\mathbb{R}^s} \mathcal{X}_Q(y) e^{-2\pi i x y} \, dy = \int_Q e^{-2\pi i x y} \, dy = \text{sinc}(x)$$

This multidimensional relationship is part of a greater theory—that of the box splines, to which Chapter 33 is devoted.

The importance of the sinc function derives largely from its role in the sampling theorem, to which we now turn. This theorem is often associated with the name Shannon because of his work [Shan]. But the theorem was known prior to Shannon's 1949 paper. For example, it was certainly known to Edmund Whittaker [Whit1] and Kotelnikov

[Kot]. A thorough historical review has been given by Higgins [Hig]. The approximation theoretic aspects of this theorem can be found in Butzer's papers, particularly [But2].

THEOREM 1 (The Sampling Theorem). *Let f be a continuous member of $L^1(\mathbb{R}^s)$ whose Fourier transform vanishes outside of the cube $Q = \left[-\frac{1}{2}, \frac{1}{2}\right]^s$. Then f is uniquely determined by its values on $\mathbb{Z}^s$, and in fact*

$$f(x) = \sum_{v \in \mathbb{Z}^s} f(v) \operatorname{sinc}(x - v) \qquad (x \in \mathbb{R}^s)$$

Proof. Since f is an L^1-function, its Fourier transform is continuous [Ru2]. By our hypotheses, $\hat{f}$ has compact support and therefore belongs to $L^2(\mathbb{R}^s)$. It follows that the Fourier series of $\hat{f}$ converges in the L^2-norm to $\hat{f}$:

$$\hat{f}(x) = \sum_{v \in \mathbb{Z}^s} c_v e^{-2\pi i v x} \qquad (x \in Q)$$

In this equation the Fourier coefficients c_v are given by

$$c_v = \int_Q \hat{f}(y) e^{2\pi i v y}\, dy$$

By the Fourier inversion formula, we have

$$c_v = \int_{\mathbb{R}^s} \hat{f}(y) e^{2\pi i v y}\, dy = f(v)$$

Using the Fourier inversion formula again yields

$$f(x) = \int_{\mathbb{R}^s} \hat{f}(y) e^{2\pi i x y}\, dy = \int_Q \hat{f}(y) e^{2\pi i x y}\, dy$$

$$= \int_Q \sum_{v \in \mathbb{Z}^s} c_v e^{-2\pi i v y} e^{2\pi i x y}\, dy$$

Since the Fourier series converges in L^2, the order of summation and integration can be reversed to give

$$f(x) = \sum_{v \in \mathbb{Z}^s} c_v \int_Q e^{-2\pi i v y} e^{2\pi i x y}\, dy$$

$$= \sum_{v \in \mathbb{Z}^s} c_v \int_Q e^{-2\pi i (v - x) y}\, dy$$

$$= \sum_{v \in \mathbb{Z}^s} f(v) \hat{\chi}_Q(v - x)$$

$$= \sum_{v \in \mathbb{Z}^s} f(v) \operatorname{sinc}(v - x)$$

$$= \sum_{v \in \mathbb{Z}^s} f(v) \operatorname{sinc}(x - v) \qquad \blacksquare$$

In order to accommodate any continuous function in $L^1(\mathbb{R}^s)$ whose Fourier transform is compactly supported, one can make a change of variable in the Sampling Theorem. The result is given next, although the proof is left to Problem 6.

THEOREM 2. *If $f \in C(\mathbb{R}^s) \cap L^1(\mathbb{R}^s)$ and $\hat{f}(x)$ vanishes for $\|x\|_\infty \geq \frac{h}{2}$ then*

$$f(x) = \sum_{v\in\mathbb{Z}^s} f(vh)\,\text{sinc}\left(\frac{x}{h} - v\right)$$

Theorem 2 is sometimes called the "interpolation theorem," because when the hypotheses are satisfied we have

$$\sum_{v\in\mathbb{Z}^s} f(vh)\,\text{sinc}(\mu - v) = f(\mu h) \qquad (\mu \in \mathbb{Z}^s)$$

(See Problem 7.)

In Chapter 36, we will define a discrete convolution

$$(f \star g)(x) = \sum_{v\in\mathbb{Z}^s} f(v)g(x - v) \qquad (x \in \mathbb{R}^s)$$

With this formalism, the equations in the Sampling Theorem and in Theorem 2 become

$$f \star \text{sinc} = f$$
$$S_h^{-1}(S_h f \star \text{sinc}) = f$$

where S_h is the operator defined by $(S_h f)(x) = f(hx)$. If f does not have compact support, these formulas can no longer be expected to reproduce f exactly. However, the discrete convolution continues to provide a good approximation to f under less restrictive conditions.

There is an alternative formulation of Theorem 2, sometimes called the "quadrature theorem."

THEOREM 3. *Let f be a continuous element of $L^1(\mathbb{R}^s)$ such that, for some $h > 0$, $\hat{f}(x)$ vanishes when $\|x\|_\infty \geq h/2$. Then*

$$f(vh) = \frac{1}{h}\int_{\mathbb{R}^s} f(x)\,\text{sinc}\left(\frac{x}{h} - v\right)dx \qquad (v \in \mathbb{Z}^s)$$

Proof. It suffices to prove the assertion when $h = 1$ and then to make the change of variable $x \to hx$. From an equation in the proof of Theorem 1,

$$f(v) = \int_{\mathbb{R}^s} \hat{f}(x)e^{2\pi ivx}\,dx$$
$$= \int_Q \hat{f}(x)e^{2\pi ivx}\,dx$$
$$= \int_Q \int_{\mathbb{R}^s} f(y)e^{-2\pi ixy}\,dy\,e^{2\pi ivx}\,dx$$
$$= \int_{\mathbb{R}^s} f(y)\int_Q e^{2\pi ix(v-y)}\,dx\,dy$$

Inverting the order of integration here is permissible because $f \in L^1(\mathbb{R}^s)$ by hypothesis and consequently the function $(x, y) \mapsto f(y)e^{2\pi i x (v-y)}$ is in $L^1(Q \times \mathbb{R}^s)$. Now, by the argument at the conclusion of Theorem 1,

$$\int_Q e^{2\pi i x (v-y)} dx = \operatorname{sinc}(y - v)$$

Hence

$$f(v) = \int_{\mathbb{R}^s} f(y) \operatorname{sinc}(y - v) \, dy \qquad \blacksquare$$

COROLLARY. *Assume that f belongs to $C(\mathbb{R}^s) \cap L^1(\mathbb{R}^s)$ and that its Fourier transform vanishes at x whenever $\|x\|_\infty > h/2$. Then $f \in L^2(\mathbb{R}^s)$, and*

$$\|f\|_2 = \left\{ h^s \sum_{v \in \mathbb{Z}^s} |f(vh)|^2 \right\}^{1/2}$$

Proof. From Theorem 2,

$$\|f\|_2^2 = \int_{\mathbb{R}^s} [f(x)]^2 \, dx = \int_{\mathbb{R}^s} \left\{ \sum_{v \in \mathbb{Z}^s} f(vh) \operatorname{sinc}\left(\frac{x}{h} - v\right) \right\}^2 dx$$

$$= \sum_{\mu \in \mathbb{Z}^s} \sum_{v \in \mathbb{Z}^s} f(vh) f(\mu h) \int_{\mathbb{R}^s} \operatorname{sinc}\left(\frac{x}{h} - v\right) \operatorname{sinc}\left(\frac{x}{h} - \mu\right) dx$$

The interchanging of the sum and the integral is justified here since the sum converges in the L^2-sense. With the substitution $y = x/h$ and the result of Problem 6, we obtain

$$\|f\|_2^2 = \sum_{\mu \in \mathbb{Z}^s} \sum_{v \in \mathbb{Z}^s} f(vh) f(\mu h) \int_{\mathbb{R}^s} \operatorname{sinc}\left(\frac{x}{h} - v + \mu\right) \operatorname{sinc}\left(\frac{x}{h}\right) dx$$

$$= \sum_{v \in \mathbb{Z}^s} [f(vh)]^2 \int_{\mathbb{R}^s} \operatorname{sinc}(y) h^s \, dy$$

$$= \sum_{v \in \mathbb{Z}^s} h^s [f(vh)]^2 \qquad \blacksquare$$

Problems

1. Prove that the support of B_n is $[-n/2, n/2]$.

2. For $x \in \mathbb{R}$, let $x_+ = \max\{x, 0\}$ and $x_- = \min\{x, 0\}$. Prove that $B_2(x) = (1 + x_- - x_+)_+$.

3. Prove that $\operatorname{sinc} \notin L^1(\mathbb{R})$ by considering the areas of triangles constructed between the graph of sinc and the horizontal axis in Figure 29.2.

4. Use the Parseval equation, $\langle \hat{f}, \hat{g} \rangle = \langle f, g \rangle$ to prove that the set of functions
$$\{x \mapsto \operatorname{sinc}(x - j) : j \in \mathbb{Z}\}$$
is orthonormal in $L^2(\mathbb{R})$.

5. Show that for $y \in \mathbb{R}^s$,

$$\int_Q e^{-2\pi i x y} \, dx = \text{sinc}(y)$$

6. Prove Theorem 2 by applying the Sampling Theorem to $x \mapsto f(hx)$.

7. Prove that if the series $\sum_{v \in \mathbb{Z}^s} f(v) \, \text{sinc}(x - v)$ converges uniformly for $x \in \mathbb{R}^s$, then

$$f(\mu) = \sum_{v \in \mathbb{Z}^s} f(v) \, \text{sinc}(\mu - v)$$

8. Prove that for $\mu \in \mathbb{Z}^s$,

$$\int_{\mathbb{R}^s} \text{sinc}(x - \mu) \, \text{sinc}(x) \, dx = \delta_{\mu 0}$$

References

[BRS] Butzer, P. L., S. Ries, and R. L. Stens. "Approximation of continuous and discontinuous functions by generalized sampling series." *J. Approx. Theory 50* (1987), 25–39.

[BSS] Butzer, P. L., W. Splettstösser, and R. L. Stens. "The sampling theorem and linear predictions in signal analysis." *Jahresber. Deutsch. Math.-Verein. 90* (1988), 1–60.

[But2] Butzer, P. L. "A survey of the Whittaker-Shannon sampling theorem and some of its extensions." *J. Math. Res. Expositions 3* (1983), 185–212.

[Chu1] Chui, C. K. *Multivariate Splines.* SIAM, Philadelphia, 1988.

[De] Deans, S. R. *The Radon Transform and Some of Its Applications.* Wiley, New York, 1983.

[DVP1] de La Vallée Poussin, C. "Sur la convergence des formules d'interpolation entre ordonnées equidistantes." *Bull. Cl. Sci. Acad. Roy. Belg. No. 4* (1908), 319–403.

[Do2] Donoghue, W. F. *Distributions and Fourier Transforms.* Pure and Applied Mathematics Series, vol. 32. Academic Press, New York, 1969.

[DMc] Dym, H., and H. P. McKean. *Fourier Series and Integrals.* Academic Press, New York, 1972.

[Foll] Folland, G. B. *Fourier Analysis and Its Applications.* Wadsworth–Brooks/Cole, Pacific Grove, CA, 1992.

[Hig] Higgins, J. R. "Five short stories about the cardinal series." *Bull. Amer. Math. Soc. 12* (1985), 45–89.

[Kot] Kotelnikov, V. A. "On the carrying capacity of the 'ether' and wire in telecommunications." *Material for the First All-Union Conference on Questions of Communication, Izd. Red. Upr.* Svyazi RKKA, Moscow, 1933. (Russian)

[LB] Lund, J., and K. L. Bowers. *SINC Methods for Quadrature and Differential Equations.* SIAM, Philadelphia, 1992.

[NWal] Nashed, M. Z., and G. G. Walter. "General sampling theorems for functions in reproducing kernel Hilbert spaces." *Math. Control Signals Systems 4* (1991), 363–390.

[OS] Oppenheim, A. V., and R. W. Schafer. *Discrete Time Signal Processing.* Prentice Hall, Englewood Cliffs, NJ, 1989.

[Ru2] Rudin, W. *Functional Analysis.* McGraw-Hill, New York, 1973.

[Sai]   Saitoh, S. *Theory of Reproducing Kernels and Its Applications*. Longmans, London, 1988.

[Shan]   Shannon, C. "Communication in the presence of noise." *Proc. IRE 37* (1949), 10–21.

[Sten]   Stenger, F. *Numerical Methods Based on Sinc and Analytic Functions*. Springer-Verlag, Berlin, 1993. Review: *Math. Intell. 18*(2), 1996, 71–73.

[Walt]   Walter, Gilbert G. *Wavelets and Other Orthogonal Systems with Applications*. CRP Press, Boca Raton, FL, 1994.

[Whit1]   Whittaker, E. T. "On the functions which are represented by the expansions of the interpolation theory." *Proc. Roy. Soc. Edinburgh 35* (1915), 181–194.

[Whit2]   Whittaker, J. M. "The Fourier theory of the cardinal function." *Proc. Math. Soc. Edinburgh 1* (1929), 169–176.

30

The Golomb-Weinberger Theory

This chapter expounds part of a theory put forth in 1958 by Golomb and Weinberger [GolW]. Our selection of material from their massive paper is highly influenced by the needs of the following Chapters 31 and 32, which concern applications of this theory.

The setting is a real Hilbert space, H, whose inner product is written $\langle \cdot , \cdot \rangle$. The problem to be studied is this: For a certain unknown element, f, in H, we have at hand partial information in the form of numerical values of some continuous linear functionals, $\phi_1(f), \phi_2(f), \ldots, \phi_n(f)$. From this inadequate information, we wish to guess what f is.

In the usual circumstances, nothing can be said with much certainty about f, for it can be any element in the linear manifold

$$(1) \qquad M = \{v \in H : \phi_i(v) = \phi_i(f) \text{ for } 1 \leq i \leq n\}$$

Still, there is a reasonable way to proceed: Take g (our guess for f) to be the point of least norm in M. We shall say more about the justification of this presently. For now, we are content to prove an elementary equation showing how well this choice of g serves as a substitute for f.

THEOREM 1. *With g and f as described above, we have*

$$(2) \qquad \|f - g\|^2 = \|f\|^2 - \|g\|^2$$

Proof. First we define the linear subspace

$$(3) \qquad V = \{v \in H : \phi_i(v) = 0, \text{ for } i = 1, 2, \ldots, n\}$$

(Notice that the manifold M is a translate of the subspace V.) Next, we prove that g is characterized by the conditions

$$(4) \qquad g \perp V, \quad \phi_i(g) = \phi_i(f) \qquad (1 \leq i \leq n)$$

To see that this is so, recall that g was chosen to minimize $\|g\|$ subject to the condition $g \in M$. This is the same as minimizing $\|g\|$ subject to the condition $f - g \in V$. Writing

$v = f - g$, we see that we are minimizing $\|f - v\|$ subject to the condition $v \in V$. This is a standard problem of best approximation, treated in Chapter 28, and it is well known that the solution, v, is characterized by the conditions $v \in V$ and $f - v \perp V$. Hence $f - g \in V$ and $g \perp V$. In particular, $g \perp f - g$, and by the Pythagorean Law,

$$\|f\|^2 = \|f - g + g\|^2 = \|f - g\|^2 + \|g\|^2 \qquad \blacksquare$$

Example. Let $H = \mathbb{R}^2$, and suppose that f is an unknown vector (ξ_1, ξ_2) such that $\xi_1 + 3\xi_2 = 20$. What is the vector g (our guess as to f)? Since V consists of vectors orthogonal to $(1, 3)$, and g is orthogonal to V, we can let $g = \lambda(1, 3)$. The other condition in Equation (3) then reads $\lambda + 9\lambda = 20$, so $\lambda = 2$ and $g = (2, 6)$. $\qquad \blacksquare$

Suppose that in addition to the information $\phi_i(f)$, we are told also that $\|f\| \leq r$. Does this additional information help in any way? Obviously not, for our "estimator," g, has minimum norm among all the elements of M. But we can now see a further justification for our choice of g by noting that it is the *center* of the set

$$C = \{v \in H : \|v\| \leq r, \phi_i(v) = \phi_i(f), \text{ for } i = 1, 2, \ldots, n\}$$

To verify this assertion, notice first that g, by its definition, is an element of C. Next, we can see that each point on the circumference of C is equidistant from g. Indeed, if $v \in C$ and $\|v\| = r$, then (because $g - v \in V$ and $g \perp V$) we have

$$\|g\|^2 = \|(g - v) + v\|^2 = \|g - v\|^2 + \|v\|^2$$

whence $\|g - v\|^2 = \|g\|^2 - \|v\|^2 = \|g\|^2 - r^2$. Since we know nothing of f except that it belongs to C, the choice of the center point of C as an approximation to f is eminently reasonable. The set C has been termed a **hypercircle** by previous workers. There is a book on the hypercircle by Synge [Syn].

Suppose now that the nature of the problem is changed so that what we want is not f itself but only $\psi(f)$, where ψ is another linear functional. Naturally, we can use $\psi(g)$ as an estimate of $\psi(f)$, and a bound on $|\psi(f) - \psi(g)|$ is required. A trivial bound is obtained from Theorem 1 as follows:

$$|\psi(f) - \psi(g)| \leq \|\psi\| \|f - g\| \leq \|\psi\| \sqrt{\|f\|^2 - \|g\|^2}$$

Can anything better be found? Yes, as one would expect, the factor $\|\psi\|$ may be much larger than necessary, for it is the largest value attained by $\psi(v)$ on the unit cell of the whole Hilbert space, H, whereas we are working with only a part of H. A better bound is obtained by observing that if V is defined as in Equation (3), then $f - g \in V$. Consequently,

$$|\psi(f) - \psi(g)| = |\psi(f - g)| \leq \|\psi|V\| \cdot \|f - g\|$$

Here $\psi|V$ denotes the restriction of ψ to V, so that

$$\|\psi|V\| = \sup_{\substack{v \in V \\ \|v\|=1}} |\psi(v)| \leq \sup_{\substack{v \in H \\ \|v\|=1}} |\psi(v)| = \|\psi\|$$

We now have two major objectives. First, can we compute the element g? Second, can we estimate $\|\psi|V\|$? In answering these questions, the Riesz Representation Theorem for Hilbert space is of significant assistance. To each bounded linear functional ψ on H, there corresponds an element $u \in H$ such that $\psi(f) = \langle f, u \rangle$ for all $f \in H$. The element u is called the **representative** of ψ in H. We will refer to the element g as defined above as the **minimal norm interpolant to** f **on** $\phi_1, \ldots, \phi_n$.

THEOREM 2. *Let $f \in H$ and let $\phi_1, \ldots, \phi_n$ be continuous linear functionals on H. Let g be the minimal norm interpolant to f on $\phi_1, \ldots, \phi_n$. Let ψ be a continuous linear functional on H with representative u. Let V be as defined in Equation (3), and let $P : H \twoheadrightarrow V$ be the orthogonal projection. Then*

$$(5) \qquad |\psi(f) - \psi(g)| \leq \|\psi|V\| \|f - g\| = \|Pu\|\sqrt{\|f\|^2 - \|g\|^2}$$

Proof. The theorem is asserting that $\|\psi|V\| = \|Pu\|$. Observe that

$$\|\psi|V\| = \sup_{\substack{v \in V \\ \|v\|=1}} |\psi(v)| = \sup_{\substack{v \in V \\ \|v\|=1}} |\langle v, u \rangle| = \sup_{\substack{v \in V \\ \|v\|=1}} |\langle v, u - Pu + Pu \rangle|$$

Because $P : H \to V$ is the orthogonal projection, $u - Pu \perp V$, and so

$$\|\psi|V\| = \sup_{\substack{v \in V \\ \|v\|=1}} |\langle v, Pu \rangle|$$

Since $Pu \in V$, $\|\psi|V\| = \|Pu\|$. ∎

The bound in Theorem 2 is best possible (see Problem 9). Our next result shows how to compute the minimal norm interpolant.

THEOREM 3. *Let $\phi_1, \ldots, \phi_n$ be continuous linear functionals on H, with representatives $u_1, \ldots, u_n$ respectively. Let $f \in H$ have minimal norm interpolant g on $\phi_1, \ldots, \phi_n$. Then $g = \sum_{j=1}^n \lambda_j u_j$, where the coefficients λ_j are chosen to solve the system of linear equations $\sum_{j=1}^n \lambda_j \langle u_i, u_j \rangle = \langle f, u_i \rangle$, $1 \leq i \leq n$.*

Proof. Let V be as defined in Equation (3). Then

$$V = \{v \in H : \phi_i(v) = 0, 1 \leq i \leq n\}$$
$$= \{v \in H : \langle v, u_i \rangle = 0, 1 \leq i \leq n\} = \bigcap_{i=1}^n u_i^\perp$$

Recall, from the proof of Theorem 1, that g is characterized by the properties $g \perp V$ and $\phi_i(g) = \phi_i(f)$ for $1 \leq i \leq n$. Since $g \in (\bigcap u_i^\perp)^\perp$, Problem 4 allows us to infer that $g \in \text{span}\{u_1, u_2, \ldots, u_n\}$. Writing $g = \sum_{j=1}^n \lambda_j u_j$, we have

$$\phi_i(f) = \phi_i(g) = \sum_{j=1}^n \lambda_j \phi_i(u_j) \qquad (1 \leq i \leq n)$$

Thus the numbers λ_j solve the system described in the statement of the theorem. ∎

The system of equations for $\lambda_1, \lambda_2, \ldots, \lambda_n$ derived above must be consistent, for we know that g exists. The coefficient matrix is the Gram matrix $\langle u_i, u_j \rangle$. This matrix is nonsingular if and only if $\{u_1, u_2, \ldots, u_n\}$ is linearly independent. It is reasonable to assume this property as a hypothesis, but the theory does not require that we do so.

Notice that the system of equations consists of "normal" equations for a least-squares approximation problem. Indeed, g (or $\sum_{j=1}^{n} \lambda_j u_j$) will be the best approximation to f in span$\{u_1, u_2, \ldots, u_n\}$.

THEOREM 4. *Let $\phi_0, \phi_1, \ldots, \phi_n$ be continuous linear functionals on H with representatives $u_0, u_1, \ldots, u_n$ respectively. Let f be an element of H having minimal norm interpolant g on $\phi_1, \ldots, \phi_n$. Let z be the best approximation to u_0 in span$\{u_1, \ldots, u_n\}$. Then*

(6) $$|\phi_0(f) - \phi_0(g)| \leq \|Pu_0\| \sqrt{\|f\|^2 - \|g\|^2} = \|u_0 - z\| \sqrt{\|f\|^2 - \|g\|^2}$$

Proof. The first inequality in Equation (6) is simply a restatement of Theorem 2. It remains to establish that $\|Pu_0\| = \|u_0 - z\|$. We begin by writing $Pu_0 = u_0 - (I - P)u_0$. Now since $P : H \to V$ is the orthogonal projection of H onto V, $(I - P)$ is the orthogonal projection of H onto $V^\perp$. Thus $(I - P)u_0$ is the best approximation to u_0 from $V^\perp$. However, Problem 4 shows that $V^\perp = \text{span}\{u_1, \ldots, u_n\}$, which completes the proof.   ∎

Now we proceed to a very special case of the theory that will be useful in the applications. We assume that there is a semi-inner product $(\cdot, \cdot)$ defined on H having kernel K and satisfying two conditions:

(7) $$\langle u, v \rangle = (u, v) + \sum_{i=1}^{r} \phi_i(u)\phi_i(v)$$

(8) $$\dim(K) = r$$

In this description, a subset of r elements from the original set $\{\phi_1, \phi_2, \ldots, \phi_n\}$ has been selected. With a possible renumbering, this subset is labeled $\{\phi_1, \phi_2, \ldots, \phi_r\}$.

As before, we have "representatives," u_i for the functionals ϕ_i. Thus, $\phi_i(v) = \langle v, u_i \rangle$ for all v in H and for $1 \leq i \leq n$. The kernel of the semi-inner product is

(9) $$K = \{u \in H : (u, u) = 0\}$$

By Problem 5, K is a subspace of H. Being finite dimensional, K is closed.

LEMMA 1. *The set $\{u_1, u_2, \ldots, u_r\}$ is an orthonormal base for K.*

Proof. If $u \in K$, then $(u, u) = 0$, and by Problem 5, $(u, v) = 0$ for all $v \in H$. Consequently,

$$\langle u, v \rangle = (u, v) + \sum_{i=1}^{r} \phi_i(u)\phi_i(v) = \sum_{i=1}^{r} \phi_i(u)\phi_i(v) = \sum_{i=1}^{r} \langle u, u_i \rangle \langle v, u_i \rangle$$

$$= \left\langle \sum_{i=1}^{r} \langle u, u_i \rangle u_i, v \right\rangle$$

It follows that $u = \sum_{i=1}^{r} \langle u, u_i \rangle u_i$. Hence $\{u_1, u_2, \ldots, u_r\}$ spans K. Since $\dim(K) = r$, that set is a basis for K. Since $u_j = \sum_{i=1}^{r} \langle u_j, u_i \rangle u_i$, we conclude that $\langle u_j, u_i \rangle = \delta_{ij}$. ∎

LEMMA 2. *The subspaces K and V are mutually orthogonal.*

Proof. The subspace V is defined in Equation (2). By Lemma 1, it suffices to prove that $u_j \perp V$ for $1 \le j \le r$. If $v \in V$, then $\phi_i(v) = 0$ for $1 \le i \le n$. Hence $\langle v, u_j \rangle = \phi_j(v) = 0$ for $1 \le j \le r$. ∎

COROLLARY. *Let $\phi_0, \phi_1, \ldots, \phi_n$ be continuous linear functionals on H with representatives $u_0, u_1, \ldots, u_n$ respectively. Let f be an element of H having minimal norm interpolant g on $\phi_1, \ldots, \phi_n$. Let z be the best approximation to u_0 in $\text{span}\{u_1, \ldots, u_n\}$. Then*

$$(10) \qquad |\phi_0(f) - \phi_0(g)| \le \|u_0 - z\| \sqrt{\|f\|^2 - \|g\|^2} \le \|u_0 - z\| \sqrt{(f,f)}$$

Proof. The first inequality in Equation (10) comes directly from Equation (6). It remains to establish that $\|f\|^2 - \|g\|^2 \le (f,f)$. From Equation (7),

$$\|f\|^2 - \|g\|^2 = \langle f, f \rangle - \langle g, g \rangle = (f,f) - (g,g) + \sum_{i=1}^{r} \{\phi_i(f)\}^2 - \sum_{i=1}^{r} \{\phi_i(g)\}^2$$

Because g is an interpolant to f on $\phi_1, \ldots, \phi_n$, we have $\phi_i(f) = \phi_i(g)$, $i = 1, \ldots, r$. Hence,

$$\|f\|^2 - \|g\|^2 = (f,f) - (g,g) \le (f,f)$$ ∎

This chapter concludes with an instructive example. Before we begin this example, it will reward us to pause for a moment and consider the role of completeness in this theory. We have assumed throughout that H is a complete space. The reason for this is that the Riesz Representation Theorem requires that hypothesis. If H is not complete, but is simply an inner-product space, then it is possible that a search for the representative of a given linear functional will generate an element which is not in H, but is in the completion of H. This is not a serious drawback, and will help to explain why the issue of completeness does not play a role in our example.

Let $u \in L^2(\mathbb{R})$ and $c \in \mathbb{R}$. Corresponding to this choice of u and c, define

$$(11) \qquad f(x) = c + \int_0^x u(t)\, dt \qquad (x \in \mathbb{R})$$

It is not difficult to prove that f is continuous. In fact, with the aid of the Cauchy-Schwarz inequality, we can write

$$(12) \qquad \begin{aligned} |f(x) - f(y)| = \left| \int_x^y u(t)\, dt \right| &\le |x - y|^{1/2} \left[\int_x^y u^2(t)\, dt \right]^{1/2} \\ &\le |x - y|^{1/2} \left[\int_{-\infty}^{\infty} u^2(t)\, dt \right]^{1/2} \end{aligned}$$

From these inequalities, we see that $f \in C(\mathbb{R})$. (Indeed, f belongs to the class Lip_α with $\alpha = 1/2$.)

Let X denote the space of all functions f that arise in the way just described. For each $f \in X$ there is an element $u \in L^2$ that produced f, and we can designate it by u_f. (The function u_f is uniquely determined by f. See Problem 15.) The constant c is, of course, $f(0)$. A semi-inner product can be defined on X as follows.

$$(f, g) = \int_{-\infty}^{\infty} u_f(t) u_g(t)\, dt \qquad (f, g \in X)$$

The kernel of this semi-inner product is $\Pi_0(\mathbb{R})$, the linear space of polynomials of degree 0 on $\mathbb{R}$. In accordance with previous theory in this chapter, a convenient inner product on X is given by

$$\langle f, g \rangle = f(0)g(0) + \int_{-\infty}^{\infty} u_f(t) u_g(t)\, dt \qquad (f, g \in X)$$

For any $x \in \mathbb{R}$, the Cauchy-Schwarz inequality yields

$$|f(x) - f(0)| = \left| \int_0^x u_f(t)\, dt \right| \le \left\{ \int_0^x |u_f(t)|^2\, dt \right\}^{1/2} \left\{ \int_0^x 1\, dt \right\}^{1/2}$$

$$\le |x|^{1/2} \left\{ \int_{-\infty}^{\infty} |u_f(t)|^2\, dt \right\}^{1/2} = \{|x| (f,f)\}^{1/2}$$

Now it follows that

$$|f(x)| \le |f(0)| + [|x| (f,f)]^{1/2} \le \sqrt{2}[|f(0)|^2 + |x| (f,f)]^{1/2}$$
$$\le \sqrt{2}(1 + |x|)^{1/2} \langle f, f \rangle^{1/2}$$

Thus the point-evaluation functional at x defined by $x^*(f) = f(x)$ for $f \in X$ is a bounded linear functional on X. Consequently, as an example of the theory we can take our functionals $\phi_1, \ldots, \phi_n$ (with respect to which the interpolant is to be constructed) to be point-evaluation functionals.

Our theory (in particular Theorem 3) emphasizes that an important task is the determination of representatives for these point-evaluation functionals. Consider the function

$$R_x(y) = \frac{1}{2}(|y| - |y - x|) + \frac{1}{2}|x| + 1 \qquad (x, y \in \mathbb{R})$$

The graph of R_x is shown in Figure 30.1. It is clear that $R_x \in X$. Now for $f \in X$ and $x \in \mathbb{R}$,

$$\langle f, R_x \rangle = f(0)R_x(0) + \int_{-\infty}^{\infty} u_f(t) u_{R_x}(t)\, dt = f(0) + \int_0^x u_f(t)\, dt = f(x)$$

Note that the above argument is valid whether $x > 0$ or $x \le 0$. Thus R_x is the representative for the point-evaluation functional x^*. Now suppose that $a_1, \ldots, a_n$ are points in $\mathbb{R}$ with $a_1 = 0$. Then the form of the minimal norm interpolant on $a_1^*, \ldots, a_n^*$ is as given in Theorem 3:

$$g(y) = \sum_{j=1}^n \lambda_j R_{a_j}(y) = \sum_{j=1}^n \lambda_j \left(\frac{1}{2}(|y| - |y - a_j|) + \frac{1}{2}|a_j| + 1 \right) \qquad (y \in \mathbb{R})$$

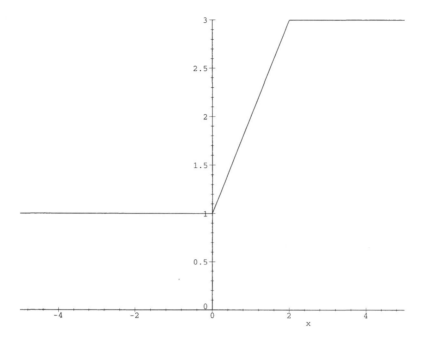

Figure 30.1 *Graph of R_x*

This interpolant can be written in an alternative form as

$$(13) \qquad g(y) = \sum_{j=1}^{n} \mu_j |y - a_j| + C \qquad (y \in \mathbb{R})$$

where $\mu_j = -\lambda_j/2$, $j = 2, \ldots, n$, $\mu_1 = (1/2)\sum_{j=2}^{n} \lambda_j$, and $C = \sum_{j=1}^{n} \lambda_j(1 + |a_j|/2)$. It is apparent from Equation (13) that the form of the interpolant is a piecewise linear function with break points at $a_1, \ldots, a_n$. Hence, g is a linear spline. If we set $A = \min_{1 \leq j \leq n} a_j$ and $B = \max_{1 \leq j \leq n} a_j$, then each function R_{a_j} is constant outside $[A, B]$, $j = 1, \ldots, n$. Hence, g is constant in $(-\infty, A)$ and (B, ∞). Thus g is the natural linear spline with knots $a_1, \ldots, a_n$.

Notice that we are really defining the relationship between u_f and f by the equation $u_f(t) = f'(t)$ for almost all $t \in \mathbb{R}$. Then X can be seen to consist of all continuous functions whose derivative is in $L^2(\mathbb{R})$. There is clearly some difficulty here, since Equation (12) shows that f is in $\mathrm{Lip}(1/2)$, and its derivative need not exist in the classical sense.

Problems

1. Let V be a closed linear subspace of a Hilbert space H. Given f in H, show that there is a unique v in V such that $\|f - v\| = \mathrm{dist}(f, V)$. Show that this element is characterized by the two conditions $v \in V$ and $f - v \perp V$.

2. Prove the following standard result in linear algebra. For a set of linear functionals $\phi_0, \phi_1, \ldots, \phi_n$ on a linear space, the condition $\phi_0 \in \mathrm{span}\{\phi_1, \phi_2, \ldots, \phi_n\}$ is equivalent to the condition $\bigcap_{i=1}^{n} \ker(\phi_i) \subset \ker(\phi_0)$.

3. By using the definition of $\|\psi|V\|$, verify the first inequality in Equation (5):
$|\psi(f) - \psi(g)| \le \|\psi|V\|\|f - g\|$.

4. Prove that for a finite set of vectors $u_1, u_2, \ldots, u_n$ in an inner-product space, $(\cap_{i=1}^{n} u_i^{\perp})^{\perp} = \text{span}\{u_1, u_2, \ldots, u_n\}$.

5. Provide a careful verification that the Cauchy–Schwarz inequality is valid in semi–inner product spaces. Then prove that in such a space the kernel $K = \{v : (v, v) = 0\}$ is a subspace.

6. Let E be a semi–inner product space. Let K be the kernel of the semi–inner product, as in the preceding problem. Define $v^{\perp} = \{u \in E : (u, v) = 0\}$. Prove that $K = \cap \{v^{\perp} : v \in E\}$ and that $K^{\perp} = E$.

7. Let E be a semi–inner product space. Let F be a subspace of E such that $(u, u) = 0$ for all u in F. Let $\text{dist}(v, F) = \inf\{(v - f, v - f)^{1/2} : f \in F\}$. Show that the norm in the quotient space E/F, defined by $\|v + F\| = \text{dist}(v, F)$, has the property $\|v + F\| = \sqrt{(v, v)}$.

8. Let $\{\phi_1, \phi_2, \ldots, \phi_n\}$ and $\{\psi_1, \psi_2, \ldots, \psi_m\}$ be two finite sets of linear functionals that span the same subspace. Prove that there exist positive constants A and B such that
$$A \sum_{i=1}^{n} [\phi_i(x)]^2 \le \sum_{j=1}^{m} [\psi_j(x)]^2 \le B \sum_{i=1}^{n} [\phi_i(x)]^2$$

9. Let Uf be the minimal norm interpolant to f. Show that there is an f_0 in H such that
$$|\psi(f_0) - \psi(Uf_0)| = \|\psi|V\|\|f_0 - Uf_0\|$$
Thus the bound in Theorem 2 is best possible.

10. Examine Problem 4. Find suitable generalizations for the case when the set of vectors u_i is infinite and for the case when the inner product is replaced by a semi-inner product.

11. Let $\{u_1, u_2, \ldots, u_n\}$ be a set of unit vectors in an inner-product space. Define a bilinear form
$$B(x, y) = \langle x, y \rangle - \sum_{i=1}^{n} \langle x, u_i \rangle \langle y, u_i \rangle$$

Prove that if the bilinear form is nonnegative definite (i.e., $B(x, x) \ge 0$ for all x), then the given set of unit vectors is orthonormal.

12. In an inner-product space, let $\{u_1, u_2, \ldots, u_n\}$ be a linearly independent set. Define
$$B(x, y) = \sum_{i=1}^{n} \langle x, u_i \rangle \langle y, u_i \rangle$$
Prove that if $B(x, y) = 0$ for all y, then x is in the linear span of $u_1, u_2, \ldots, u_n$.

13. Show that there is no loss of generality in assuming that $a_1 = 0$ in the interpolation problem at the end of the chapter. Show that for a general choice of a_1 and suitable choice for the semi-inner product, we have
$$R_x(y) = \frac{1}{2}(|y - a_1| - |y - x|) + \frac{1}{2}|x - a_1| + 1 \qquad (x, y \in \mathbb{R})$$

14. Repeat the example at the end of the chapter, using the class of functions X whose second derivative is in $L^2(\mathbb{R})$. These functions can be written

$$f(x) = f(0) + xf'(0) + \int_0^x u_f(t)\,dt$$

where $u_f \in L^2(\mathbb{R})$. Show that $X \subset C^{(1)}(\mathbb{R})$. Show that the equation

$$\langle f, g \rangle = f(0)g(0) + f(1)g(1) + \int_0^x u_f(t)u_g(t)\,dt \qquad (f, g \in X)$$

defines an inner product on X. Show that for each x in $\mathbb{R}$, x^* is a bounded, linear functional on X. Calculate R_x. Show that the minimal norm interpolant is a natural cubic spline.

15. For $u \in L^2(\mathbb{R})$ and $c \in \mathbb{R}$, define f by Equation (11). Prove that the map $(u, c) \mapsto f$ is injective.

16. Prove Theorem 2 by establishing and using the fact that an orthogonal projection is self-adjoint: $\langle Px, y \rangle = \langle x, Py \rangle$.

References

[GolW] Golomb, M., and H. F. Weinberger. "Optimal approximation and error bounds." In *On Numerical Approximation,* ed. by R. E. Langer, University of Wisconsin Press, Madison, WI, 1959, pp. 117–190.

[Syn] Synge, J. L. *The Hypercircle in Mathematical Physics.* Cambridge University Press, Cambridge, UK, 1957.

31
Hilbert Function Spaces and Reproducing Kernels

In the previous chapter, the discussion focused on the general interpolation problem in a Hilbert space H. That is, bounded linear functionals $\phi_1, \ldots, \phi_n$ were prescribed, and from the values of $\phi_1(f), \ldots, \phi_n(f)$ we sought to approximate the element f in H. We wish now to consider the case where f is a function and $\phi_1, \ldots, \phi_n$ are point-evaluation functionals. Because we want to consider these point-evaluation functionals as bounded linear functionals on H, we make the following definition.

Definition 1. A Hilbert space H will be called a **Hilbert function space** if the elements of H are complex-valued functions on a set X, and for each x in X, there exists a positive constant c_x such that $|f(x)| \leq c_x \|f\|$ for all f in H. When we wish to make explicit the role of X, we will say that H is a Hilbert function space over X.

Note that the familiar space $L^2[a, b]$ is not a Hilbert function space; indeed, point evaluation is not defined. Each function f in this space is actually an equivalence class of functions equal to each other almost everywhere. The "value" at a point is ambiguous, since any point has measure zero.

THEOREM 1. *Let H be a Hilbert function space over X. For each x in X, there exists a unique element $K_x \in H$ such that $\langle f, K_x \rangle = f(x)$ for all $f \in H$.*

Proof. The Riesz Representation Theorem for Hilbert space asserts the existence and uniqueness of the element K_x. ∎

Definition 2. Let H be a Hilbert function space over X. Let x be a point in X. The mapping $K : X \times X \rightarrow \mathbb{C}$ defined by $K(x, y) = K_x(y)$ is called the **reproducing kernel** of H.

The reproducing kernel is so called because it has the potential of reproducing each function in $H : \langle f, K_x \rangle = f(x)$ for all f in H and all x in X. The standard definition of the "sections" of a function of two variables is being used here. Thus, $K_a(b) = K(a, b) = K^b(a)$. In other notation, we have $K_a = K(a, \cdot)$ and $K^b = K(\cdot, b)$. Using x^* for the point-evaluation functional at x, we have $x^*(f) = \langle f, K_x \rangle$ for all f in H and for all x in X.

THEOREM 2. *Let H be a Hilbert function space over X with reproducing kernel K. Then*

    a. $K(x, y) = \langle K_x, K_y \rangle$ *for all* $x, y \in X$

    b. $K(x, y) = \overline{K(y, x)}$ *for all* $x, y \in X$

    c. $K(x, x) \geq 0$ *for all* $x \in X$

    d. *If, for some point* x_0, $K(x_0, x_0) = 0$ *then for all* f *in* H, $f(x_0) = 0$

    e. $|K(x, y)| \leq \sqrt{K(x, x)} \sqrt{K(y, y)}$ *for all* $x, y \in X$

Proof. Part **a.** Since $K_x \in H$ for each $x \in X$, it follows that $K(x, y) = \langle K_x, K_y \rangle$ for all $x, y \in X$.

Part **b.** For $x, y \in X$, using (a), we obtain

$$K(x, y) = \langle K_x, K_y \rangle = \overline{\langle K_y, K_x \rangle} = \overline{K(y, x)}$$

Part **c.** From (a), for all $x \in X$,

$$K(x, x) = \langle K_x, K_x \rangle \geq 0$$

Part **d.** If $K(x_0, x_0) = 0$, then from (c), $K_{x_0} = 0$. Hence for each $f \in H$,

$$f(x_0) = \langle f, K_{x_0} \rangle = \langle f, 0 \rangle = 0$$

Part **e.** Using the Cauchy-Schwarz inequality and (a), we obtain

$$|K(x, y)|^2 = |\langle K_x, K_y \rangle|^2 \leq \langle K_x, K_x \rangle \langle K_y, K_y \rangle$$
$$= K(x, x) K(y, y) \qquad\blacksquare$$

In Chapter 12, we introduced the idea of a positive definite function on a linear space. The following definition generalizes this concept slightly.

Definition 3. Let X be a set and F a complex-valued function on $X \times X$. The function F is said to be **positive definite** if for any n and for any n points $x_1, \ldots, x_n$ in X, the matrix $A_{ij} = F(x_i, x_j)$ is nonnegative definite, that is,

$$u^* A u = \sum_{i,j=1}^{n} \overline{u}_i u_j A_{ij} \geq 0$$

for all $u \in \mathbb{C}^n$. If $u^* A u > 0$ whenever the points $x_1, \ldots, x_n$ are distinct and $u \neq 0$, then F is said to be **strictly positive definite.**

THEOREM 3. *Let H be a Hilbert function space over X with reproducing kernel K. Then K is a positive definite function. If for any n and any distinct points $x_1, \ldots, x_n$ in X, the corresponding set of point-evaluation functionals is linearly independent over H, then K is strictly positive definite.*

Proof. Select $n \in \mathbb{N}$ and points $x_1, \ldots, x_n$ in X. Let $u_1, \ldots, u_n$ be arbitrary complex numbers. Then by Theorem 2(a),

$$
\sum_{i,j=1}^{n} \bar{u}_i u_j K(x_i, x_j) = \sum_{i,j=1}^{n} \bar{u}_i u_j \langle K_{x_i}, K_{x_j} \rangle
$$

(1)
$$
= \left\langle \sum_{i=1}^{n} \bar{u}_i K_{x_i}, \sum_{j=1}^{n} \bar{u}_j K_{x_j} \right\rangle
$$

$$
= \left\| \sum_{i=1}^{n} \bar{u}_i K_{x_i} \right\|^2 \geq 0
$$

Now suppose that K is not strictly positive definite. Then there exist distinct points $x_1, \ldots, x_n$ in X and a nonzero vector u in $\mathbb{C}^n$ such that

$$
\sum_{i,j=1}^{n} \bar{u}_i u_j K(x_i, x_j) = 0
$$

It now follows from Equation (1) that $\sum_{i=1}^{n} \bar{u}_i K_{x_i} = 0$. Thus, for any $f \in H$,

$$
\sum_{i=1}^{n} u_i f(x_i) = \sum_{i=1}^{n} u_i \langle f, K_{x_i} \rangle = \left\langle f, \sum_{i=1}^{n} \bar{u}_i K_{x_i} \right\rangle = 0
$$

This shows that the set of point-evaluation functionals $x_1^*, \ldots, x_n^*$ is linearly dependent. ∎

If H is a Hilbert function space, then any closed subspace of H is a Hilbert function space, and so possesses a reproducing kernel. The next result shows how to obtain this kernel from the reproducing kernel for H.

THEOREM 4. *Let H be a Hilbert space, and let E be a closed subspace of H that is itself a Hilbert function space over the set X with reproducing kernel G. Let P be the orthogonal projection of H onto E. For each f in H,*

$$
(Pf)(x) = \langle f, G_x \rangle \qquad (x \in X)
$$

Furthermore, if H is a Hilbert function space over X with reproducing kernel K, then for each x in X, $G_x = PK_x$.

Proof. Let f be any element of H. Then $f - Pf \perp E$. Hence, for each x in X,

$$
\langle f, G_x \rangle = \langle f - Pf, G_x \rangle + \langle Pf, G_x \rangle = \langle Pf, G_x \rangle = (Pf)(x)
$$

For the second part of the theorem, let $x, y \in X$. Then by the first part,

$$
(PK_x)(y) = \langle K_x, G_y \rangle = \overline{\langle G_y, K_x \rangle} = \overline{G(y, x)}
$$

Now by applying Theorem 2(b), we obtain

$$
(PK_x)(y) = \overline{G(y, x)} = G(x, y). \qquad \blacksquare
$$

COROLLARY 1. *Let H be a Hilbert function space over X with reproducing kernel K.*

> **a.** *If E and F are closed subspaces of H such that E ⊥ F and E ⊕ F = H, then K = L + M, where L and M are the reproducing kernels of E and F respectively.*
> **b.** *If E is a closed subspace of H with reproducing kernel L, then L(x, x) ≤ K(x, x) for all x ∈ X.*

Proof. Part **a.** Let P be the orthogonal projection of H onto E. Then I − P is the orthogonal projection of H onto F. For each x in X,

$$K_x = PK_x + (I - P)K_x = L_x + M_x$$

by Theorem 4.

Part **b.** Using Theorem 2(b) and Theorem 4, we arrive at

$$L(x, x) = \langle L_x, L_x \rangle = \|L_x\|^2 = \|PK_x\|^2$$
$$\leq \|K_x\|^2 = \langle K_x, K_x \rangle = K(x, x) \qquad \blacksquare$$

THEOREM 5. *Let H be a separable Hilbert function space over X. Let $\{u_n\}_{n=1}^\infty$ be an orthonormal basis for H. The reproducing kernel for H is given by*

$$K(x, y) = \sum_{k=1}^\infty \overline{u_k(x)} u_k(y) \qquad (x, y \in X)$$

Proof. Fix $n \in \mathbb{N}$, and let E denote the span of $\{u_1, \ldots, u_n\}$. Then E is a Hilbert function space over X. Set

$$G(x, y) = \sum_{k=1}^n \overline{u_k(x)} u_k(y) \qquad (x, y \in X)$$

Then G is the reproducing kernel for E. To verify this, fix x in X and f in E. Then there exist $\alpha_1, \ldots, \alpha_n$ such that $f = \sum_{j=1}^n \alpha_j u_j$. Now

$$\langle f, G_x \rangle = \sum_{k=1}^n u_k(x)\langle f, u_k \rangle = \sum_{k=1}^n u_k(x)\left\langle \sum_{j=1}^n \alpha_j u_j, u_k \right\rangle = \sum_{k=1}^n \alpha_k u_k(x) = f(x)$$

Keeping x fixed in X, observe that, by Corollary 1,

$$\sum_{k=1}^n |u_k(x)|^2 = G(x, x) \leq K(x, x)$$

Thus $\sum_{k=1}^\infty |u_k(x)|^2 < \infty$. Now the partial sums of $\sum_{k=1}^\infty u_k(x)u_k$ form a Cauchy sequence in H, since for $j > i$,

$$\left\| \sum_{k=1}^j u_k(x)u_k - \sum_{k=1}^i u_k(x)u_k \right\|^2 = \left\| \sum_{k=i}^j u_k(x)u_k \right\|^2 = \sum_{k=i}^j |u_k(x)|^2$$

We therefore conclude that $\sum_{k=1}^{\infty} u_k(x)u_k$ lies in H. Take f in H. Then $f = \sum_{k=1}^{\infty} \langle f, u_k \rangle u_k$, and so

$$\left\langle f, \sum_{i=1}^{\infty} \overline{u_i(x)}\,u_i \right\rangle = \left\langle \sum_{k=1}^{\infty} \langle f, u_k \rangle u_k, \sum_{i=1}^{\infty} \overline{u_i(x)}\,u_i \right\rangle$$

$$= \sum_{i,k=1}^{\infty} \langle f, u_k \rangle \langle u_k, u_i \rangle u_i(x)$$

$$= \sum_{k=1}^{\infty} \langle f, u_k \rangle u_k(x)$$

$$= f(x)$$

Because the reproducing kernel for H is unique, we conclude that

$$K(x, y) = \sum_{k=1}^{\infty} \overline{u_k(x)}\,u_k(y) \qquad (x, y \in X) \qquad \blacksquare$$

So far, we have worked with a Hilbert space. What happens when the underlying space is not complete? To answer this question, let E be a linear space whose elements are complex-valued functions on a set X. Let $\langle \cdot , \cdot \rangle$ be an inner product on E with associated norm $\| \cdot \|$. Suppose that for each $x \in X$ there exists a constant $c_x > 0$ such that $|f(x)| \le c_x \|f\|$ for all $f \in E$. Let H denote the completion of E. Let $\{f_n\}$ be a Cauchy sequence in E with limit $f \in H$. The first question is whether f can be regarded as a mapping from X to $\mathbb{C}$. To answer this question in the affirmative, we must assign a meaningful value to $f(x)$ for each $x \in X$. Note that since

$$|f_n(x) - f_m(x)| \le c_x \|f_n - f_m\| \qquad (x \in X)$$

$\{f_n(x)\}$ is a Cauchy sequence of complex numbers for each $x \in X$. Thus we may define $g : X \to \mathbb{C}$ by $g(x) = \lim_{n \to \infty} f_n(x)$ for x in X. Now let $x^* : E \to \mathbb{C}$ denote the point-evaluation functional at x. Then $\|x^*\| \le c_x$ and so $x^* \in E^*$. Thus x^* has a unique continuous linear extension to H. This extension is defined at f (as given above) by the equation

$$x^*(f) = \lim_{n \to \infty} x^*(f_n) = \lim_{n \to \infty} f_n(x) = g(x)$$

Thus we see that $f(x) = g(x)$ for $x \in X$. This establishes the following result.

LEMMA 1. *Let E be a linear space whose elements are complex-valued functions on a set X. Let $\langle \cdot , \cdot \rangle$ be an inner product on E with associated norm $\| \cdot \|$. Suppose that to each $x \in X$ there corresponds a number c_x such that $|f(x)| \le c_x \|f\|$ for all $f \in E$. Then the completion of E with respect to the norm $\| \cdot \|$ is a Hilbert function space.*

Chapter 32 presents a specific application of the previous theory. These applications have a rather special nature, which will be considered in the rest of this chapter. We also want to make the error estimates of the previous chapter more explicit. In the remainder of the chapter, X will denote an arbitrary set, and $\mathbb{C}^X$ will denote the set of all complex-valued mappings on X. We begin with a result in linear algebra.

LEMMA 2. *Let G be an n-dimensional linear space of complex-valued functions defined on a set X. Then there exist points $x_1, \ldots, x_n$ in X such that the corresponding set of point-evaluation functionals $\{x_1^*, \ldots, x_n^*\}$ is a basis for G^*.*

Proof. Let $\{g_1, \ldots, g_n\}$ be a basis for G. Then the matrix $(g_j(x))$ has column rank n. (This matrix has n columns and one row for each x in X.) It therefore has row rank n, and consequently, there exist points $x_1, \ldots, x_n$ such that $\det(g_j(x_i)) \neq 0$. (See Problem 8 in Chapter 1, page 8.) The lemma also follows from Auerbach's Theorem, proved in Chapter 6, page 42. In that theorem, take B to be the set of all point-evaluation functionals. ∎

Let us assume that E is a subspace of $\mathbb{C}^X$ and $(\cdot, \cdot)$ is a semi-inner product on E with null space N, where $\dim N = m$. By Lemma 2, there exist $x_1, \ldots, x_m \in X$ such that if $p \in N$ and $p(x_i) = 0$ for $i = 1, \ldots, m$ then $p = 0$. Define $\langle \cdot, \cdot \rangle$ by the equation

$$(2) \qquad \langle f, g \rangle = \sum_{i=1}^{m} f(x_i)\overline{g(x_i)} + (f, g) \qquad (f, g \in E)$$

It is elementary to see that Equation (2) defines an inner product on E. Given $x \in X$, let x^* denote the point-evaluation functional at x. Then the set of functionals $x_1^*, \ldots, x_m^*$ is linearly independent over N. Hence, there exist $p_1, \ldots, p_m$ in N such that $p_i(x_j) = \delta_{ij}$ for $i, j = 1, \ldots, m$. Note that for $1 \leq k, j \leq m$, $(p_j, p_k) = 0$, since these elements are in the null space of the semi-inner product. Consequently,

$$\langle p_j, p_k \rangle = \sum_{i=1}^{m} p_j(x_i)\overline{p_k(x_i)} + (p_j, p_k) = p_j(x_k) = \delta_{jk}$$

From this we see that $\{p_1, \ldots, p_m\}$ is an orthonormal base for N. Let $L : E \longrightarrow N$ denote the orthogonal projection. Then

$$(3) \qquad Lf = \sum_{i=1}^{m} \langle f, p_i \rangle p_i \qquad (f \in E)$$

Note that the projection L has the somewhat unusual feature that it is an interpolation operator. For any j in the range $1 \leq j \leq m$,

$$(Lf)(x_j) = \sum_{i=1}^{m} \langle f, p_i \rangle p_i(x_j) = \sum_{i=1}^{m} \langle f, p_i \rangle \delta_{ij}$$

$$= \langle f, p_j \rangle = \sum_{i=1}^{m} f(x_i)\overline{p_j(x_i)} + (f, p_j) = f(x_j)$$

This establishes the property $\langle f, p_j \rangle = f(x_j)$, for $j = 1, 2, \ldots, m$ and for $f \in E$. Consequently, the projection can be written down without referring to the semi-inner product or the inner product on E:

$$(4) \qquad Lf = \sum_{i=1}^{m} f(x_i) p_i \qquad (f \in E)$$

Definition 4. Let E be a subspace of $\mathbb{C}^X$ and let $(\,\cdot\,,\,\cdot)$ be a semi-inner product on E with null space N, where $\dim N = m$. Let $\langle\,\cdot\,,\,\cdot\rangle$ denote the inner product on E defined by Equation (2). The pair $\big(E,(\,\cdot\,,\,\cdot)\big)$ will be called a pre-Hilbert function space if $(H, \langle\,\cdot\,,\,\cdot\rangle)$ is a Hilbert function space, where H denotes the completion of E with respect to the norm defined by the inner product.

It would be helpful if we were able to recognize whether $(E,(\,\cdot\,,\,\cdot))$ is a pre-Hilbert space without resorting to the inner product as defined in Equation (2), or the completion. The next two results explore this possibility.

> **THEOREM 6.** *Let E be a subspace of $\mathbb{C}^X$ and let $(\,\cdot\,,\,\cdot)$ be a semi-inner product on E with finite–dimensional null space N. Let H denote the Hilbert space defined in Definition 4. Let $P : H \twoheadrightarrow N$ be a projection. If, for each x in X, there exists $c_x > 0$ such that*
> $$|(f - Pf)(x)| \le c_x\sqrt{(f,f)} \qquad (f \in E)$$
> *then $(E,(\,\cdot\,,\,\cdot))$ is a pre-Hilbert function space.*

Proof. For each f in E and x in X,
$$|f(x)| \le |(Pf)(x)| + |(f - Pf)(x)| \le \|P\|\,\|f\| + c_x|f| \le (\|P\| + c_x))\|f\|$$
Lemma 1 shows that the completion of E with respect to $\|\cdot\|$ is a Hilbert function space. An appeal to Definition 4 confirms that $(E,(\,\cdot\,,\,\cdot))$ is a pre-Hilbert function space. ∎

Note that Theorem 6 does not achieve our stated objective. We have to know that P is a bounded linear operator from H onto N, and this involves knowledge of the completion of E with respect to the norm associated with the inner product given in Equation (2). However, one projection from H onto N is the orthogonal projection L defined prior to Definition 4. Using this projection, we obtain the following immediate consequence of Theorem 6.

> **COROLLARY 2.** *Let E be a subspace of $\mathbb{C}^X$, and let $(\,\cdot\,,\,\cdot)$ be a semi-inner product on E having a finite-dimensional kernel N. Let x_i and p_i be as defined just after Lemma 2. Assume that for each $x \in X$, there exists $c_x > 0$ such that*
> $$\left| f(x) - \sum_{i=1}^{m} f(x_i)p_i(x) \right| \le c_x\sqrt{(f,f)}$$
> *for all $f \in E$. Then $(E,(\,\cdot\,,\,\cdot))$ is a pre-Hilbert space.*

We now go on to investigate the form of the reproducing kernel in a pre-Hilbert function space.

> **THEOREM 7.** *Let $(H, (\,\cdot\,,\,\cdot))$ be a pre-Hilbert function space over X. Let the kernel of $(\,\cdot\,,\,\cdot)$ be N, where $\dim N = m$. Let $\langle\,\cdot\,,\,\cdot\rangle$ be defined as in Equation (2). Let $(N^{\perp}, \langle\,\cdot\,,\,\cdot\rangle)$ have reproducing kernel G and let*

$\{v_1, \ldots, v_m\}$ *be an orthonormal basis for* N. *Then* $(H, \langle \cdot , \cdot \rangle)$ *has a reproducing kernel* K *given by*

$$K(x, y) = \sum_{i=1}^{m} \overline{v_i(x)} v_i(y) + G(x, y) \qquad (x, y \in X)$$

Proof. We can express H as the orthogonal sum of N and $N^\perp$. Corollary 1 shows that the reproducing kernel for H is the sum of the reproducing kernels for N and $N^\perp$. Theorem 5 shows that the reproducing kernel for N is $(x, y) \mapsto \sum_{i=1}^{m} \overline{v_i(x)} v_i(y)$ for $x, y \in X$, and this concludes the proof. ∎

It is clear from Theorem 7 that the major task in discovering the reproducing kernel for $(H, \langle \cdot , \cdot \rangle)$ is to determine G. We require two things of G. First, that $G_x \in N^\perp$ for every $x \in X$. Second, that

$$\langle f, G_x \rangle = (f, G_x) = f(x)$$

for all $x \in X$ and $f \in N^\perp$. It turns out that one need not be so precise in specifying G. In fact, the first property (that $G_x \in N^\perp$ for every $x \in X$) can be dropped.

> **COROLLARY 3.** *Adopt the setting and notation of Theorem 7, but let* $G : X \times X \to \mathbb{C}$ *be such that* $G_x \in H$ *for each* $x \in X$ *and*
>
> $$\langle f, G_x \rangle = (f, G_x) = f(x) \qquad (f \in N^\perp, x \in X)$$
>
> *Let* $p_1, \ldots, p_m$ *be as defined prior to Definition 4. Then the reproducing kernel for* $(H, \langle \cdot , \cdot \rangle)$ *is*
>
> $$K(x, y) = \sum_{i=1}^{m} \overline{p_i(x)} p_i(y) + G(x, y) - \sum_{i=1}^{m} G(x, x_i) p_i(y) \qquad (x, y \in X)$$

Proof. For any $x, y \in X$, define $G' : X \times X \to \mathbb{C}$ by

$$G'(x, y) = G(x, y) - \sum_{i=1}^{m} G(x, x_i) p_i(y)$$

$$= G(x, y) - \sum_{i=1}^{m} \langle G_x, p_i \rangle p_i(y)$$

$$= G(x, y) - (LG_x)(y)$$

This shows that G'_x is in $N^\perp$ for each $x \in X$. For any $f \in N^\perp$,

$$\langle f, G'_x \rangle = (f, G'_x) = (f, G_x) = f(x)$$

Hence G' is the reproducing kernel for $N^\perp$. Since $\{p_1, \ldots, p_m\}$ is an orthonormal basis for N, the result now follows from Theorem 7. ∎

Our final results in this chapter make the error estimates of Chapter 30 (Theorem 2 or the Corollary) considerably more explicit in the setting of a pre-Hilbert space. By virtue of Lemma 1, it is more convenient to work in the Hilbert function space $(H, \langle \cdot , \cdot \rangle)$.

Let $\mathcal{A} = \{x_1, \ldots, x_k\}$ be a set of points in X, where $k \geq m = \dim N$. We assume that $x_1, \ldots, x_k$ are still chosen so that $\{x_1^*, \ldots, x_k^*\}$ is linearly independent over N. Given any $f \in H$, we will let Uf denote the minimal norm interpolant to f on $\mathcal{A}$. The results of Chapter 30 enable us to estimate the quantity $|f(x) - (Uf)(x)|$ for any $x \in X$ and $f \in H$. Indeed, if K is the reproducing kernel for $(H, \langle \cdot\,, \cdot \rangle)$, then the Corollary in Chapter 30 (page 227) shows that

(5)
$$|f(x) - (Uf)(x)| \leq \|K_x - PK_x\| \sqrt{(f,f)}$$

Because P is an orthogonal projection, PK_x is the best approximation to K_x from the span of $\{K_{x_1}, \ldots, K_{x_k}\}$. Usually, the number k is large, which makes this best approximation very difficult to compute. This difficulty is circumvented by observing that since PK_x is the best approximation to K_x, any other approximation from the span of $\{K_{x_1}, \ldots, K_{x_k}\}$ will provide an upper bound.

THEOREM 8. *In the error estimate of Equation (5),*

$$\|K_x - PK_x\| \leq \left(K(x, x) - \sum_{i=1}^{k} |K(x, x_i)|^2 \right)^{1/2}$$

Proof. We have, by the remarks preceding this theorem,

$$\|K_x - PK_x\| \leq \|K_x - LK_x\|$$

where L is as in Equation (4). For any $x \in X$,

$$\begin{aligned}
\|K_x - LK_x\|^2 &= \langle K_x - LK_x, K_x - LK_x \rangle \\
&= \langle K_x - LK_x, K_x \rangle \\
&= K(x, x) - [LK_x](x) \\
&= K(x, x) - \sum_{i=1}^{k} K(x, x_i) p_i(x)
\end{aligned}$$

Theorem 7 shows that for $1 \leq j \leq k$,

$$K(x, x_j) = \sum_{i=1}^{k} \overline{p_i(x)} p_i(x_j) + G(x, x_j)$$

where G is the reproducing kernel for $N^\perp$. Since G_x lies in $N^\perp$, it follows that $PG_x = \sum_{i=1}^{k} G(x, x_i) p_i = 0$ and so $G(x, x_i) = 0$ for $x \in X$ and $i = 1, \ldots, k$. Thus $K(x, x_i) = \overline{p_i(x)}$, $i = 1, \ldots, k$ and so

$$\begin{aligned}
\|K_x - LK_x\|^2 &= K(x, x) - \sum_{i=1}^{k} K(x, x_i) \overline{K(x, x_i)} \\
&= K(x, x) - \sum_{i=1}^{k} |K(x, x_i)|^2 \qquad\blacksquare
\end{aligned}$$

Of course, it may happen that the kernel of the semi-inner product $(\cdot\,,\,\cdot)$ is trivial, so that we have a genuine inner product on H. In this case Theorem 8 yields the upper bound $K(x, x)$, which is likely to be very much poorer than the upper bound of Equation (5). Our last error estimate works around this difficulty in a way that is computable in many applications. Lemma 3 smooths the way to the proof.

LEMMA 3. *Let $x_1, \ldots, x_k$ be points in X and $\alpha_1, \ldots, \alpha_k$ be real numbers such that $p(x) = \sum_{i=1}^{k} \alpha_i p(x_i)$ for all $p \in N$. Let $f_1, \ldots, f_k$ be bounded, complex-valued functions on X. Define $M : X \times X \to \mathbb{C}$ by*

$$M(u, v) = \sum_{i=1}^{k} f_i(u)p_i(v) \qquad (u, v \in X)$$

Then

$$M(x, x) - \sum_{i=1}^{k} \alpha_i M(x_i, x) - \sum_{j=1}^{k} \alpha_j M(x, x_j) + \sum_{i,j=1}^{k} \alpha_i \alpha_j M(x_i, x_j) = 0$$

Proof. Note first that for any $u \in X$,

$$\sum_{j=1}^{k} \alpha_j M(u, x_j) = \sum_{j=1}^{k} \alpha_j \sum_{s=1}^{k} f_s(u)p_s(x_j) = \sum_{s=1}^{k} f_s(u) \sum_{j=1}^{k} \alpha_j p_s(x_j)$$

$$= \sum_{s=1}^{k} f_s(u)p_s(x) = M(u, x)$$

Using this relationship twice, we obtain

$$M(x, x) - \sum_{i=1}^{k} \alpha_i M(x_i, x) - \sum_{j=1}^{k} \alpha_j M(x, x_j) + \sum_{i,j=1}^{k} \alpha_i \alpha_j M(x_i, x_j)$$

$$= M(x, x) - \sum_{i=1}^{k} \alpha_i M(x_i, x) - M(x, x) + \sum_{i=1}^{k} \alpha_i \sum_{j=1}^{k} \alpha_j M(x_i, x_j)$$

$$= -\sum_{i=1}^{k} \alpha_i M(x_i, x) + \sum_{i=1}^{k} \alpha_i M(x_i, x)$$

$$= 0 \qquad\qquad\qquad\qquad\qquad\qquad\qquad\qquad\qquad\qquad\blacksquare$$

THEOREM 9. *Adopt the setting and notation of Theorem 7, but let $G : X \times X \to \mathbb{C}$ be such that $G_x \in H$ for each $x \in X$ and*

$$\langle f, G_x \rangle = (f, G_x) = f(x) \qquad (f \in N^{\perp}, x \in X)$$

Let $\alpha_1, \ldots, \alpha_k$ be real numbers such that $p(x) = \sum_{i=1}^{k} \alpha_i p(x_i)$ for all $p \in N$. Set

$$\mathcal{P}(x) = G(x, x) - \sum_{i=1}^{k} \alpha_i \big(G(x_i, x) + G(x, x_i) \big) + \sum_{i,j=1}^{k} \alpha_i \alpha_j G(x_i, x_j)$$

Then

$$|f(x) - (Uf)(x)| \le (\mathcal{P}(x))^{1/2}\sqrt{(f,f)}$$

for all $f \in H$.

Proof. Fix $x \in X$. Then $\sum_{i=1}^{k} \alpha_i K_{x_i}$ lies in the span of the set of functions $\{K_{x_1}, \ldots, K_{x_k}\}$. Thus by the corollary in Chapter 30 (page 225), and the remarks preceding Theorem 8,

$$|f(x) - (Uf)(x)| \le \left\| K_x - \sum_{i=1}^{k} \alpha_i K_{x_i} \right\| \sqrt{(f,f)}$$

for all $f \in H$. Now,

$$\left\| K_x - \sum_{i=1}^{k} \alpha_i K_{x_i} \right\|^2 = \left\langle K_x - \sum_{i=1}^{k} \alpha_i K_{x_i}, \ K_x - \sum_{i=1}^{k} \alpha_i K_{x_i} \right\rangle$$

$$= \left\langle K_x - \sum_{i=1}^{k} \alpha_i K_{x_i}, \ K_x \right\rangle - \sum_{j=1}^{k} \alpha_j \left\langle K_x - \sum_{i=1}^{k} \alpha_i K_{x_i}, \ K_{x_j} \right\rangle$$

$$= K(x, x) - \sum_{i=1}^{k} \alpha_i K(x_i, x) - \sum_{j=1}^{k} \alpha_j K(x, x_j) + \sum_{i,j=1}^{k} \alpha_i \alpha_j K(x_i, x_j)$$

From Theorem 7,

$$K(x, y) = \sum_{s=1}^{m} \overline{p_s(x)} p_s(y) + F(x, y) \qquad (y \in X)$$

where F is the reproducing kernel for $N^\perp$. Set

$$M(u, v) = \sum_{s=1}^{m} \overline{p_s(u)} p_s(v) \qquad (u, v \in X)$$

It follows from Lemma 3 that

$$(6) \qquad M(x, x) - \sum_{i=1}^{k} \alpha_i M(x_i, x) - \sum_{j=1}^{k} \alpha_j M(x, x_j) + \sum_{i,j=1}^{k} \alpha_i \alpha_j M(x_i, x_j) = 0$$

Hence,

$$\left\| K_x - \sum_{i=1}^{k} \alpha_i K_{x_i} \right\|^2 = F(x, x) - \sum_{i=1}^{k} \alpha_i \big(F(x_i, x) + F(x, x_i) \big) + \sum_{i,j=1}^{k} \alpha_i \alpha_j F(x_i, x_j)$$

Corollary 3 shows that for $u, v \in X$,

$$F(u, v) = G(u, v) - \sum_{s=1}^{m} \langle G_u, p_s \rangle p_s(v) = G(u, v) - \sum_{s=1}^{m} G(u, x_s) p_s(v)$$

Set

$$Q(u, v) = \sum_{s=1}^{m} G(u, x_s) p_s(v) \qquad (u, v \in X)$$

Another application of Lemma 3 shows that

$$Q(x, x) - \sum_{i=1}^{k} \alpha_i Q(x_i, x) - \sum_{j=1}^{k} \alpha_j Q(x, x_j) + \sum_{i,j=1}^{k} \alpha_i \alpha_j Q(x_i, x_j) = 0$$

Thus,

$$\left\| K_x - \sum_{i=1}^{k} \alpha_i K_{x_i} \right\|^2 = G(x, x) - \sum_{i=1}^{k} \alpha_i (G(x_i, x) + G(x, x_i)) + \sum_{i, j=1}^{k} \alpha_i \alpha_j G(x_i, x_j) \quad \blacksquare$$

Problems

1. Show that Equation (2) defines an inner product on E as asserted.

2. A function $F : X \times X \to \mathbb{C}$ is called **conditionally positive definite of order** m if whenever $x_1, \ldots, x_n \in X$ and $u_1, \ldots, u_n \in \mathbb{C}$ are such that $\sum_{i=1}^{n} u_i p(x_i) = 0$ for all $p \in \Pi_{m-1}$, then $\sum_{i,j=1}^{n} \overline{u}_i u_j F(x_i, x_j) \geq 0$. Show that if $N = \Pi_{m-1}$, then G as defined in Theorem 7 is conditionally positive definite of order m.

3. Let H be a Hilbert function space over X with reproducing kernel K. For u in X, let $H_u = \{f \in H : f(u) = 0\}$. Show that the reproducing kernel for H_u is

   $$G(x, y) = K(x, y) - K(x, u)K(u, y)/K(u, u)$$

4. Generalize the previous example to the case when

   $$H_Y = \{f \in H : f(y) = 0, \text{for all } y \in Y\}$$

   where Y is any finite set in X.

5. Let H be a Hilbert function space over X with reproducing kernel K. Let ψ be a bounded linear functional on H. Calculate the reproducing kernel for the null space of ψ.

6. Develop a suitable generalization to the previous example along the lines of Exercise 4.

7. Let H be a Hilbert function space over X with reproducing kernel K. Let ψ be a continuous linear functional on H. Show that the representative for ψ is r where $r(x) = \overline{\psi(K_x)}$ for all $x \in X$.

8. Let H be a Hilbert function space over X with reproducing kernel K. Let A be a bounded linear operator on H. Show that

   $$(Af)(x) = \langle f, A^* K_x \rangle \qquad (f \in H, x \in X)$$

   (Note here that A^* is the adjoint of A. Its action is defined by $A^* \psi = \psi \circ A$ for all bounded linear functionals ψ on H.)

9. Let H be a Hilbert function space over X with reproducing kernel K. Let ψ be a bounded linear functional on H. Prove that

   $$\|\psi\|^2 \leq \psi \otimes \psi(K)$$

   Here, a tensor product of two functionals, $\psi \otimes \phi$ acts on a function of two variables, say F, as follows: first let $f(x) = \phi(F_x)$; then $\psi \otimes \phi(F) = \psi(f)$. In an abuse of language, one says that ψ acts on F as a function of the second argument, and ψ acts on F as a function of the first argument.

10. The function G described in Corollary 3 is not unique. Describe the set of all $G : X \times X \rightarrow \mathbb{C}$ that satisfy the conditions of Corollary 3.

11. Show that Lemma 3 is valid when

$$M(u, v) = \sum_{i=1}^{m} p_i(u) f_i(v) \qquad (u, v \in X)$$

12. This problem continues the example given at the end of Chapter 30 (page 227). Let H be the class of functions from $\mathbb{R}$ to $\mathbb{R}$ whose first derivative is in $L^2(\mathbb{R})$. These functions can be written

$$f(x) = f(0) + \int_0^x u_f(t)\, dt$$

where $u_f \in L^2(\mathbb{R})$. A suitable inner product is

$$\langle f, g \rangle = f(0)g(0) + \int_{-\infty}^{\infty} u_f(t) u_g(t)\, dt \qquad (f, g \in X)$$

Show that for any x in $\mathbb{R}$ and $f \in H$,

$$|f(x)| \le \sqrt{2}(1 + \sqrt{x})\sqrt{\langle f, f \rangle}$$

13. This example continues Problem 14 in Chapter 30, page 231. Let H be the class of functions from $\mathbb{R}$ to $\mathbb{R}$ whose second derivative is in $L^2(\mathbb{R})$. These functions can be written

(7) $$f(x) = f(0) + xf'(0) + \int_0^x \int_0^y u_f(t)\, dt\, dy$$

where $u_f \in L^2(\mathbb{R})$. A suitable inner product is

$$\langle f, g \rangle = f(0)g(0) + f(1)g(1) + \int_{-\infty}^{\infty} u_f(t) u_g(t)\, dt \qquad (f, g \in X)$$

Show that each such f is in $C^{(1)}(\mathbb{R})$. Note that it helps to observe that the derivative of f is easily computable from Equation (7). Show that

$$|f'(0)| \le |f(0)| + |f(1)| + \left(\int_{-\infty}^{\infty} (u_f(t))^2\, dt \right)^{1/2}$$

Show that for any x in $\mathbb{R}$, there exists a number c_x' such that $|f'(x)| \le c_x'\sqrt{\langle f, f \rangle}$, for all $f \in H$. Show also that there is a number c_x such that $|f(x)| \le c_x\sqrt{\langle f, f \rangle}$.

References

[Aron] Aronszajn, N. "Theory of reproducing kernels." *Trans. Amer. Math. Soc.* **68** (1950), 337–404.

[Att]   Atteia, M. *Hilbertian Kernels and Spline Functions.* North-Holland, Amsterdam, 1992.

[Berg]   Bergman, Stefan. *The Kernel Function and Conformal Mapping.* Mathematical Surveys, vol. 5. Amer. Math. Soc., Providence, RI, 1950. 2nd (rev.) ed., 1970.

[GP]   Goffman, C., and G. Pedrick. *First Course in Functional Analysis.* 2nd ed., Chelsea, New York. Now available from Amer. Math. Soc.

[NWal]   Nashed, M. Z., and G. G. Walter. "General sampling theorems for functions in reproducing kernel Hilbert spaces." *Math. Control Signals Systems 4* (1991), 363–390.

[RS]   Reed, M., and B. Simon. *Methods of Modern Mathematical Physics,* Vol. I, *Functional Analysis.* 2nd ed., Academic Press, New York, 1980.

[Sai]   Saitoh, S. *Theory of Reproducing Kernels and Its Applications.* Longmans, London, 1988.

[Shap1]   Shapiro, H. S. *Topics in Approximation Theory.* Lecture Notes in Mathematics, vol. 187, Springer-Verlag, New York, 1971.

[Shap2]   Shapiro, H. S. *Smoothing and Approximation of Functions.* Van Nostrand, New York, 1969.

32
Spherical Thin-Plate Splines

In this chapter, an important application of Chapters 31 and 30 is pursued. We intend to examine Hilbert spaces H that are subspaces of $C(\mathbb{S}^{s-1})$. Here $\mathbb{S}^{s-1}$ is the $(s-1)$-dimensional unit sphere in $\mathbb{R}^s$. That is, $\mathbb{S}^{s-1} = \{x \in \mathbb{R}^s : \|x\|_2 = 1\}$. Historically, this is not the way the variational theory of radial basis functions developed. Early work by Duchon [Du] focused on subspaces of $C(\mathbb{R}^s)$. This theory is best understood with the aid of techniques from distribution theory. The corresponding theory for $\mathbb{S}^{s-1}$ is more elementary. The early development came from Wahba [Wah1]. Subsequently, Freeden and co-workers added considerably to the theory, producing many papers in the process; see [Free] and [FGS] for example. The development in this chapter is based on Levesley, Light, Ragozin, and Sun [LLRS].

It is necessary, first of all, to understand a little about spherical harmonics. Sources for this topic include [Mu], [See], and [SW]. Let $\Pi_k(\mathbb{R}^s)$ denote the space of all polynomials of total degree at most k on $\mathbb{R}^s$. Then $\Pi_k(\mathbb{S}^{s-1})$ denotes the restriction of these polynomials to the sphere $\mathbb{S}^{s-1}$. If $p \in \Pi_k(\mathbb{S}^{s-1})$, then we can write

$$p(x) = \sum \{c_\alpha x^\alpha : \alpha \in \mathbb{Z}_+^s, \ |\alpha| \le k\} \qquad (x \in \mathbb{S}^{s-1})$$

The set of monomials x^α is obviously linearly dependent on $\mathbb{S}^{s-1}$, because $\|x\|^2 - 1 = 0$. Consequently, in the above representation of p, the coefficients c_α are not uniquely determined. To restore unicity, we can demand that $c_\alpha = 0$ if $\alpha_1 > 1$. Alternatively, we can set

$$\Gamma = \{\alpha = (\alpha_1, \ldots, \alpha_s) \in \mathbb{Z}_+^s : |\alpha| \le k \text{ and } \alpha_1 = 0 \text{ or } \alpha_1 = 1\}$$

Then

$$p(x) = \sum_{\alpha \in \Gamma} c_\alpha x^\alpha$$

and the c_α, $\alpha \in \Gamma$, are uniquely determined by p.

There is a natural inner product on $C(S^{s-1})$ defined by

(1) $$[f, g] = \int_{S^{s-1}} fg\, d\omega(S^{s-1}) \qquad (f, g \in C(S^{s-1}))$$

Here ω is the usual Lebesgue measure on S^{s-1}, so that for example, $\omega(S^1) = 2\pi$.

Let $\Pi_{k-1}^{\perp}(S^{s-1})$ denote the orthogonal complement of $\Pi_{k-1}(S^{s-1})$ in $\Pi_k(S^{s-1})$ with respect to the inner product given by Equation (1). For the purposes of this definition, Π_{-1} consists of 0 alone. The dimension of $\Pi_{k-1}^{\perp}(S^{s-1})$ will be denoted by d_k. (The dependence of d_k on s is suppressed in the notation.) The set of **spherical harmonics** $\{Y_j^{(k)} : j = 1, \dots, d_k\}$ can be defined to be any orthonormal basis for $\Pi_{k-1}^{\perp}(S^{s-1})$. Then any p in $\Pi_k(S^{s-1})$ can be written as

$$p = \sum_{i=0}^{k} \sum_{j=1}^{d_i} a_{ij} Y_j^{(i)}$$

The numbers d_k, $k = 0, 1, \dots$, are computable, and a formula for them can be found in Müller [Mu]. However, the specific value of d_k is not important for our designs. It is clear that $d_0 = 1$, irrespective of the value of s. For $s = 2$, the spherical harmonics are connected to the trigonometric functions sin and cos . See Problem 1 for details. The following addition formula can be found in Müller's monograph [Mu]:

(2) $$\sum_{j=1}^{k} Y_j^{(k)}(x) Y_j^{(k)}(y) = \omega_{s-1}^{-1} d_k P_k(xy) \qquad (x, y \in S^{s-1}, k = 0, 1, \dots)$$

Here xy denotes the inner product of x and y, and P_k is a certain Gegenbauer polynomial of degree k. These polynomials are mutually orthogonal on $[-1, 1]$ with respect to the weight function w given by

$$w(t) = (1 - t^2)^{(s-3)/2} \qquad (t \in [-1, 1])$$

The normalization used in Chapter 17 is adopted here; that is, $P_k(1) = 1$.

Let β be a sequence $[\beta_0, \beta_1, \dots]$ such that, for some integer k, we have $\beta_i = 0$ for $0 \le i < k$, $\beta_i > 0$ for $i \ge k$, and $\sum_{i \ge k} \beta_i^{-2} d_i < \infty$. Corresponding to any such β, there is a space X_β, defined by

$$X_\beta = \left\{ \sum_{i=0}^{\infty} \sum_{j=1}^{d_i} a_{ij} Y_j^{(i)} : \sum_{i=k}^{\infty} \beta_i^2 \sum_{j=1}^{d_i} a_{ij}^2 < \infty \right\}$$

Every f in X_β can be expressed in the form

$$f = \sum_{i=0}^{\infty} \sum_{j=1}^{d_i} a_{ij}(f) Y_j^{(i)}$$

where $\sum_{i=0}^{\infty} \beta_i^2 \sum_{j=1}^{d_i} a_{ij}^2(f) < \infty$. Note that we are here considering a formal orthogonal series for f, and are not making any assertions about its convergence. The equation

$$(f, g) = \sum_{i=k}^{\infty} \beta_i^2 \sum_{j=1}^{d_i} a_{ij}(f) a_{ij}(g) \qquad (f, g \in X_\beta)$$

defines a semi-inner product on X_β. Lemma 1, below, shows that the series defining (f, g) does indeed converge absolutely. This bilinear form has as its kernel the set

$$N = \left\{ \sum_{i=0}^{k-1} \sum_{j=1}^{d_i} a_{ij} Y_j^{(i)} \; : \; a_{ij} \in \mathbb{R} \right\}$$

It is clear that $N = \Pi_{k-1}(\mathcal{S}^{s-1})$ and that $\dim N = d_0 + \cdots + d_{k-1}$. We shall adopt the notation $|f|_\beta = \sqrt{(f,f)}$ for $f \in X_\beta$ and define $m := \dim N$.

LEMMA 1. *If $\sum_{i=k}^\infty \beta_i^{-2} d_i < \infty$, then X_β is a subspace of $C(\mathcal{S}^{s-1})$. If f is an element of X_β having the form $f = \sum_{i=k}^\infty \sum_{j=1}^{d_i} a_{ij} Y_j^{(i)}$, then for all $x \in \mathcal{S}^{s-1}$, $|f(x)| \le \omega_{s-1}^{-1/2} A |f|_\beta$, where $A^2 := \sum_{i=k}^\infty \beta_i^{-2} d_i$.*

Proof. Given f as above, let

$$g_i = \sum_{j=1}^{d_i} a_{ij} Y_j^{(i)} \qquad (i = k, k+1, \dots)$$

The Cauchy-Schwarz inequality and an application of Equation (2) lead to

$$|g_i(x)| \le \left(\sum_{j=1}^{d_i} a_{ij}^2 \right)^{1/2} \left(\sum_{j=1}^{d_i} \left(Y_j^{(i)}(x) \right)^2 \right)^{1/2}$$

$$= \omega_{s-1}^{-1/2} \left(\sum_{j=1}^{d_i} a_{ij}^2 \right)^{1/2} d_i^{1/2} \left(P_i(|x|^2) \right)^{1/2}$$

Since $P_i(|x|^2) = P_i(1) = 1$, we have

$$\|g_i\|_\infty \le \omega_{s-1}^{-1/2} d_i^{1/2} \left(\sum_{j=1}^{d_i} a_{ij}^2 \right)^{1/2}$$

Using the Cauchy-Schwarz inequality again leads to

$$|f(x)| \le \sum_{i=k}^\infty \|g_i\|_\infty \le \omega_{s-1}^{-1/2} \sum_{i=k}^\infty \beta_i \left(\sum_{j=1}^{d_i} a_{ij}^2 \right)^{1/2} d_i^{1/2} \beta_i^{-1}$$

$$\le \omega_{s-1}^{-1/2} \left(\sum_{i=k}^\infty \beta_i^2 \sum_{j=1}^{d_i} a_{ij}^2 \right)^{1/2} \left(\sum_{i=k}^\infty d_i \beta_i^{-2} \right)^{1/2}$$

$$= A \omega_{s-1}^{-1/2} \sqrt{(f,f)} = A \omega_{s-1}^{-1/2} |f|_\beta$$

This establishes the required inequality. The Weierstrass M-test now shows that f belongs to $C(\mathcal{S}^{s-1})$. Since every element of X_β is the sum of a polynomial and a function f of the type considered above, we see that $X_\beta \subset C(\mathcal{S}^{s-1})$. ∎

A consequence of Lemma 1 is that we may now follow the construction in Chapter 31, pages 235 and 236, which leads to an inner product on X_β. Take $x_1, \dots, x_m$ in

$\mathcal{S}^{s-1}$ such that if $p \in N$ and $p(x_i) = 0$ for $i = 1, \ldots, m$, then $p = 0$. Such a choice is possible by Lemma 2 of Chapter 31. A set with this property will be said to be **total** with respect to N. The inner product on X_β is now defined by

$$(3) \qquad \langle f, g \rangle := \sum_{i=1}^{m} f(x_i)g(x_i) + (f, g) \qquad (f, g \in X_\beta)$$

We will use $\|\cdot\|$ to denote the norm on X_β that is derived from this inner product. One might expect now to use the theory of pre-Hilbert spaces to see that the completion of X_β is a Hilbert function space. However, part of this theory will not be needed, since it will turn out that X_β is already complete.

> **LEMMA 2.** *If $\sum_{i=k}^{\infty} \beta_i^{-2} d_i < \infty$, then there exists a constant C such that $|f(x)| \le C\|f\|$, for all f in X_β and for all x in $\mathcal{S}^{s-1}$.*

Proof. Let $A^2 = \sum_{i=k}^{\infty} \beta_i^{-2} d_i$. We intend to apply Theorem 6 of Chapter 31, using a suitable projection P. Let $f \in X_\beta$; then f has the form $f = \sum_{i=0}^{\infty} \sum_{j=1}^{d_i} a_{ij} Y_j^{(i)}$. Set $Pf = \sum_{i=0}^{k-1} \sum_{j=1}^{d_i} a_{ij} Y_j^{(i)}$. We have to show that P is a projection from X_β onto N. The only question is whether P is bounded. An application of Lemma 1 shows that for every x in $\mathcal{S}^{s-1}$,

$$|(f - Pf)(x)| \le \omega_{s-1}^{-1/2} A |f - Pf|_\beta = \omega_{s-1}^{-1/2} A |f|_\beta$$

Now, because Pf is in N,

$$\|Pf\| = \left(\sum_{i=1}^{m} ((Pf)(x_i))^2 \right)^{1/2} \le \left(\sum_{i=1}^{m} ((f - Pf)(x_i))^2 \right)^{1/2} + \left(\sum_{i=1}^{m} (f(x_i))^2 \right)^{1/2}$$

$$\le \sqrt{m} \omega_{s-1}^{-1/2} A |f|_\beta + \left(\sum_{i=1}^{m} (f(x_i))^2 \right)^{1/2}$$

$$\le \sqrt{2}\left(1 + \sqrt{m} \omega_{s-1}^{-1/2} A\right) \left(\sum_{i=1}^{m} (f(x_i))^2 + |f|_\beta^2 \right)^{1/2}$$

$$= \sqrt{2}\left(1 + \sqrt{m} \omega_{s-1}^{-1/2} A\right) \|f\|$$

Thus P is a bounded linear operator. An application of Theorem 6 of Chapter 31 (page 238) shows now that $(X_\beta, |\cdot|_\beta)$ is a pre-Hilbert function space. It therefore follows that the constant C exists as claimed in the statement of the lemma. ∎

The next issue is the completeness of X_β. Let ℓ_β denote the linear space of all real sequences $a = [a_{ij} : 1 \le j \le d_i, i \ge k]$ such that $\sum_{i=k}^{\infty} \beta_i^2 \sum_{j=1}^{d_i} a_{ij}^2 < \infty$. If $a \in \ell_\beta$, then the equation

$$(4) \qquad \|a\| = \left(\sum_{i=k}^{\infty} \beta_i^2 \sum_{j=1}^{d_i} a_{ij}^2 \right)^{1/2}$$

defines a norm on ℓ_β. The space ℓ_β is then a weighted version of ℓ^2 and thus is a Hilbert space.

LEMMA 3. *Let the mapping* $J : X_\beta \rightarrow \ell_\beta$ *be defined by* $Jf = a$, *where* $f = \sum_{i=0}^{\infty} \sum_{j=1}^{d_i} a_{ij} Y_j^{(i)}$ *and* $a = [a_{ij}]_{j=1, i=k}^{d_i, \infty}$. *Then* $|f|_\beta = \|Jf\|$ *for all* $f \in X_\beta$, *and* J *maps* X_β *onto* ℓ_β.

Proof. The proof is elementary. ∎

THEOREM 1. *The space* $(X_\beta, \|\cdot\|)$ *is a Hilbert function space.*

Proof. Let $[f_n]$ be a Cauchy sequence in X_β. From the inequality

$$\|Jf_i - Jf_j\| = |f_i - f_j|_\beta \le \|f_i - f_j\|$$

we see that $[Jf_n]$ is a Cauchy sequence in ℓ_β. Since ℓ_β is complete, $Jf_n \rightarrow b$ for some $b \in \ell_\beta$. Since J is "onto" (i.e., surjective), there exists $g \in X_\beta$ such that $Jg = b$. Moreover, $|f_n - g|_\beta = \|Jf_n - b\| \rightarrow 0$ as $n \rightarrow \infty$. Now let $L : X_\beta \rightarrow N$ be the orthogonal projection. Then $\|Lf_i - Lf_j\| \le \|f_i - f_j\|$, showing that $[Lf_n]$ is a Cauchy sequence in N. Since N is finite-dimensional, it is complete in any norm. Hence, there exists $p \in N$ such that $\|Lf_n - p\| \rightarrow 0$ as $n \rightarrow \infty$. Now, from the inequality

$$\|f_n - g - p + Lg\| \le \|f_n - g - Lf_n + Lg\| + \|Lf_n - p\|$$
$$= \|f_n - g - L(f_n - g)\| + \|Lf_n - p\|$$
$$= |f_n - g|_\beta + \|Lf_n - p\|$$

we see that $\|f_n - g - p + Lg\| \rightarrow 0$ as $n \rightarrow \infty$. Thus $f_n \rightarrow g + p - Lg$ as $n \rightarrow \infty$, showing that X_β is complete. From Lemma 2, we already know that $(X_\beta, \|\cdot\|_\beta)$ is a pre-Hilbert function space. We have just shown that this space is complete, and so it is indeed a Hilbert function space. ∎

Our next goal is to calculate the reproducing kernel for X_β. We have already selected $\{x_1, \ldots, x_m\}$ to be total with respect to N. In accordance with Chapter 31, we now choose $p_1, \ldots, p_m$ in N such that $p_i(x_j) = \delta_{ij}$. The projection L defined by $Lf = \sum_{i=1}^{m} f(x_i) p_i$ is the orthogonal projection from X_β onto N (see Chapter 31, page 237). We intend to calculate the reproducing kernel for X_β with the help of Theorem 5 in Chapter 31.

LEMMA 4. *Assume that* $\sum_{i=k}^{\infty} \beta_i^{-2} d_i < \infty$. *Let* L *be the orthogonal projection of* X_β *onto* N. *Then the reproducing kernel for* $N^\perp$ *is the function* G *defined for all* x *and* y *in* $\mathbf{S}^{s-1}$ *by the equation*

$$G(x, y) = \sum_{i=k}^{\infty} \beta_i^{-2} \sum_{j=1}^{d_i} \left(Y_j^{(i)}(x) - (LY_j^{(i)})(x) \right) \left(Y_j^{(i)}(y) - (LY_j^{(i)})(y) \right)$$

Proof. Set $Z_j^{(i)} = Y_j^{(i)} - LY_j^{(i)}$, $j = 1, \ldots, d_i$, $i \ge k$. Then

$$\langle Z_j^{(i)}, Z_q^{(p)} \rangle = \sum_{\nu=1}^{m} Z_j^{(i)}(x_\nu) Z_q^{(p)}(x_\nu) + (Z_j^{(i)}, Z_q^{(p)}) = \langle Y_j^{(i)}, Y_q^{(p)} \rangle$$

Because $Z_j^{(i)} = Y_j^{(i)} - LY_j^{(i)}$, and because $(Lf)(x_\nu) = f(x_\nu)$ for all $f \in X_\beta$ and $1 \le \nu \le m$, we have

$$\langle Z_j^{(i)}, Z_q^{(p)} \rangle = \left(Y_j^{(i)} - LY_j^{(i)}, \; Y_q^{(p)} - LY_q^{(p)} \right) = \left(Y_j^{(i)}, Y_q^{(p)} \right)$$

It now follows from the form of the semi-inner product that $\langle Z_j^{(i)}, Z_q^{(p)} \rangle = \beta_i^2$ if $i = p \geq k$ and $1 \leq j = q \leq d_i$. Otherwise, $\langle Z_j^{(i)}, Z_q^{(p)} \rangle$ is zero. Thus the sequence $\{Z_j^{(i)} : 1 \leq j \leq i, k \leq i\}$ is an orthonormal sequence in X_β. Moreover,

$$N^\perp = \{(I - L)f \; : \; f \in X_\beta\} = \left\{ \sum_{i=0}^{\infty} \sum_{j=1}^{d_i} a_{ij}(I - L)Y_j^{(i)} \; : \; \sum_{i=k}^{\infty} \beta_i^2 \sum_{j=1}^{d_i} a_{ij}^2 < \infty \right\}$$

$$= \left\{ \sum_{i=k}^{\infty} \sum_{j=1}^{d_i} \beta_i a_{ij} \beta_i^{-1} Z_j^{(i)} \; : \; \sum_{i=k}^{\infty} \beta_i^2 \sum_{j=1}^{d_i} a_{ij}^2 < \infty \right\}$$

Thus $\{Z_j^{(i)}/\beta_i \; : \; 1 \leq j \leq d_i, \, i \geq k\}$ is an orthonormal basis for $N^\perp$, and the result now follows from Theorem 5 of Chapter 31. ∎

THEOREM 2. *Assume that $\sum_{i=k}^{\infty} \beta_i^{-2} d_i < \infty$. Define G by the equation in the statement of the preceding lemma. Then the reproducing kernel for X_β is the function K defined by the equation*

$$K(x, y) = \sum_{i=1}^{m} p_i(x) p_i(y) + G(x, y) \qquad (x, y \in \mathbf{S}^{s-1})$$

Proof. The result is an immediate consequence of Theorem 7 in Chapter 31, pages 238 and 239. ∎

The function G in Theorem 2 can be cast into a simpler form by an application of Equation (2), (the addition formula for spherical harmonics). This leads to one of the main results of this chapter.

THEOREM 3. *Assume that $\sum_{i=k}^{\infty} \beta_i^{-2} d_i < \infty$. Define Φ by the equation*

$$\Phi(x, y) := \omega_{s-1}^{-1} \sum_{i=k}^{\infty} \beta_i^{-2} d_i P_i(xy) \qquad (x, y \in \mathbf{S}^{s-1})$$

where P_i is the i-th Gegenbauer polynomial appearing in Equation (2). Then the reproducing kernel K for X_β is given by

$$K = \sum_{i=1}^{m} p_i \otimes p_i + (I - L \otimes I)(I - I \otimes L)\Phi.$$

Proof. Theorem 2 shows that $K = G + \sum_{i=1}^{m} p_i \otimes p_i$. The function G can be transformed as follows:

$$G(x, y) = \sum_{i=k}^{\infty} \beta_i^{-2} \sum_{j=1}^{d_i} \left(Y_j^{(i)}(x) - (LY_j^{(i)})(x) \right)\left(Y_j^{(i)}(y) - (LY_j^{(i)})(y) \right)$$

$$= \sum_{i=k}^{\infty} \beta_i^{-2} \sum_{j=1}^{d_i} \left((I - L \otimes I)(I - I \otimes L)(Y_j^{(i)} \otimes Y_j^{(i)}) \right)(x, y)$$

$$= \left((I - L \otimes I)(I - I \otimes L) \sum_{i=k}^{\infty} \beta_i^{-2} \sum_{j=1}^{d_i} Y_j^{(i)} \otimes Y_j^{(i)} \right)(x, y)$$

The final components of the preceding equation can also be transformed by an application of Equation (2):

$$\left(\sum_{i=k}^{\infty} \beta_i^{-2} \sum_{j=1}^{d_i} Y_j^{(i)} \otimes Y_j^{(i)} \right)(x, y) = \sum_{i=k}^{\infty} \beta_i^{-2} \sum_{j=1}^{d_i} Y_j^{(i)}(x) Y_j^{(i)}(y)$$

$$= \sum_{i=k}^{\infty} \beta_i^{-2} \omega_{s-1}^{-1} d_i P_i(xy)$$

$$= \Phi(x, y) \qquad \blacksquare$$

The function Φ appearing in Theorem 3 is a radial function on $\mathcal{S}^{s-1}$. Thus, $\Phi(x, y)$ depends only on the geodesic distance $\mathrm{Arccos}(xy)$ between x and y. There are some close links now with Chapters 15, 16, and 17. The following generalizes the definition of a positive definite function on the sphere $\mathcal{S}^{s-1}$ as given in Chapter 17, page 123.

Definition. A function F in $C[-1, 1]$ is said to be **conditionally positive definite** of order k on $\mathcal{S}^{s-1}$ if the inequality

$$\sum_{i,j=1}^{n} c_i c_j F(x_i x_j) \geq 0$$

is valid whenever

a. $x_1, \ldots, x_n$ are distinct points in $\mathcal{S}^{s-1}$
b. $c_1, \ldots, c_n$ are real numbers such that the functional $\sum_{\nu=1}^{n} c_\nu x_\nu^*$ annihilates all spherical harmonics $Y_j^{(i)}$ having degree i less than k.

If, in addition, $\sum_{i,j=1}^{n} c_i c_j F(x_i x_j) > 0$ whenever at least one of the $c_1, \ldots, c_n$ is nonzero, then F is said to be **strictly conditionally positive definite** of order k on $\mathcal{S}^{s-1}$.

THEOREM 4. *Suppose that $\sum_{i=k}^{\infty} \beta_i^{-2} d_i < \infty$ and set $\gamma_i = \beta_i^{-2} d_i \omega_{s-1}^{-1}$, $i = k, k+1, \ldots$. Let $\phi = \sum_{i=k}^{\infty} \gamma_i P_i$, where P_i is the i-th Gegenbauer polynomial appearing in Equation (2). Then ϕ is strictly conditionally positive definite of order k on $\mathcal{S}^{s-1}$.*

Proof. Let $x_1, \ldots, x_n$ be distinct points in $\mathcal{S}^{s-1}$ and let $c_1, c_2, \ldots, c_n$ be real numbers. Then Equation (2) yields

$$\sum_{i,j=1}^{n} c_i c_j \phi(x_i x_j) = \sum_{i,j=1}^{n} c_i c_j \sum_{\nu=k}^{\infty} \gamma_\nu P_\nu(x_i x_j)$$

$$= \sum_{i,j=1}^{n} c_i c_j \sum_{\nu=k}^{\infty} \beta_\nu^{-2} \sum_{\mu=1}^{d_\nu} Y_\mu^{(\nu)}(x_i) Y_\mu^{(\nu)}(x_j)$$

$$= \sum_{\nu=k}^{\infty} \beta_\nu^{-2} \sum_{\mu=1}^{d_\nu} \left(\sum_{i=1}^{n} c_i Y_\mu^{(\nu)}(x_i) \right)^2 \geq 0$$

If the above inequality is an equality, then (since $\beta_v^{-2} > 0$ for $v = k, k + 1, \ldots$) we must have $\sum_{i=1}^{n} c_i Y_\mu^{(v)}(x_i) = 0$ when $1 \le \mu \le d_v$ and $v = k, k + 1, \ldots$. If we also assume that $\sum_{i=1}^{n} c_i Y_\mu^{(v)}(x_i) = 0$ for $1 \le \mu \le d_\mu$ and $k = 0, \ldots, k - 1$, then we see that the functional that maps $f \in C(\mathbb{S}^{s-1})$ to $\sum_{i=1}^{n} c_i f(x_i)$ annihilates every spherical harmonic. We there-fore conclude that this functional is the zero functional. That is, $c_1 = \cdots = c_n = 0$. Since $\phi \in C[-1, 1]$, it follows that ϕ is strictly conditionally positive definite of order k on $\mathbb{S}^{s-1}$. ∎

We now want to discuss the minimal norm interpolant. Our approach is identical to that in Chapter 31. We take $\mathcal{A} = \{x_1, \ldots, x_n\}$ to be a set of points in $\mathbb{S}^{s-1}$ for which $n \ge m = \dim N$. We assume still that $\{x_1, \ldots, x_m\}$ is total with respect to N. The inner product on X_β will continue to be defined by Equation (3). Given any $f \in X_\beta$, we will let Uf denote the minimal norm interpolant to f on $\mathcal{A}$. By Theorem 3 of Chapter 30, page 225,

$$Uf = \sum_{i=1}^{n} \lambda_i K(x_i, \cdot)$$

where $\lambda_1, \ldots, \lambda_n$ satisfy $\sum_{i=1}^{n} \lambda_i K(x_i, x_j) = f(x_j), j = 1, \ldots, n$. Here K is the reproduc-ing kernel for X_β given by Theorem 3, page 251. The minimal norm interpolant can be written in an alternative form given in the next theorem.

THEOREM 5. *Let $\{x_1, \ldots, x_n\}$ be a set of points in $\mathbb{S}^{s-1}$ that is total with respect to $\Pi_{k-1}(\mathbb{S}^{s-1})$. Given $f \in X_\beta$, let Uf be the minimal norm inter-polant to f on the given set of points. Let ϕ be as given in Theorem 4. Then Uf has the form*

$$(Uf)(x) = \sum_{i=1}^{n} \mu_i \phi(x_i x) + \sum_{v=0}^{k-1} \sum_{j=1}^{d_v} \varepsilon_{vj} Y_j^{(v)}(x), \qquad (x \in \mathbb{S}^{s-1})$$

The parameters μ_i and ε_{vj} are uniquely determined by these equations

$$(5) \quad (Uf)(x_\sigma) = \sum_{i=1}^{n} \mu_i \phi(x_i x_\sigma) + \sum_{v=0}^{k-1} \sum_{j=1}^{d_v} \varepsilon_{vj} Y_j^{(v)}(x_\sigma) = f(x_\sigma) \qquad (\sigma = 1, \ldots, n)$$

$$(6) \qquad \sum_{i=1}^{n} \mu_i Y_j^{(v)}(x_i) = 0 \qquad (j = 1, 2, \ldots, d_v, v = 0, 1, \ldots, k - 1)$$

Proof. We begin with $Uf = \sum_{i=1}^{n} \lambda_i K(x_i, \cdot)$. From Theorem 3,

$$K = \Phi - (L \otimes I)\Phi - (I \otimes L)\Phi + (L \otimes L)\Phi$$

Recalling that $Lf = \sum_{i=1}^{m} f(x_i)p_i$, we can write

$$[(I \otimes L)\Phi](u, v) = \sum_{i=1}^{m} \phi(u x_i)p_i(v) \qquad (u, v \in \mathbb{S}^{s-1})$$

and

$$[(L \otimes L)\Phi](u, v) = \sum_{j=1}^{m} \sum_{i=1}^{m} \phi(x_j x_i)p_i(v)p_j(u) \qquad (u, v \in \mathbb{S}^{s-1})$$

It follows that for each $x \in \mathbb{S}^{s-1}$, $[(I \otimes L)\Phi](x, \cdot) \in N$ and $[(L \otimes L)\Phi](x, \cdot) \in N$. We define $q_1, \ldots, q_n \in N$ by

$$q_i(v) = [(I \otimes L)\Phi](x_i, v) + [(L \otimes L)\Phi](x_i, v) \qquad (v \in \mathbb{S}^{s-1})$$

Then, for $x \in \mathbb{S}^{s-1}$ and $i = 1, \ldots, n$,

$$K(x_i, x) = \phi(x_i x) - [(L \otimes I)\Phi](x_i, x) - q_i(x)$$

$$= \phi(x_i x) - \sum_{j=1}^{n} \phi(x_j x)p_j(x_i) - q_i(x)$$

Thus, for any $x \in \mathbb{S}^{s-1}$,

$$(Uf)(x) = \sum_{i=1}^{n} \lambda_i K(x_i, \cdot) = \sum_{i=1}^{n} \lambda_i \left(\phi(xx_i) - \sum_{j=1}^{m} \phi(x_j x)p_j(x_i) \right) - \sum_{i=1}^{n} \lambda_i q_i(x)$$

$$= \sum_{i=1}^{n} \mu_i \phi(x_i x) + \sum_{v=0}^{k-1} \sum_{j=1}^{d_v} \varepsilon_{vj} Y_j^{(v)}(x)$$

Here we have written the function $\sum_{i=1}^{n} \lambda_i q_i$, which is an element of N, in terms of the basis $\{Y_j^{(v)} : j = 1, \ldots, d_v, v = 0, \ldots, k - 1\}$. Moreover,

$$\mu_i = \begin{cases} \lambda_i - \sum_{j=1}^{n} \lambda_j p_i(x_j), & i = 1, \ldots, m \\ \lambda_i, & i = m + 1, \ldots, n \end{cases}$$

Thus Uf has the required form. To show that Uf is uniquely determined by Equations (5) and (6), suppose that

(7) $$\sum_{i=1}^{n} \mu_i \phi(x_i x_\sigma) + \sum_{v=0}^{k-1} \sum_{j=1}^{d_v} \varepsilon_{vj} Y_j^{(v)}(x_\sigma) = 0, \qquad (\sigma = 1, \ldots, n)$$

and

(8) $$\sum_{i=1}^{n} \mu_i Y_j^{(v)}(x_i) = 0, \qquad (j = 1, \ldots, d_v, v = 0, 1, \ldots, k - 1)$$

We have to show that Equations (7) and (8) have only the trivial solution. Multiplying the Equation (7) by μ_σ and summing over σ, we see that

$$\sum_{i,\sigma=1}^{n} \mu_i \mu_\sigma \phi(x_i x_\sigma) + \sum_{v=0}^{k-1} \sum_{j=1}^{d_v} \varepsilon_{vj} \sum_{\sigma=1}^{n} \mu_\sigma Y_j^{(v)}(x_\sigma) = 0$$

Using Equation (8) leads to

$$\sum_{i,\sigma=1}^{n} \mu_i \mu_\sigma \phi(x_i x_\sigma) = 0$$

Because ϕ is conditionally positive definite of order k (by Theorem 4), it follows that $\mu_1 = \mu_2 = \cdots = \mu_n = 0$. Equation (7) now reads

$$\sum_{v=0}^{k-1} \sum_{j=1}^{d_v} \varepsilon_{vj} Y_j^{(v)}(x_\sigma) = 0 \qquad (\sigma = 1, \ldots, n)$$

Since $\{x_1, \ldots, x_n\}$ is total with respect to N, it follows that $\varepsilon_{vj} = 0$, for $1 \le j \le d_v$ and $0 \le v \le k - 1$. ∎

The last few results in this chapter give estimates of $|f(x) - (Uf)(x)|$, the error in approximating f by its minimal norm interpolant.

LEMMA 5. *Let $x_1, \ldots, x_n$ be distinct points in $\mathbb{S}^{s-1}$. Fixing x in $\mathbb{S}^{s-1}$, choose $\alpha_1, \alpha_2, \ldots, \alpha_n$ so that $p(x) = \sum_{i=1}^{n} \alpha_i p(x_i)$ for all $p \in N$. For $f \in X_\beta$, let Uf be the minimal norm interpolant to f on $x_1, \ldots, x_n$, as given in Theorem 5. Then*

$$|f(x) - (Uf)(x)| \le \left\{ \phi(1) - 2\sum_{i=1}^{n} \alpha_i \phi(xx_i) + \sum_{i,j=1}^{n} \alpha_i \alpha_j \phi(x_i x_j) \right\}^{1/2} |f|_\beta$$

Proof. Let K be the reproducing kernel for X_β. By Theorem 3,

$$K = (I - L \otimes I)(I - I \otimes L)\Phi + \sum_{i=1}^{m} p_i \otimes p_i$$

For any $f \in N^\perp$ and $u \in \mathbb{S}^{s-1}$,

$$\left\langle f, K(u, \cdot) \right\rangle = \left\langle f, \Phi(u, \cdot) - [(L \otimes I)\Phi](u, \cdot) \right.$$

$$\left. - [(I \otimes L)\Phi](u, \cdot) + [(L \otimes L)\Phi](u, \cdot) + \sum_{i=1}^{m} p_i(u)p_i \right\rangle$$

Since $[(I \otimes L)\Phi](x, \cdot)$, $[(L \otimes L)\Phi](x, \cdot)$, and $\sum_{i=1}^{m} p_i(x)p_i$ all belong to N,

$$\left\langle f, K(x, \cdot) \right\rangle = \left\langle f, \Phi(x, \cdot) - [(L \otimes I)\Phi](x, \cdot) \right\rangle \qquad (f \in N^\perp)$$

This shows that in Theorem 9 of Chapter 31, we may take

$$G(u, \cdot) = \Phi(u, \cdot) - [(L \otimes I)\Phi](u, \cdot) \qquad (u \in \mathbb{S}^{s-1})$$

That same theorem shows that $|f(x) - (Uf)(x)|^2 \le \Theta(x)|f|_\beta$, where

$$\Theta(x) = G(x, x) - \sum_{i=1}^{n} \alpha_i (G(x_i, x) + G(x, x_i)) + \sum_{i,j=1}^{n} \alpha_i \alpha_j G(x_i, x_j)$$

Now,

$$[(L \otimes I)\Phi](u,v) = \sum_{\sigma=1}^{m} \phi(x_\sigma u) p_\sigma(v) \qquad (u,v \in \mathcal{S}^{s-1})$$

and therefore Problem 11 of Chapter 31 shows that

$$[(L \otimes I)\Phi](x,x) = \sum_{i=1}^{n} \alpha_i \{ [(L \otimes I)\Phi](x_i,x) + [(L \otimes I)\Phi](x,x_i) \}$$

$$- \sum_{i,j=1}^{n} \alpha_i \alpha_j [(L \otimes I)\Phi](x_i,x_j)$$

Thus,

$$\Theta(x) = \Phi(x,x) - \sum_{i=1}^{n} \alpha_i (\Phi(x_i,x) + \Phi(x,x_i)) + \sum_{i,j=1}^{n} \alpha_i \alpha_j \Phi(x_i,x_j)$$

$$= \phi(1) - 2 \sum_{i=1}^{n} \alpha_i \phi(xx_i) + \sum_{i,j=1}^{n} \alpha_i \alpha_j \phi(x_i x_j) \qquad \blacksquare$$

It is natural to have expectations about the error bound given in Theorem 5. If the points $x, x_1, \ldots, x_n$ are close together, then $|f(x) - (Uf)(x)|$ should be small. This is apparent from the form of the bound, since in this case $x_i x_j$ and xx_i are close to 1, and the fact that $\sum_{i=1}^{n} \alpha_i = 1$ means that the term whose square root is taken will be small. The natural way to investigate this is by means of the Taylor series.

THEOREM 6. *Let $\{x_1, \ldots, x_n\}$ be a subset of $\mathcal{S}^{s-1}$ that is total with respect to $\Pi_{k-1}(\mathcal{S}^{s-1})$. Fixing $x \in \mathcal{S}^{s-1}$, let $\alpha_1, \ldots, \alpha_n$ be real numbers such that $p(x) = \sum_{i=1}^{n} \alpha_i p(x_i)$ for all $p \in \Pi_{v-1}(\mathcal{S}^{s-1})$, where v is a fixed integer satisfying $v \geq k$. For $f \in X_\beta$, let the minimal norm interpolant on $x_1, \ldots, x_n$ be given (as in Theorem 5) by*

$$(Uf)(y) = \sum_{i=1}^{n} \mu_i \phi(x_i y) + p(y) \qquad (y \in \mathcal{S}^{s-1})$$

where $\mu_1, \ldots, \mu_n \in \mathbb{R}$, $p \in \Pi_{k-1}(\mathcal{S}^{s-1})$ and $\phi \in C^{(v)}[1 - h_0, 1]$, for some h_0 in $(0, 1)$. Assume that $1 - x_i x_j < h < h_0$ whenever $\alpha_i \alpha_j \neq 0$ and $i, j = 1, \ldots, n$. Assume that $1 - xx_i < h < h_0$ whenever $\alpha_i \neq 0$ and $i = 1, \ldots, n$. Then there exists a number C independent of x, h and f such that

$$|f(x) - (Uf)(x)| \leq C h^{v/2} \left(\sum_{i=1}^{n} |\alpha_i| \left(\sum_{j=1}^{n} |\alpha_j| + 2 \right) \right)^{1/2} |f|_\beta \qquad (f \in X_\beta)$$

Proof. Since $v \geq k$, it follows that $p(x) = \sum_{i=1}^{n} \alpha_i p_i(x)$ for all $p \in \Pi_{k-1}(\mathcal{S}^{s-1})$. Hence the error estimate of Lemma 5 applies. By expanding ϕ as a Taylor series about 1, one concludes that there exists ξ_{ij} and ξ_i in the interval $(1 - h_0, 1)$ such that

$$\sum_{i,j=1}^{n} \alpha_i \alpha_j \phi(x_i x_j) - 2 \sum_{i=1}^{n} \alpha_i \phi(x x_i)$$

$$= \sum_{i,j=1}^{n} \alpha_i \alpha_j \left\{ \sum_{\sigma=0}^{\nu-1} \frac{(1-x_i x_j)^{\sigma}}{\sigma!} \phi^{(\sigma)}(1) + \frac{(1-x_i x_j)^{\nu}}{\nu!} \phi^{(\nu)}(\xi_{ij}) \right\}$$

$$- 2 \sum_{i=1}^{n} \alpha_i \left\{ \sum_{\sigma=0}^{\nu-1} \frac{(1-x x_i)^{\sigma}}{\sigma!} \phi^{(\sigma)}(1) + \frac{(1-x x_i)^{\nu}}{\nu!} \phi^{(\nu)}(\xi_i) \right\}$$

Set $p_\sigma(y) = \sum_{j=1}^{n} \alpha_j (1 - x_j y)^{\sigma}$ and $q_\sigma(y) = (1 - xy)^{\sigma}$ for $y \in \mathbb{S}^{s-1}$ and $\sigma = 0, \dots, n-1$. Then, since p_σ and q_σ belong to $\Pi_{\nu-1}$ for $\sigma = 0, \dots, \nu-1$, we have

$$\sum_{i,j=1}^{n} \alpha_i \alpha_j \phi(x_i x_j) - 2 \sum_{i=1}^{n} \alpha_i \phi(x x_i)$$

$$= \sum_{\sigma=0}^{\nu-1} \frac{\phi^{(\sigma)}(1)}{\sigma!} \left(\sum_{i=1}^{n} \alpha_i [p_\sigma(x_i) - 2 q_\sigma(x_i)] \right)$$

$$+ \sum_{i=1}^{n} \alpha_i \left(\sum_{j=1}^{n} \alpha_j \frac{(1-x_i x_j)^{\nu}}{\nu!} \phi^{(\nu)}(\xi_{ij}) - 2 \frac{(1-x_i x)^{\nu}}{\nu!} \phi^{(\nu)}(\xi_i) \right)$$

$$= \sum_{\sigma=0}^{\nu-1} \frac{\phi^{(\sigma)}(1)}{\sigma!} \left(p_\sigma(x) - 2 q_\sigma(x) \right)$$

$$+ \sum_{i=1}^{n} \alpha_i \left(\sum_{j=1}^{n} \alpha_j \frac{(1-x_i x_j)^{\nu}}{\nu!} \phi^{(\nu)}(\xi_{ij}) - 2 \frac{(1-x_i x)^{\nu}}{\nu!} \phi^{(\nu)}(\xi_i) \right)$$

Note that if $\sigma \neq 0$, then $q_\sigma(x) = 0$ for all $x \in \mathbb{S}^{s-1}$, whereas $q_0(x) = 1$. Similarly,

$$p_\sigma(x) = \sum_{i=1}^{n} \alpha_j (1 - x_j x)^{\sigma} = \sum_{j=1}^{n} \alpha_j q_\sigma(x_j) = q_\sigma(x)$$

Hence $p_\sigma(x) = 1$ if $\sigma = 0$ and is zero otherwise. Therefore,

$$\left| \phi(1) - 2 \sum_{i=1}^{n} \alpha_i \phi(x x_i) + \sum_{i,j=1}^{n} \alpha_i \alpha_j \phi(x_i x_j) \right|$$

$$= \left| \phi(1) - \phi(1) + \sum_{i=1}^{n} \alpha_i \left(\sum_{j=1}^{n} \alpha_j \frac{(1-x_i x_j)^{\nu}}{\nu!} \phi^{(\nu)}(\xi_{ij}) - 2 \frac{(1-x_i x)^{\nu}}{\nu!} \phi^{(\nu)}(\xi_i) \right) \right|$$

$$\leq \max_{z \in [1-h_0,\, 1]} |\phi^{(\nu)}(z)| \frac{h^{\nu}}{\nu!} \left| \sum_{i=1}^{n} |\alpha_i| \left(\sum_{j=1}^{n} |\alpha_j| + 2 \right) \right|$$

The desired result is established by putting

$$C = \frac{1}{\nu!} \max_{t \in [1-h_0, 1]} |\phi^{(\nu)}(t)| \qquad \blacksquare$$

Although the number C in Theorem 6 is independent of h, it is by no means clear that the quantity $\sum_{i=1}^{m} |\alpha_i|$ enjoys the same property. This is in fact the case, but the proof is quite difficult. A proof can be found in [GoL]. However, one special case of Theorem 6 is elementary.

COROLLARY 1. *Let k be 0 or 1. Let the minimal norm interpolant to an element f in X on $x_1, \ldots, x_n$ be given by*

$$(Uf)(y) = \sum_{i=1}^{n} \lambda_i \phi(yx_i) + p(y) \qquad (y \in \mathcal{S}^{s-1})$$

where $\lambda_1, \ldots, \lambda_n \in \mathbb{R}, p \in \Pi_{k-1}(\mathcal{S}^{s-1})$ and $\phi \in C^{(1)}[1 - h_0, 1]$ for some $0 < h_0 < 1$. Suppose $1 - x_1 x < h < h_0$. Then there exists a number C independent of h, f, and x such that

$$|f(x) - (Uf)(x)| \le C\sqrt{h}|f|_\beta$$

Proof. Take $\nu = 1$ in Theorem 6. Then $p(x) = p(x_1)$ for all $p \in \Pi_0(\mathcal{S}^{s-1})$, and so we may take $\alpha_1 = 1$ and $\alpha_2 = \cdots = \alpha_n = 0$. The error estimate of Theorem 6 shows that there is a constant C' such that

$$|f(x) - (Uf)(x)| \le \sqrt{3}C'\sqrt{h}|f|_\beta \qquad (f \in X_\beta)$$

Taking $C = \sqrt{3}C'$ completes the proof. $\qquad \blacksquare$

A very important question is whether the minimal norm interpolant converges to f as the interpolation points fill out $\mathcal{S}^{s-1}$ in some sense. The natural measure of distance between two points y and z in $\mathcal{S}^{s-1}$ is the geodesic distance $d(y, z) = \text{Arccos}(yz)$. This definition coincides with that of Equation (2) in Chapter 17. Given a set $\mathcal{A} \subset \mathcal{S}^{s-1}$ and a point $x \in \mathcal{S}^{s-1}$, the quantity $\text{dist}(x, \mathcal{A})$ is defined in the usual way as $\inf\{d(x, a) : a \in \mathcal{A}\}$. The density of $\mathcal{A}$ in $\mathcal{S}^{s-1}$ is then defined by the number $\sup\{\text{dist}(x, \mathcal{A}) : x \in \mathcal{S}^{s-1}\}$.

COROLLARY 2. *Let k be 0 or 1. There exists a positive number δ such that for any set of points $\mathcal{A} = \{x_1, \ldots, x_n\}$ in $\mathcal{S}^{s-1}$ of density less than δ, the minimal norm interpolant to f on $\mathcal{A}$ is of the form*

$$(Uf)(y) = \sum_{i=1}^{n} \lambda_i \phi(yx_i) + p(y) \qquad (y \in \mathcal{S}^{s-1})$$

where $\lambda_i \in \mathbb{R}$ and $p \in \Pi_{k-1}(\mathcal{S}^{s-1})$. If $\phi \in C^{(1)}[1 - h, 1]$ for some h in the interval $(0, 1)$, then we may take $\delta = \sqrt{2h}$, and the operator U will satisfy $\|f - Uf\|_\infty \le Ch|f|_\beta$ for all f in X_β. Here C depends only on h.

Proof. The minimal norm interpolant has the form given in the statement of the corollary whenever $\mathcal{A}$ is a total set of points with respect to $\Pi_{k-1}(S^{s-1})$. (This is the conclusion of Theorem 5.) Since $k \leq 1$, the set $\Pi_{k-1}(S^{s-1})$ is either the set consisting solely of the zero polynomial, or $\Pi_0(S^{s-1})$. Thus the requirement that $\mathcal{A}$ be total is met as long as $\mathcal{A}$ is nonempty. For the second part, take $\delta = \sqrt{2h}$. Since $\mathcal{A}$ is of density δ in S^{s-1}, there exists $a \in \mathcal{A}$ such that $d(x, a) < \delta$. This means that $\text{Arccos}(xa) < \delta$. Thus

$$1 - xa = 1 - \cos\big(\text{Arccos}(xa)\big) < 1 - \cos(\delta) < 1 - \left(1 - \frac{\delta^2}{2}\right) = \frac{\delta^2}{2}$$

Since $\delta = \sqrt{2h}$, it follows that $\delta^2/2 = h$. Corollary 1 now shows that there exists C' such that

$$\big| f(x) - (Uf)(x)\big| \leq C'\sqrt{\delta^2/2}\,|f|_k \qquad (f \in X_\beta)$$

Because x was chosen only to lie in S^{s-1}, the choice $c = C'/\sqrt{2}$ completes the proof. ∎

Problems

1. The following fact was used several times in this chapter: If $L : X_\beta \rightarrow N$ is the orthogonal projection and if f is an element of X having the form $f = \sum_{i=0}^{\infty} \sum_{j=1}^{d_i} a_{ij} Y_j^{(i)}$, then $Lf = \sum_{i=0}^{\infty} \sum_{j=1}^{d_i} a_{ij} L Y_j^{(i)}$. Does this last series converge? Is its sum equal to Lf?

2. Show that for each $x \in S^{s-1}$, the function $f = \sum_{i=k}^{\infty} \beta_k^{-2} \sum_{j=1}^{d_i} Y_j^{(k)}(x) Y_j^{(k)}$ is in X_β, provided that $\sum_{i=k}^{\infty} \beta_k^{-2} d_k < \infty$.

3. Show that p_k and q_k, as defined in the proof of Theorem 6, lie in $\Pi_{k-1}(S^{s-1})$.

4. Prove that there exists a positive h (depending on k and s) such that every set of density less than h in S^{s-1} is total with respect to $\Pi_{k-1}(S^{s-1})$.

References

[Du]   Duchon, J. "Splines minimizing rotation-invariant semi-norms in Sobolev spaces." In *Constructive Theory of Functions of Several Variables*, ed. by W. Schempp and K. Zeller. Lecture Notes in Math. vol. 571, Springer-Verlag, New York, 1976, pp. 85–100.

[Free]   Freeden, W. "Spherical spline interpolation." *J. Comput. Appl. Math 11* (1984), 367–375.

[FGS]   Freeden, W., T. Gervens, and M. Schreiner. *Constructive Approximation on the Sphere.* Oxford Univ. Press, 1998.

[GoL]   Golitschek, M., and W. Light. "Interpolation by polynomials and radial basis functions on spheres." *Constr. Approx.* (to appear).

[LLRS]   Levesley, J., W. Light, D. Ragozin, and Xingping Sun. "Variational theory for interpolation on spheres." In *IDOMAT Conference 1998*.

[Mu]   Müller, Claus. *Spherical Harmonics.* Lecture Notes in Mathematics, vol. 17. Springer-Verlag, Berlin, 1966.

[See]   Seeley, R. T. "Spherical harmonics." *Amer. Math. Monthly 73* (Part II), (1966), 115–121.

[SW]   Stein, E. M., and G. Weiss. *Introduction to Fourier Analysis on Euclidean Spaces.* Princeton Univ. Press, Princeton, NJ, 1971.

[Wah1]   Wahba, G. "Spline interpolation and smoothing on the sphere." *Society for Industrial and Applied Mathematics J. Sci. Stat. Comput. 2* (1981), 5–16. Ibid. 4 (1982), 385–386.

33

Box Splines

There are several ways of generalizing univariate splines to a multivariate setting. One of these procedures leads to "box splines," which are the multivariate analogue of B-splines (in one variable). Box splines were introduced by de Boor and DeVore in [BD]. The development in this chapter is drawn from de Boor, Höllig, and Riemenschneider [BHR], Chui [Chu1], and Höllig [Ho].

We work with functions of s real variables, and the box splines will be defined on $\mathbb{R}^s$. Let A be an $s \times n$ matrix, where $n \geq s$. The columns of A are labeled as $d_1, d_2, \ldots, d_n$ so that

$$A = [d_1, d_2, \ldots, d_n]$$

These columns are sometimes called "direction vectors," and A would then be referred to as the "direction set." As an ordered set, A may contain repetitions. It is always assumed, however, that $n \geq s$ and that $[d_1, d_2, \cdots d_s]$ is linearly independent (hence, a basis for $\mathbb{R}^s$). For each m such that $1 \leq m \leq n$, we write

(1) $$A_m = [d_1, d_2, \ldots, d_m]$$

Thus, $A_n = A$. We can interpret the vectors $d_1, d_2, \ldots, d_m$ as the columns of an $s \times m$ matrix, and we freely use this idea, referring to A_m as a matrix or as an ordered set. In particular, A_s is an $s \times s$ nonsingular matrix. Its determinant is denoted by det (A_s).

For each natural number m we let Q_m be the cube $[0, 1)^m$. Thus, the elements of Q_m are m-tuples $(\lambda_1, \lambda_2, \ldots, \lambda_m)$ in which each component satisfies the inequality $0 \leq \lambda_i < 1$. The set $A_m(Q_m)$ is the image of Q_m under the linear map A_m. We have

$$A_m(Q_m) = \left\{ \sum_{i=1}^{m} \lambda_i d_i \ : \ 0 \leq \lambda_i < 1 \right\}$$

Figure 33.1 *Sets $A_m(Q_m)$*

Figure 33.1 shows $A_m(Q_m)$ for these three matrices:

$$\begin{bmatrix} 1 & 0 \\ 0 & 1 \end{bmatrix} \quad \begin{bmatrix} 1 & 0 & 0 \\ 0 & 1 & 1 \end{bmatrix} \quad \begin{bmatrix} 1 & 0 & 1 \\ 0 & 1 & 1 \end{bmatrix}$$

The reader can think of $A_m(Q_m)$ as a "box" associated with the matrix A_m.

Assume that the columns of A have been ordered so that $\det(A_s) > 0$. The **box spline** associated with A_s is the function

(2)
$$B(x \mid A_s) = \begin{cases} (\det(A_s))^{-1} & \text{if } x \in A_s(Q_s) \\ 0 & \text{otherwise} \end{cases}$$

The box splines associated with the direction sets $A_{s+1}, A_{s+2}, \ldots, A_n$ are defined recursively by the equation

(3) $$B(x \mid A_m) = \int_0^1 B(x - td_m \mid A_{m-1})\, dt \qquad (m = s+1, s+2, \ldots, n)$$

Notice that $B(\cdot \mid A_s)$ is a scalar multiple of the characteristic function of the box $A_s(Q_s)$.

Example 1. Let $s = 1$ and $A = [1, 1, \ldots, 1]$. Then $A_1(Q_1)$ is the interval $[0, 1)$, and $B(\cdot \mid A_1)$ is its characteristic function. It is a spline of degree 0 having knots 0 and 1. The next box spline is

$$B(x \mid A_2) = \int_0^1 B(x - t \mid A_1)\, dt \qquad (x \in \mathbb{R})$$

This is a piecewise linear function supported on the interval $[0, 2]$ and having knots at 0, 1, and 2. One can continue in this manner to see that $B(\cdot \mid A_n)$ is the B-spline of degree $n - 1$ based on the knots $0, 1, \ldots, n$. (See Problem 1.)

Example 2. Let $s = 2$ and $A = [e_1, e_2, e_2]$, where $e_1 = (1, 0)$ and $e_2 = (0, 1)$. In this case, the box spline $B(\cdot \mid A_2)$ is the characteristic function of $A_2(Q_2)$, and this box is the square $[0, 1)^2$. An easy calculation leads to

$$B(x \mid A_3) = \int_0^1 B(x - te_2 \mid A_2)\, dt = B_0(\xi_1)B_1(\xi_2)$$

where $x = (\xi_1, \xi_2)$ and B_i denotes the B-spline based on knots $0, 1, \ldots, i + 1$. Thus, $B(\cdot \mid A_3) = B_0 \otimes B_1$, a tensor product of B-splines. The support of $B(\cdot \mid A_3)$ is shown in the middle sketch of Figure 33.1. In each square patch, the box spline is a linear function, and there is continuity across the shared boundary.

Example 3. Let $s = 2$ and $A = [e_1, e_2, e_1 + e_2]$. As in Example 2, $B(\cdot \mid A_2)$ is the characteristic function of the square $[0, 1)^2$. We borrow a result (Theorem 1 below) to the effect that $B(x \mid A_n) = 0$ for $x \notin A_n(Q_n)$. For this example, $A_3(Q_3)$ is shown as the third sketch in Figure 33.1. One verifies the correctness of the figure by writing

$$A_3(Q_3) = \{\lambda_1 e_1 + \lambda_2 e_2 + \lambda_3(e_1 + e_2) \ : \ 0 \le \lambda_i < 1 \, , \, 1 \le i \le 3\}$$

With $\lambda_3 = 0$, we obtain every point in $[0, 1)^2$. From each of these points, we obtain additional points by adding $\lambda_3(e_1 + e_2)$.

The first theorem establishes some elementary properties of box splines.

> **THEOREM 1.** *Let A be an $s \times n$ matrix, where $n \ge s$. Let $d_1, d_2, \ldots, d_n$ be the columns of A. Assume that $\det(A_s) > 0$. The following assertions are valid.*
>
> (i) *The function $B(\cdot \mid A)$ is nonnegative on $\mathbb{R}^s$.*
> (ii) *If $x = \sum_{i=1}^n \lambda_i d_i$, with $0 < \lambda_i < 1$, then $B(x \mid A) > 0$.*
> (iii) *The support of $B(\cdot \mid A)$ is the closure of $A(Q_n)$.*

Proof. (i) From the definition of $B(x \mid A_s)$ in Equation (2) we have $B(x \mid A_s) \ge 0$ at once. Now proceed by induction, using the recursive definition in Equation (3). The integrands in Equation (3) will be nonnegative.

(ii) Again, the definition in Equation (2) shows that $B(x \mid A_s) > 0$ for $x \in A_s(Q_s)$. For an inductive proof, suppose that for some m in the range $\{s, s + 1, \ldots, n - 1\}$ it is known that $B(x \mid A_m) > 0$ when $x = \sum_{i=1}^m \lambda_i d_i$ and $0 < \lambda_i < 1$. To prove the same assertion for $m + 1$, use Equation (3) to write

(4)
$$B(x \mid A_{m+1}) = \int_0^1 B(x - td_{m+1} \mid A_m) \, dt$$

Assume that x is fixed and is of the form $x = \sum_{i=1}^{m+1} \lambda_i d_i$, with $0 < \lambda_i < 1$. Since $m + 1 > s$, we can express d_{m+1} in the form $d_{m+1} = \sum_{i=1}^m \alpha_i d_i$, employing suitable coefficients α_i. Hence,

$$x - td_{m+1} = \sum_{i=1}^m \lambda_i d_i + \lambda_{m+1} d_{m+1} - td_{m+1}$$

$$= \sum_{i=1}^m \lambda_i d_i + (\lambda_{m+1} - t)\sum_{i=1}^m \alpha_i d_i$$

$$= \sum_{i=1}^m [\lambda_i + (\lambda_{m+1} - t)\alpha_i]d_i = \sum_{i=1}^m \lambda_i^*(t)d_i$$

where $\lambda_i^*(t) = \lambda_i + (\lambda_{m+1} - t)\alpha_i$, for $1 \le i \le m$. Notice that $0 < \lambda_i^*(\lambda_{m+1}) < 1$. By the continuity of λ_i^* there is an open interval I contained in $(0, 1)$ such that for t in I we have $0 < \lambda_i^*(t) < 1$ for $1 \le i \le m$. By the induction hypothesis, $B(x - td_{m+1} \mid A_m) > 0$ when $t \in I$. Consequently, the integral in Equation (4) is positive.

(iii) By the definition in Equation (2), $B(x \mid A_s)$ is zero when x is not of the form $x = \sum_{i=1}^{s} \lambda_i d_i$ with $0 \le \lambda_i \le 1$. As an induction hypothesis, assume that for some m in the range $s \le m \le n - 1$ we know that $B(x \mid A_m) = 0$ whenever x is not of the form $\sum_{i=1}^{m} \lambda_i d_i$ $(0 \le \lambda_i \le 1)$. In order to prove the same assertion for $m + 1$, let x be any point not of the form $\sum_{i=1}^{m+1} \lambda_i d_i$ $(0 \le \lambda_i \le 1)$. Since

$$B(x \mid A_{m+1}) = \int_0^1 B(x - t d_{m+1} \mid A_m)\, dt$$

it suffices to prove that the integrand is 0 for t in $[0, 1]$. Suppose on the contrary that for some t_0 in this range we have $B(x - t_0 d_{m+1} \mid A_m) \ne 0$. By the induction hypothesis, $x - t_0 d_{m+1} = \sum_{i=1}^{m} \lambda_i d_i$ for suitable $\lambda_i \in [0, 1]$. This manifestly contradicts our choice of x. $\blacksquare$

In the next result we use the interpretation of A_n as an $s \times n$ matrix, in accordance with remarks made previously.

LEMMA 1. *Suppose A_n is such that $\det(A_s) > 0$ and let $Q_n = [0, 1)^n$. If $f \in C(\mathbb{R}^s)$ then,*

$$\int_{\mathbb{R}^s} B(x \mid A_n) f(x)\, dx = \int_{Q_n} f(A_n y)\, dy$$

Proof. We use induction on n, starting with $n = s$. With the substitution $x = A_s y$, we have $dx = |\det(A_s)|\, dy = \det(A_s)\, dy$ and

$$\int_{\mathbb{R}^s} B(x \mid A_s) f(x)\, dx = \frac{1}{\det(A_s)} \int_{A_s(Q_s)} f(x)\, dx$$

$$= \frac{1}{\det(A_s)} \int_{Q_s} f(A_s y) \det(A_s)\, dy = \int_{Q_s} f(A_s y)\, dy$$

Now suppose that the result is true for some m, where $s \le m < n$. We use the recursive definition in Equation (3) and the Fubini theorem to write

$$\int_{\mathbb{R}^s} B(x \mid A_{m+1}) f(x)\, dx = \int_{\mathbb{R}^s} \int_0^1 B(x - t d_{m+1} \mid A_m) f(x)\, dt\, dx$$

$$= \int_0^1 \int_{\mathbb{R}^s} B(x - t d_{m+1} \mid A_m) f(x)\, dx\, dt$$

$$= \int_0^1 \int_{\mathbb{R}^s} B(u \mid A_m) f(u + t d_{m+1})\, du\, dt$$

By the induction hypothesis,

$$\int_{\mathbb{R}^s} B(x \mid A_{m+1}) f(x)\, dx = \int_0^1 \int_{Q_m} f(A_m u + t d_{m+1})\, du\, dt$$

$$= \int_{Q_{m+1}} f(A_{m+1} y)\, dy$$

In the last step, let $y = (u_1, u_2, \ldots, u_m, t)^T$, and interpret the product of a matrix with a column vector as a linear combination of the columns of the matrix. ∎

THEOREM 2. *Suppose A_n is such that $\det(A_s) > 0$. The Fourier transform of $B(\cdot \mid A_n)$ is given by*

$$B(\cdot \mid A_n)^\wedge(y) = \prod_{j=1}^{n} \exp(-\pi i y d_j) \frac{\sin(\pi y d_j)}{\pi y d_j}$$

Proof. In Lemma 1, let $f(x) = \exp(-2\pi i x y)$. Then

$$B(\cdot \mid A_n)^\wedge(y) = \int_{\mathbb{R}^s} B(x \mid A_n) \exp(-2\pi i x y)\, dx = \int_{Q_n} \exp(-2\pi i A_n u y)\, du$$

$$= \prod_{j=1}^{n} \int_0^1 \exp(-2\pi i u_j d_j y)\, du_j$$

$$= \prod_{j=1}^{n} \left[\frac{\exp(-2\pi i u_j d_j y)}{-2\pi i d_j y} \right]_0^1$$

$$= \prod_{j=1}^{n} \frac{\exp(-2\pi i d_j y) - 1}{-2\pi i d_j y}$$

$$= \prod_{j=1}^{n} \frac{\exp(-\pi i d_j y)[\exp(\pi i d_j y) - \exp(-\pi i d_j y)]}{2\pi i d_j y}$$

$$= \prod_{j=1}^{n} \frac{\exp(-\pi i d_j y)}{\pi d_j y} \sin(\pi d_j y)$$

∎

Note that the the Fourier transform of $B(\cdot \mid A)$ is well-defined, independent of the ordering of the vectors d_j, and is in $L^2(\mathbb{R}^s)$. This gives us the possibility to define $B(\cdot \mid A)$ via its Fourier transform. Such a definition would be independent of the ordering of A, and independent of the requirement that $\det(A_s) > 0$. Problem 11 investigates the effect of this alternative definition, when $\det(A_s) < 0$. We will take advantage of this fact so as to talk about $B(\cdot \mid A)$ where $\det(A_s)$ is not positive, but $\operatorname{span}(A) = \mathbb{R}^s$. In this case, we will compute $B(\cdot \mid A)$ via the recursive definition in Equation (3) by first rearranging A so that $\det(A_s) > 0$. Theorem 2 establishes that the box spline is independent of any rearrangement of columns that may have been made to bring about the condition $\det(A_s) > 0$.

LEMMA 2. *Let f be a bounded, compactly supported function on $\mathbb{R}^s$. Suppose $\hat{f}(v) = 0$ for all $v \in \mathbb{Z}^s \setminus \{0\}$. Then $\sum_{v \in \mathbb{Z}^s} f(x + v) = \hat{f}(0)$ for all $x \in \mathbb{R}^s$.*

Proof. Since f is bounded and has compact support, $f \in L^1(\mathbb{R}^s)$. Thus $\hat{f} \in C(\mathbb{R}^s)$, and so $\hat{f}(v)$ certainly exists for each $v \in \mathbb{Z}^s$. Define F by $F(x) = \sum_{v \in \mathbb{Z}^s} f(x + v)$. This definition is a valid one, since for each $x \in \mathbb{R}^s$ the sum that defines F contains only finitely many nonzero terms. The function F is bounded, and also integer-periodic, that is, $F(x + \mu) = F(x)$ for all $x \in \mathbb{R}^s$ and $\mu \in \mathbb{Z}^s$. Let $Q = [0, 1)^s$. Then

$$\int_Q |F(x)|\, dx = \int_Q \left| \sum_{v \in \mathbb{Z}^s} f(x + v) \right| dx$$

$$\leq \sum_{v \in \mathbb{Z}^s} \int_Q |f(x + v)|\, dx$$

$$= \sum_{v \in \mathbb{Z}^s} \int_{Q+v} |f(y)|\, dy$$

$$= \int_{\mathbb{R}^s} |f(y)|\, dy$$

We infer that $F \in L^1(Q)$. Now F has Fourier coefficients given by

$$c_\mu = \int_Q F(x)\, e^{-2\pi i x \mu}\, dx$$

$$= \sum_{v \in \mathbb{Z}^s} \int_{v+Q} f(y) e^{-2\pi i (y-v)\mu}\, dy$$

$$= \int_{\mathbb{R}^s} f(y) e^{-2\pi i y \mu}\, dy$$

$$= \hat{f}(\mu) \qquad (\mu \in \mathbb{Z}^s)$$

By our hypotheses on $\hat{f}$, $c_\mu = 0$ for all $\mu \in \mathbb{Z}^s$ except possibly $\mu = 0$. Thus $F - c_0 \in L^1(Q)$, and has all Fourier coefficients zero. We conclude, by the same argument as that found at the end of the proof in Lemma 5 of Chapter 35 (page 296), that $F(x) = c_0$ for almost all $x \in Q$. Because F is bounded,

$$F(x) = c_0 = \int_{\mathbb{R}^s} f(y)\, dy = \hat{f}(0)$$

for all $x \in \mathbb{R}^s$. ∎

THEOREM 3. *If A is a nonsingular $s \times n$ matrix having integer entries, then the corresponding box spline satisfies $\sum_{v \in \mathbb{Z}^s} B(x - v \mid A) = 1$.*

Proof. From Theorem 1, $B(\cdot \mid A)$ is bounded and has compact support. Furthermore, by Theorem 2,

$$B(\cdot \mid A)^\wedge(v) = \prod_{j=1}^n \exp(-\pi i v d_j) \frac{\sin(\pi v d_j)}{\pi v d_j} = 0$$

whenever $v \in \mathbb{Z}^s$ and $v \neq 0$. The box spline therefore satisfies the hypotheses of Lemma 2. Hence,

$$\sum_{v \in \mathbb{Z}^s} B(x - v \mid A) = B(\cdot \mid A)^\wedge(0) = 1$$ ∎

LEMMA 3. *Let f be a member of $C^r(\mathbb{R})$ that is a piecewise polynomial of degree at most k having knots $x_1, x_2, \ldots, x_n$. Define $g(t) = \int_t^{t+1} f(x)\, dx$. Then g belongs to $C^{r+1}(\mathbb{R})$ and is a piecewise polynomial of degree at most $k + 1$, having knots $x_1, x_2, \ldots, x_n, x_1 - 1, x_2 - 1, \ldots, x_n - 1$.*

Proof. The elementary constituents of this result are as follows.

1. If $f \in C^r(\mathbb{R})$ and $g(t) = \int_a^t f(x)\, dx$, then $g \in C^{r+1}(\mathbb{R})$. The proof is immediate: $g'(t) = f(t)$, so $g' \in C^r$ and $g \in C^{r+1}$.

2. If $f \in C^r(\mathbb{R})$ and $h(t) = \int_a^{t+b} f(x)\, dx$, then $h \in C^{r+1}$. Again, the proof is obvious, for $h(t) = g(t + b)$, where g is as defined in (1).

3. If $f \in C^r(\mathbb{R})$ and $\phi(t) = \int_{t+a}^{t+b} f(x)\, dx$, then $\phi \in C^{r+1}$. This follows from (2) because ϕ is the difference of two functions having the form of h in (2).

4. If f is a piecewise polynomial of degree at most k having knot set X, then g, as in (1) above, is a piecewise polynomial of degree at most $k + 1$ having knot set X. To verify this let I be an open interval (b, c) containing no knots. Let p be the polynomial such that $f \mid I = p \mid I$. Let $q' = p$, so that $q \in \Pi_{k+1}$. For $t \in I$

$$
g(t) = \int_a^t f(x)\, dx = \int_a^b f(x)\, dx + \int_b^t f(x)\, dx
$$

$$
= C + \int_b^t p(x)\, dx = C + q(t) - q(b)
$$

Thus $g \mid I \in \Pi_{k+1}$.

5. If g and h are piecewise polynomials having knot sets X and Y respectively, then $g + h$ is a piecewise polynomial having knot set $X \cup Y$. ∎

LEMMA 4. *Fixing x in $\mathbb{R}^s$ and $m > s$, we let*

$$
f(t) = B(x - td_m \mid A_{m-1}) \qquad g(t) = B(x - td_m \mid A_m) \qquad (t \in \mathbb{R})
$$

If f is a piecewise polynomial of degree at most k having a knot at t_0, then g is a piecewise polynomial of degree at most $k + 1$ with knots at t_0 and $t_0 - 1$. If $f \in C^r$ then $g \in C^{r+1}$. We interpret C^{-1} to be the space of piecewise continuous functions.

Proof. From the definition in Equation (3),

$$
g(t) = B(x - td_m \mid A_m)
$$

$$
= \int_0^1 B(x - td_m - \tau d_m \mid A_{m-1})\, d\tau
$$

$$
= \int_t^{t+1} B(x - \sigma d_m \mid A_{m-1})\, d\sigma
$$

$$
= \int_t^{t+1} f(\sigma)\, d\sigma
$$

The proof is completed by using Lemma 3. ∎

THEOREM 4. *Let A be an $s \times n$ matrix such that the rank of A_s is s. Let Λ be the set of n-tuples $(\lambda_1, \lambda_2, \ldots, \lambda_n)$ such that $0 \le \lambda_i \le 1$ for each i and at most $s - 1$ of the λ_i are in the open interval $(0, 1)$. Then in each connected component of $\mathbb{R}^s \backslash A(\Lambda)$, the box spline $B(\cdot \mid A)$ is a polynomial of degree at most $n - s$.*

Proof. For each m in $\{s, s + 1, \ldots, n\}$, let Λ_m be the set of m-tuples $(\lambda_1, \ldots, \lambda_m)$ in $[0, 1]^m$ such that at most $s - 1$ of the λ_i satisfy $0 < \lambda_i < 1$. Note that $\Lambda_n = \Lambda$. The proof is by induction on m, starting with $m = s$. The box spline $B(\cdot \mid A_s)$ is a scalar multiple of the characteristic function of $A_s(Q_s)$, where

$$A_s(Q_s) = \left\{ \sum_{i=1}^{s} \lambda_i d_i \ : \ 0 \leq \lambda_i < 1 \right\}$$

However, $A_s(\Lambda_s)$ is the set of points $\sum_{i=1}^{s} \lambda_i d_i$ in which all coefficients satisfy $0 \leq \lambda_i \leq 1$ and at least one λ_j satisfies $\lambda_j(1 - \lambda_j) = 0$. A typical point of $A_s(\Lambda_s)$ will have either the form

$$\sum_{\substack{i=1 \\ i \neq j}}^{s} \lambda_i d_i \qquad \text{or} \qquad d_j + \sum_{\substack{i=1 \\ i \neq j}}^{s} \lambda_i d_i$$

These points form the boundary of $A_s(Q_s)$. The connected components of $\mathbb{R}^s \backslash A_s(\Lambda_s)$ are the interior of $A_s(Q_s)$, where $B(\cdot \mid A_s)$ has a constant nonzero value, and the exterior of $A_s(Q_s)$, where $B(\cdot \mid A_s)$ is zero.

   Now suppose that the theorem has been proved for the integer m, where $s \leq m \leq n - 1$. We consider $y \in \mathbb{R}^s \backslash A_{m+1}(\Lambda_{m+1})$. Observe first that the set $A_m(\Lambda_m) \cup (A_m(\Lambda_m) + d_m)$ consists of all points in $\mathbb{R}^s$ of the form $\sum_{i=1}^{m} \lambda_i d_i + \lambda_{m+1} d_{m+1}$ such that $0 \leq \lambda_i \leq 1$ for $i = 1, 2, \ldots, m$, at most $s - 1$ of these λ_i lie in $(0, 1)$, and $\lambda_{m+1} = 0$ or $\lambda_{m+1} = 1$. This shows that $A_m(\Lambda_m) \cup \left(A_m(\Lambda_m) + d_m\right) \subset A_{m+1}(\Lambda_{m+1})$. Hence $y \notin A_m(\Lambda_m)$ and $y - d_{m+1} \notin A_m(\Lambda_m)$. Therefore, we can find a neighborhood U of y such that U and $U - d_{m+1}$ do not intersect $A_m(\Lambda_m)$. Since $\text{span}(A_s) = \mathbb{R}^s$, we may write $y = \sum_{i=1}^{m+1} \mu_i d_i$ where $\mu_i \in \mathbb{R}$. We can then assume U has the form

$$U = \left\{ \sum_{i=1}^{m+1} v_i d_i \ : \ |v_i - \mu_i| < \varepsilon \text{ for } i = 1, 2, \ldots, m + 1 \right\}$$

for some $\varepsilon > 0$. Let x be a point in U of the form $x = \sum_{i=1}^{m} v_i d_i + \mu_{m+1} d_{m+1}$, where $|v_i - \mu_i| < \varepsilon$ for $i = 1, 2, \ldots, m$. Then $t \mapsto B(x - td_{m+1} \mid A_m)$ has no knots in the ranges $|t| < \varepsilon$ or $|t - 1| < \varepsilon$, and so is a polynomial of degree at most $m - s$ in each of these ranges. An application of Lemma 4 shows that $t \mapsto B(x - td_{m+1} \mid A_{m+1})$ has no knots in $|t| < \varepsilon$, and so in this range is a polynomial of degree at most $m - s + 1$. It follows that for any x in U, the directional derivative of $B(\cdot \mid A_{m+1})$ at x of order $m - s + 2$ in the direction d_{m+1} is zero. Since $B(\cdot \mid A_{m+1})$ is independent of the ordering of the columns of A_{m+1}, we conclude that all directional derivatives of $B(\cdot \mid A_{m+1})$ of order $m - s + 2$ in the directions $d_1, d_2, \ldots, d_{m+1}$ are zero at each point in U. Because these directions span $\mathbb{R}^s$, we conclude that $B(\cdot \mid A_{m+1})$ is a polynomial in U. (See Problem 12.) Lemma 4 also shows that $t \mapsto B(x - td_{m+1} \mid A_{m+1})$ is a piecewise polynomial of degree at most one higher than the degree of the piecewise polynomial $t \mapsto B(x - td_{m+1} \mid A_m)$. The inductive hypothesis now shows that the degree of $B(\cdot \mid A_{m+1})$ in U is at most $m - s + 1$. ∎

Example 4. Let $A = (e_1, e_2, e_1 - e_2)$, where $e_1 = (1, 0)^T$ and $e_2 = (0, 1)^T$. The set $A(\Lambda)$ is shown in Figure 33.2. In the figure, the numbered line segments are all described by $\lambda(e_1 - e_2) + \mu_1 e_1 + \mu_2 e_2$, where $\lambda \in [0, 1]$ and $\mu_1, \mu_2 \in \{0, 1\}$. For example, the segment 2 is of the form $\lambda(e_1 - e_2) + e_1$.

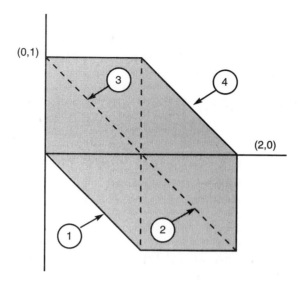

Figure 33.2

LEMMA 5. *Let f be a real-valued function defined on $\mathbb{R}^s$, and let $\{u_1, u_2, \ldots, u_s\}$ be a basis for $\mathbb{R}^s$. For $x \in \mathbb{R}^s$ and $1 \le i \le s$, define $g_{x,i}(t) = f(x + tu_i)$. If the family $\{g_{x,i} : x \in \mathbb{R}, 1 \le i \le s\}$ is equicontinuous at 0, then f is uniformly continuous.*

Proof. If x is any point in $\mathbb{R}^s$, we can write $x = \sum_{i=1}^{s} \lambda_i(x)u_i$, where the λ_i are continuous linear functionals. Define a modulus of continuity:

$$\omega(\delta) = \sup_{x \in \mathbb{R}^s} \ \sup_{1 \le i \le s} \ \sup_{|t| \le \delta} |g_{x,i}(0) - g_{x,i}(t)|$$

By hypothesis, $\omega(\delta) \downarrow 0$ as $\delta \downarrow 0$. Let x and y be any two points in $\mathbb{R}^s$. Write $x - y = \sum_{i=1}^{s} t_i u_i$ where $t_i = \lambda_i(x - y)$. Put $x_0 = x$ and $x_i = x_{i-1} + t_i u_i$ so that $x_s = y$. Then

$$\begin{aligned}
|f(x) - f(y)| &\le |f(x_0) - f(x_1)| + |f(x_1) - f(x_2)| + \cdots + |f(x_{s-1}) - f(x_s)| \\
&\le \omega(|t_1|) + \omega(|t_2|) + \cdots + \omega(|t_s|) \\
&\le \omega(\|\lambda_1\| \|x - y\|) + \omega(\|\lambda_2\| \|x - y\|) + \cdots + \omega(\|\lambda_s\| \|x - y\|) \\
&\le s\omega(c\|x - y\|) \qquad\qquad \blacksquare
\end{aligned}$$

Definition. For any $s \times n$ matrix A of rank s, let

$$\rho(A) = \min \{\#W : W \subset A, \operatorname{span}(A \backslash W) \neq \mathbb{R}^s\}$$

THEOREM 5. *If $\rho(A) \ge k \ge 2$ then $B(\cdot \mid A) \in C^{k-2}(\mathbb{R}^s)$.*

Proof. We use induction on k, starting with $k = 2$. Let A be an $s \times n$ matrix for which $\rho(A) \ge 2$. It is to be proved that $B(\cdot \mid A) \in C(\mathbb{R}^s)$. We write $A \backslash d_i$ instead of the more cumbersome $A \backslash \{d_i\}$. For each i, $\operatorname{span}(A \backslash d_i) = \mathbb{R}^s$ because $\rho(A) \ge 2$. By Lemma 4, the function $t \mapsto B(x - td_i \mid A \backslash d_i)$ is a compactly supported piecewise polynomial, for each $x \in \mathbb{R}^s$ and for $1 \le i \le n$. Define $g_{xi}(t) = B(x - td_i \mid A)$. Then $\{g_{xi} \; : \; x \in \mathbb{R}^s,$ $1 \le i \le n\}$ is equicontinuous at 0. By Lemma 5, $B(\cdot \mid A) \in C(\mathbb{R}^s)$.

Now suppose that the theorem has been proved for $k = 2, 3, 4, \ldots, m$. We then prove it for $k = m + 1$. Let A be a matrix for which $\rho(A) \ge m + 1$. As before, $\text{span}(A \backslash d_i) = \mathbb{R}^s$ for each i. We assert that $\rho(A \backslash d_i) \ge m$. To prove this, suppose that $\rho(A \backslash d_j) \le m - 1$ for some j. There exists a set $W \subset A \backslash d_j$ such that $\#W \le m - 1$ and $\text{span}((A \backslash d_j) \backslash W) \ne \mathbb{R}^s$. Let $U = W \cup \{d_j\}$. Then $\#U \le \#W + 1 \le m$. Since $\text{span}(A \backslash U) \ne \mathbb{R}^s$, we conclude that $\rho(A) \le m$, contrary to our assumptions. Since $\rho(A \backslash d_i) \ge m$, the induction hypothesis implies that $B(\cdot \mid A \backslash d_i) \in C^{m-2}(\mathbb{R}^s)$. It follows that the function $t \mapsto B(x - td_i \mid A \backslash d_i)$ belongs to $C^{m-2}(\mathbb{R})$ for all x and i. By Lemma 4, $B(x - td_i \mid A) \in C^{m-1}(\mathbb{R})$. Hence $D^{m-1}B(x - td_i \mid A) \in C(\mathbb{R})$. (Here, D denotes the operation of differentiation with respect to t.) From this it follows that all partial derivatives $D^\alpha B(\cdot \mid A)$ for $|\alpha| \le m - 1$ are continuous, and that $B(\cdot \mid A) \in C^{m-1}(\mathbb{R}^s)$. The details of this part of the proof are explored in Problem 5. ∎

THEOREM 6. *Let A be an $s \times n$ integer matrix of rank s, such that $\rho(A) \ge 2$. For each positive h, define an operator L_h by*

$$(L_h f)(x) = \sum_{v \in \mathbb{Z}^s} f(vh) B\left(\frac{x}{h} - v \mid A\right) \qquad (x \in \mathbb{R}^s, f \in C(\mathbb{R}^s))$$

There is a constant C, depending only on A, such that

$$\|f - L_h f\| \le Ch \max_{|\alpha|=1} \|D^\alpha f\|_\infty$$

for all $f : \mathbb{R}^s \to \mathbb{R}$ such that $D^\alpha f$ is continuous when $|\alpha| = 1, \alpha \in \mathbb{Z}_+^s$.

Proof. Let S_h be the scaling operator defined by $(S_h f)(x) = f(hx)$. Let L be defined by

$$(Lf)(x) = \sum_{v \in \mathbb{Z}^s} f(v) B(x - v \mid A) \qquad (x \in \mathbb{R}^s)$$

In the notation of Chapter 36, we can write $Lf = f \star B(\cdot \mid A)$. Now apply Theorem 2 of Chapter 36 (page 318) to complete the proof. ∎

To illustrate Theorem 5, we describe the box spline first constructed by Zwart [Zw]. Let $A = \{e_1, e_2, e_1 + e_2, e_1 - e_2\}$, where $e_1 = (1, 0)^T$ and $e_2 = (0, 1)^T$. By Theorem 1, the support of $B(\cdot \mid A)$ is

$$A(Q_4) = \{\lambda_1 e_1 + \lambda_2 e_2 + \lambda_3 (e_1 + e_2) + \lambda_4 (e_1 - e_2) \; : \; 0 \le \lambda_i \le 1\}$$
$$= \{(\lambda_1 + \lambda_3 + \lambda_4, \lambda_2 + \lambda_3 - \lambda_4) \; : \; 0 \le \lambda_i \le 1\}$$

The set $A(Q_4)$ is shown in Figure 33.3. There are 30 line segments on which the knots lie. These can be described by 4 expressions, in which $0 \le \lambda \le 1$ and $\mu_i \in \{0, 1\}$.

$$\lambda e_1 + \mu_1 e_2 + \mu_2 (e_1 + e_2) + \mu_3 (e_1 - e_2)$$
$$\lambda e_2 + \mu_1 e_1 + \mu_2 (e_1 + e_2) + \mu_3 (e_1 - e_2)$$
$$\lambda (e_1 + e_2) + \mu_1 e_1 + \mu_2 e_2 + \mu_3 (e_1 - e_2)$$
$$\lambda (e_1 - e_2) + \mu_1 e_1 + \mu_2 e_2 + \mu_3 (e_1 + e_2)$$

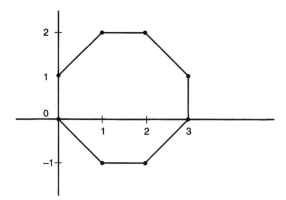

Figure 33.3

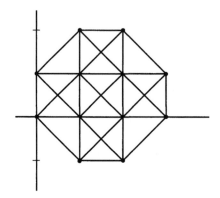

Figure 33.4

The knot lines partition $\mathbb{R}^2$ as shown in the Figure 33.4. The degree of the polynomial pieces in the triangles is $\#A - s = 4 - 2 = 2$. The continuity class of $B(\cdot \mid A)$ is $C^1(\mathbb{R}^2)$. This follows from Theorem 5 and the observation that $\rho(A) = 3$. (Any pair of elements in A spans $\mathbb{R}^2$, and thus $\rho(A) \geq 3$.)

This box spline was also studied by Powell and Sabin [PS].

Problems

1. Continue the work begun in Example 1. Prove that the box spline $x \mapsto B(x \mid A_n)$ has support $[0, n]$, that it is a piecewise polynomial of degree $n - 1$, and that its knots are $0, 1, \ldots, n$.

2. Let A be the $2 \times 2n$ matrix whose columns are $e_1, e_2, e_1, e_2, \ldots, e_1, e_2$. Show that the box spline $B(x \mid A)$ is the tensor product $B_{n-1} \otimes B_{n-1}$. That is, if $x = (\xi_1, \xi_2) \in \mathbb{R}^2$ then $B(x, A) = B_{n-1}(\xi_1)B_{n-1}(\xi_2)$.

3. Let $s = 1$, $n = 2$, $A = [1, 2]$. Prove that on the intervals $(-\infty, 0)$, $(0, 1)$, $(1, 2)$, $(2, 3)$, $(3, \infty)$, $B(x \mid A_2)$ has these values: 0, x, 1, $2 - x$, 0.

4. Prove that $B(\cdot \mid A)$ is a piecewise polynomial of exact degree $n - s$ on its support. *Hint:* Suitable changes will be needed in Lemmas 2 and 3.

5. Let D_v be the directional derivative in direction v. Prove that if A is a nonsingular matrix then
$$D_v(f \circ A) = (D_{Av} f) \circ A$$

6. (Continuation) Prove that $D_u D_v (f \circ A) = (D_{Au} D_{Av} f) \circ A$.

7. (Continuation) Prove that $D_{u_1}^{\alpha_1} \cdots D_{u_n}^{\alpha_n}(f \circ A) = \left(D_{Au_1}^{\alpha_1} \cdots D_{Au_n}^{\alpha_n} f\right) \circ A$.

8. Prove that if $\tilde{A}$ is an $s \times m$ submatrix of A then $\rho(\tilde{A}) \leq \rho(A)$.

9. Draw the set $A(\Lambda)$ for the box splines $B(\cdot \mid A)$ on $\mathbb{R}^2$, where
 a. $A = \{e_1, e_2, e_2, e_2\}$
 b. $A = \{e_1, e_2, e_1 + e_2, e_2 - e_1\}$
 c. $A = \{e_1, e_1 + e_2, e_1 - 2e_2\}$

10. Let $e_1, \ldots, e_s$ be the usual basis for $\mathbb{R}^s$. Let A be an $s \times n$ matrix with each column being one of these basis vectors. Show that $B(\cdot \mid A)$ is a tensor product of univariate B-splines.

11. Suppose A is an $s \times s$ matrix with $\det(A) < 0$. Define $B(\cdot \mid A_s)$ by means of its Fourier transform as given in Theorem 2. Show that $B(\cdot \mid A_s)$ must be $1/|\det(A_s)|$ times the characteristic function of $A_s(Q_s)$, and is therefore a nonnegative function.

12. Let $d_1, d_2, \ldots, d_s$ span $\mathbb{R}^s$. Suppose that for some $k \in \mathbb{N}$ the directional derivatives $D_{d_i}^k f$ are 0 for $i = 1, 2, \ldots, s$. Show that f is a polynomial of coordinate degree at most $k - 1$. A polynomial p has coordinate degree $k - 1$ if it has the form $p(x) = \sum_{\alpha \in \Omega} c_\alpha x^\alpha$, where
$$\Omega = \{\alpha \in \mathbb{Z}^s \ : \ \alpha = (\alpha_1, \alpha_2, \ldots, \alpha_s), \text{ and } 0 \leq \alpha_i \leq k - 1, i = 1, 2, \ldots, s\}$$

References

[BHR]   de Boor, C., K. Höllig, and S. Riemenschneider. *Box Splines*. Springer-Verlag, New York, 1993.

[BD]   de Boor, C., and R. DeVore. "Approximation by smooth multivariate splines." *Trans. Amer. Math. Soc.* 276 (1983), 775–788.

[Chu1]   Chui, Charles K. *Multivariate Splines*. SIAM, Philadelphia, 1988.

[DM3]   Dahmen, W., and C. Micchelli. "Translates of multivariate splines." *Linear Alg. Appl. 52* (1983), 217–234.

[Ho]   Höllig, K. "A remark on multivariate B-splines." *J. Approx. Theory 33* (1982), 119–125.

[Nea]   Neamtu, M. *A contribution to the theory and practice of multivariate splines.* Unpublished doctoral dissertation, University of Twente, 1991.

[PSa]   Powell, M. J. D., and M. A. Sabin. "Piecewise quadratic approximation on triangles." *Asso. for Comp. Mach. Trans. on Math. Software 3* (1977), 316–325.

[Zw]   Zwart, P. B. "Multivariate splines with non-degenerate partitions." *SIAM J. Numer. Anal. 10* (1973), 665–673.

34
Wavelets, I

We are going to interpret wavelets as a tool for signal analysis. Classical Fourier theory is also available for that purpose, and we shall begin our discussion by reviewing the basics of that topic. This will help in understanding the objectives of wavelet theory.

Fourier theory deals with functions f in the space $L^2[0, 1]$. These functions are extended to all of $\mathbb{R}$ by integer-periodicity; they are Lebesgue-measurable and square-integrable. Formally,

1. $$f(x + n) = f(x) \qquad (x \in \mathbb{R}, n \in \mathbb{Z})$$

2. $$\int_0^1 |f(x)|^2 \, dx < \infty$$

As a matter of precision, f stands for an equivalence class of such functions, two functions being declared equivalent to each other if they differ only on a set of measure zero.

The **Fourier coefficients** of such a function f are defined by the equation

(1) $$c_k = c_k(f) = \int_0^1 f(x) e^{-2\pi i k x} \, dx \qquad (k \in \mathbb{Z})$$

The **Fourier series** of f is then

(2) $$f \sim \sum_{k=-\infty}^{\infty} c_k e^{2\pi i k x} \qquad (x \in \mathbb{R})$$

In (2) we have written $\sim$ simply to indicate that there is a formal series associated with f. The sense in which the series represents f is not as one might expect, but rather

(3) $$\lim_{n,m\to\infty} \int_0^1 \left| f(x) - \sum_{k=-m}^{n} c_k e^{2\pi i k x} \right|^2 dx = 0$$

It is convenient to define $e_k(x) = e^{2\pi ikx}$. The Fourier series is an orthonormal expansion, since

(4)
$$\langle e_k, e_m \rangle = \int_0^1 e_k(t)\overline{e_m(t)}\, dt = \delta_{km}$$

Parseval's identity shows how the "energy" of the function f is encoded in its Fourier series:

(5)
$$\left\{ \int_0^1 |f(x)|^2 \, dx \right\}^{1/2} = \left\{ \sum_{k=-\infty}^{\infty} |c_k|^2 \right\}^{1/2}$$

The function e_k is "rapidly varying" on $[0, 1]$ if $|k|$ is large, and "slowly varying" if k is small. It is usual to call k the "frequency" and to refer to high and low frequencies instead of rapid and slow variation. Thus, in the Fourier expansion (2) the decomposition of f into high and low frequencies is very clear.

In signal processing, the variable x usually represents time, and the signal f could be encoding the playing of an orchestra, for example. If over a relatively short interval of time the orchestra plays something different, then every coefficient in the Fourier series expansion must be recomputed. It is not possible for a *local* change in the function (that is, one which takes place over a short time interval) to be reflected in a change of only a small number of Fourier coefficients. This is one drawback of the Fourier theory that wavelet theory seeks to overcome.

Observe that the Fourier series can be constructed from the single function $e(x) = e^{2\pi ix}$. Each function e_k is obtained from e by "dilation:" $e_k(x) = e(kx)$. Hence, the Fourier expansion in Equation (2) if truncated is seen to be a linear combination of integer dilates of the single function e. This function is often called a "sinusoidal wave," and the term $c_k e_k$ in the series (2) is responsible for frequency information at the frequency k.

Wavelet theory also begins with a single function, ψ. However, there is no longer a restriction to periodic signals. Accordingly, we assume that $\psi \in L^2(\mathbb{R})$, so that ψ is Lebesgue measurable and $\int_{-\infty}^{\infty} |\psi(t)|^2 \, dt < \infty$.

We also give up, as being too restrictive, the idea of using only linear combinations of dilates of ψ to recover signals. Instead, both dilates and translates are used. The set of functions employed is given by the definition

(6)
$$\psi_{kj}(x) = 2^{k/2} \psi(2^k x - j) \qquad (j, k \in \mathbb{Z})$$

An ideal situation would be to have an orthonormal set in Equation (6); that is,

(7)
$$\langle \psi_{kj}, \psi_{ni} \rangle = \int_{-\infty}^{\infty} \psi_{kj}(x)\psi_{ni}(x)\, dx = \delta_{kn}\delta_{ji} \qquad (k, n, i, j \in \mathbb{Z})$$

The coefficient $2^{k/2}$ in Equation (6) has been inserted to help achieve this. The following lemma shows how inner products behave in this formalism.

LEMMA 1. *Let f and g be two functions in $L^2(\mathbb{R})$. Then, for all n, m, i, j in $\mathbb{Z}$,*

(8)
$$\langle f_{ni}, g_{nj} \rangle = \langle f_{mi}, g_{mj} \rangle$$

Proof. We have

$$\langle f_{ni}, g_{nj}\rangle = \int 2^{n/2} f(2^n t - i) 2^{n/2} g(2^n t - j)\, dt$$

In this integral, make the substitution $t = 2^{m-n}x$, getting

$$\int 2^{n/2} f(2^m x - i) 2^{n/2} g(2^m x - j) 2^{m-n}\, dx = \int 2^{m/2} f(2^m x - i) 2^{m/2} g(2^m x - j)\, dx$$

$$= \langle f_{mi}, g_{mi}\rangle \qquad \blacksquare$$

Thus if ψ has unit norm, then so does each ψ_{kj}.

Definition 1. An **orthonormal wavelet** is a function ψ in $L^2(\mathbb{R})$ such that the family $\{\psi_{kj} : k, j \in \mathbb{Z}\}$, as defined in Equation (6), is orthonormal, as in Equation (7).

   In the wavelet literature, more general definitions will be encountered. In particular, there are weaker definitions in which orthonormality is relaxed. There are also wavelets in spaces $L^2(\mathbb{R}^s)$ for $s > 1$.
   The wavelet series for a function f in $L^2(\mathbb{R})$, based on a given orthonormal wavelet ψ, is

$$(9) \qquad f \sim \sum_{k \in \mathbb{Z}} \sum_{j \in \mathbb{Z}} c_{kj} \psi_{kj}$$

in which the coefficients are given by

$$(10) \qquad c_{kj} = \langle f, \psi_{kj}\rangle = \int_{-\infty}^{\infty} f(x) \psi_{kj}(x)\, dx \qquad (k, j \in \mathbb{Z})$$

As in the classical Fourier series, the symbol $\sim$ in (9) signifies only that each f in $L^2(\mathbb{R})$ has associated with it a formal series. We hope to arrange matters so that the double series converges to f in the metric of $L^2(\mathbb{R})$. In that case, the orthonormal family is an orthonormal base for $L^2(\mathbb{R})$. Of course, it may turn out that too much is being demanded of ψ. But surely, dear reader, you do not expect *that*. As a matter of fact, an example is immediately at hand:

Example 1. The **Haar function** is defined by the equation

$$(11) \qquad \mathfrak{h}(x) = \begin{cases} 1 & \text{if } 0 \le x < 1/2 \\ -1 & \text{if } 1/2 \le x < 1 \\ 0 & \text{otherwise} \end{cases}$$

THEOREM 1. *The Haar function defined in Equation (11) has the orthonormality property expressed in Equation (7), and is therefore an orthonormal wavelet.*

Proof. Notice that $\mathfrak{h}$ has norm 1, and that, by Equation (8), the same is true of each $\mathfrak{h}_{kj}$. By the same change of variable used in arriving at Equation (8), we have

$$(12) \qquad \int \mathfrak{h}_{kj}(x)\mathfrak{h}_{ni}(x)\,dx = 2^{-k/2}2^{n/2}\int \mathfrak{h}(y)\mathfrak{h}(2^{n-k}(y+j)-i)\,dy$$

If $k = n$ in this equation, one concludes that

$$\int \mathfrak{h}_{kj}(x)\mathfrak{h}_{ki}(x)\,dx = \int \mathfrak{h}(y)\mathfrak{h}(y+j-i)\,dy = \delta_{0,\,j-i} = \delta_{ij}$$

Here we have noted that $\mathfrak{h}(y) \neq 0$ if and only if $0 \leq y < 1$, while $\mathfrak{h}(y+j-i) \neq 0$ if and only if $i-j \leq y < i-j+1$. These intervals are disjoint from each other unless $i = j$.

To complete the proof, we can assume $k \neq n$. By symmetry it suffices to consider the case $n > k$. Let $p = n - k$. By Equation (12), we see that we must prove

$$\int \mathfrak{h}(y)\mathfrak{h}(2^p y + q)\,dy = 0 \qquad (p \in \mathbb{N}, \quad q \in \mathbb{Z}, \quad p > 0)$$

Taking into account Equation (11), we see that it suffices to prove

$$\int_0^{1/2} \mathfrak{h}(2^p y + q)\,dy - \int_{1/2}^1 \mathfrak{h}(2^p y + q)\,dy = 0$$

With elementary changes of variables, we arrive at the following equation to be proved:

$$(13) \qquad \int_a^b \mathfrak{h}(x)\,dx - \int_b^c \mathfrak{h}(x)\,dx = 0 \qquad (a = q, \quad b = a + 2^{p-1}, \quad c = a + 2^p)$$

Surprisingly, both integrals in Equation (13) are zero. Let us examine the first. It will certainly be zero if $a \geq 1$ or $b \leq 0$ because in these cases the interval $[a, b]$ does not overlap the support of $\mathfrak{h}$. Assume therefore that $a < 1$ and $b > 0$. Since $p > 0$, both a and b are integers. Hence $a \leq 0$ and $b \geq 1$. Thus $[a, b]$ contains the support of $\mathfrak{h}$, and

$$\int_a^b \mathfrak{h}(x)\,dx = \int_0^1 \mathfrak{h}(x)\,dx = 0$$

The second integral in Equation (13) is treated in the same way. ∎

Further properties of the Haar wavelet will be developed later in this chapter. It stands as an important example, illustrating many features of the general theory. However, its discontinuous nature is a serious defect in many applications. Thus, we embark on a quest for further examples.

If $\phi \in L^2(\mathbb{R})$ and if the family $\{\phi_{0j} : j \in \mathbb{Z}\}$ is orthonormal, then $\{\phi_{nj} : j \in \mathbb{Z}\}$ is orthonormal, for each $n \in \mathbb{Z}$, by Lemma 1. We put

$$(14) \qquad V_n = \text{closure}\Big(\text{span}\{\phi_{nj} : j \in \mathbb{Z}\}\Big) \qquad (n \in \mathbb{Z})$$

where the closure is in the topology of $L^2(\mathbb{R})$. We have, by Lemma 2, below,

$$(15) \qquad V_n = \left\{ \sum_{j \in \mathbb{Z}} c_j \phi_{nj} \; : \; c \in \ell^2(\mathbb{Z}) \right\}$$

LEMMA 2. *Let $\{u_n\}_1^\infty$ be an orthonormal sequence in a Hilbert space. The closure of the linear span of $\{u_n\}_1^\infty$ is the set of all points $\sum_{n=1}^\infty c_n u_n$, where c runs over ℓ^2.*

Proof. Let V be the closure of the linear span of $\{u_n\}$, and let U be the set of all $\sum c_n u_n$ with $c \in \ell^2$. If $c \in \ell^2$, the series $\sum c_n u_n$ converges, because its partial sums form a Cauchy sequence:

$$\left\| \sum_{k=1}^m c_k u_k - \sum_{k=1}^n c_k u_k \right\|^2 = \left\| \sum_{k=n+1}^m c_k u_k \right\|^2 = \sum_{k=n+1}^m c_k^2 .$$

Since finite sums $\sum_1^n c_k u_k$ belong to V, we have also $\sum_1^\infty c_k u_k \in V$. This proves that $U \subset V$.

Now we observe that U is a closed subspace, being isometrically isomorphic to the complete space ℓ^2. Since $u_n \in U$, we have also span $\{u_n\} \subset U$ and closure$\big($span $\{u_n\}\big)$ $\subset U$. Hence $V \subset U$. ∎

A generalization of Lemma 2 is given in Chapter 35 (Lemma 1 on page 287).

If we assume, in addition, that $\phi \in V_1$, then for a suitable $a \in \ell^2(\mathbb{Z})$ we will have

$$(16) \qquad \phi(t) = \sum a_j \phi(2t - j)$$

Such an equation is called a **two-scale relation.** It is an example of a "functional equation."

LEMMA 3. *Let ϕ be an element of $L^2(\mathbb{R})$ such that $\{\phi_{0j} : j \in \mathbb{Z}\}$ is orthonormal and such that the two-scale equation (16) is valid. Then for all n and m in $\mathbb{Z}$,*

$$(17) \qquad \sum_j [a_{n-2j} a_{m-2j} + (-1)^{n+m} a_{1-n+2j} a_{1-m+2j}] = 2\delta_{nm}$$

Proof. If $n + m$ is odd, the expression on the left in Equation (17) becomes

$$\sum_j a_{n-2j} a_{m-2j} - \sum_j a_{1-n+2j} a_{1-m+2j}$$

In the first sum, let $j = -i$. In the second sum, let $j = i + (n + m - 1)/2$. The result is

$$\sum_i a_{n+2i} a_{m+2i} - \sum_i a_{m+2i} a_{n+2i} = 0 = 2\delta_{nm}$$

If $n + m$ is even, the expression on the left of Equation (17) is

$$\sum_j a_{n-2j} a_{m-2j} + \sum_j a_{1-n+2j} a_{1-m+2j}$$

In the first sum, let $j = -i$. In the second sum, let $j = i + (m + n)/2$. We get

$$\sum_i a_{n+2i} a_{m+2i} + \sum_i a_{m+2i+1} a_{n+2i+1} = \sum_j a_{n+j} a_{m+j} = \sum_i a_i a_{i+m-n}$$

Since $\{\phi_{1k} : k \in \mathbb{Z}\}$ is orthonormal, the Parseval equation applies to give

$$(18) \qquad \sum_i a_i a_{i+m-n} = \left\langle \sum_i a_i \phi_{1i}, \sum_i a_{i+m-n} \phi_{1i} \right\rangle$$

By the two-scale equation, we have

$$\sum_i a_i \phi_{1i}(t) = \sum_i a_i 2^{1/2} \phi(2t - i) = 2^{1/2} \phi(t)$$

Similarly,

$$\sum_i a_{i+m-n} \phi_{1i}(t) = \sum_i a_{i+m-n} 2^{1/2} \phi(2t - i)$$

$$= \sum_j a_j 2^{1/2} \phi(2t - j + m - n)$$

$$= 2^{1/2} \sum_j a_j \phi\left(2\left(t - \frac{n-m}{2}\right) - j\right)$$

$$= 2^{1/2} \phi\left(t - \frac{n-m}{2}\right) = 2^{1/2} \phi_{0, (n-m)/2}(t)$$

Finally, the right side of Equation (18) becomes

$$\left\langle 2^{1/2} \phi_{00}, 2^{1/2} \phi_{0, (n-m)/2} \right\rangle = 2\delta_{nm} \qquad \blacksquare$$

LEMMA 4. *Let ϕ be an element of $L^2(\mathbb{R})$ such that $\{\phi_{0j} : j \in \mathbb{Z}\}$ is orthonormal. Assume that ϕ satisfies the two-scale equation (16). Define ψ by the equation*

$$\psi(t) = \sum_k (-1)^k a_{1-k} \phi(2t - k)$$

Then

(19)
$$\phi_{1n} = 2^{-1/2} \sum_j [a_{n-2j}\phi_{0j} + (-1)^n a_{1-n+2j}\psi_{0j}]$$

Proof. By the two-scale equation, $\phi(t) = \sum_i a_i \phi(2t - i)$. Therefore

$$\phi(t - j) = \sum_i a_i \phi(2(t - j) - i) = \sum_i 2^{-1/2} a_i 2^{1/2} \phi(2t - 2j - i)$$

In other terms, $\phi_{0j} = \sum_i 2^{-1/2} a_i \phi_{1,\,2j+i}$. A similar equation holds for ψ, namely

$$\psi_{0j} = \sum_i (-1)^i 2^{-1/2} a_{1-i} \phi_{1,\,2j+i}$$

The expression on the right in Equation (19) can now be written as

$$2^{-1/2} \sum_j \left[a_{n-2j} \sum_i 2^{-1/2} a_i \phi_{1,\,i+2j} + (-1)^n a_{1-n+2j} \sum_i 2^{-1/2}(-1)^i a_{1-i}\phi_{1,\,i+2j} \right]$$

Change the summation index i to m by the equation $m = i + 2j$, obtaining

$$2^{-1} \sum_m \sum_j [a_{n-2j} a_{m-2j} + (-1)^{n+m} a_{1-n+2j} a_{1-m+2j}]\phi_{1m}$$

According to the preceding lemma, the sum over j is $2\delta_{nm}$. Hence the entire expression reduces to ϕ_{1n}. ∎

We now come to the principal theorem of this chapter. It shows how each ϕ in $L^2(\mathbb{R})$ that possesses the two properties hypothesized in Lemma 4 leads directly to a wavelet ψ. An important feature of wavelet theory is the idea of an orthogonal decomposition. We will use the notation $U = V \oplus^{\perp} W$ to signify that each element $u \in U$ can be written uniquely as $u = v + w$, where $v \in V$, $w \in W$, and $V \perp W$.

THEOREM 2. *Let ϕ be an element of $L^2(\mathbb{R})$ for which $\{\phi(\cdot - n) : n \in \mathbb{Z}\}$ is orthonormal. Assume that ϕ satisfies a two-scale relation*

(20)
$$\phi(t) = \sum_k a_k \phi(2t - k) \qquad (a \in \ell^2(\mathbb{Z}))$$

Define ψ by the equation

(21)
$$\psi(t) = \sum_j (-1)^j a_{1-j} \phi(2t - j)$$

Then ψ is an orthonormal wavelet. Furthermore, $V_1 = V_0 \oplus^{\perp} W_0$, where W_0 is the space generated by $\{\psi_{0n} : n \in \mathbb{Z}\}$.

Proof. First we prove that for all n and m, $\phi_{0n} \perp \psi_{0m}$:

$$
\begin{aligned}
\langle \phi_{0n}, \psi_{0m} \rangle &= \int \phi(t-n)\psi(t-m)\,dt \\
&= \int \sum_k a_k \phi(2t-2n-k) \sum_j (-1)^j a_{1-j} \phi(2t-2m-j)\,dt \\
&= \sum_k \sum_j (-1)^j a_k a_{1-j} \int \phi(x-2n-k)\phi(x-2m-j)\,\frac{1}{2}\,dx \\
&= \frac{1}{2} \sum_k \sum_j (-1)^j a_k a_{1-j} \delta_{2n+k,\,2m+j} \\
&= \frac{1}{2} \sum_k (-1)^{2n+k-2m} a_k a_{1-2n-k+2m} \\
&= \frac{1}{2} \sum_k (-1)^k a_k a_{p-k} \qquad (p = 2m - 2n + 1) \\
&= \frac{1}{4}\left[\sum_k (-1)^k a_k a_{p-k} + \sum_j (-1)^j a_j a_{p-j} \right] \\
&= \frac{1}{4}\left[\sum_k (-1)^k a_k a_{p-k} + \sum_k (-1)^{p-k} a_{p-k} a_k \right] \\
&= \frac{1}{4}\left[\sum_k (-1)^k a_k a_{p-k} - \sum_k (-1)^k a_{p-k} a_k \right] = 0
\end{aligned}
$$

The orthogonality relation just proved establishes that $W_0 \perp V_0$. Equations (20) and (21) imply that ϕ and ψ belong to V_1. The same is true of their integer translates. Hence $V_0 + W_0 \subset V_1$.

The next step in the proof is to establish that $\{\psi_{0n} : n \in \mathbb{Z}\}$ is orthonormal. A quick calculation confirms that

(22) $\phi_{0n} = 2^{-1/2} \sum_k a_k \phi_{1,\,k+2n} \qquad \psi_{0n} = 2^{-1/2} \sum_k (-1)^k a_{1-k} \phi_{1,\,k+2n}$

Hence,

$$
\delta_{nm} = \langle \phi_{0n}, \phi_{0m} \rangle = \frac{1}{2}\left\langle \sum_k a_k \phi_{1,\,k+2n}, \sum_j a_j \phi_{1,\,j+2m} \right\rangle
$$

(23)
$$
= \frac{1}{2} \sum_k \sum_j a_k a_j \delta_{k+2n,\,j+2m}
$$

$$
= \frac{1}{2} \sum_k a_k a_{k+2n-2m}
$$

Now, a similar calculation using Equations (22) and (23) yields

$$\langle \psi_{0n}, \psi_{0m} \rangle = \frac{1}{2} \sum_k (-1)^k (-1)^{k+2n-2m} a_{1-k} a_{1-2n+2m-k}$$

$$= \frac{1}{2} \sum_j a_j a_{j+2m-2n} = \delta_{mn}$$

The next step in the proof is to establish that $V_1 \subset V_0 + W_0$. By Lemma 4, $\phi_{1n} \in V_0 + W_0$. It follows that

$$\text{span}\{\phi_{1n} : n \in \mathbb{Z}\} \subset V_0 + W_0$$

and that

$$V_1 \equiv \text{closure}(\text{span}\{\phi_{1n} : n \in \mathbb{Z}\}) \subset \text{closure}(V_0 + W_0)$$

By Lemma 5 (below), $V_0 + W_0$ is closed, since V_0 and W_0 are closed and orthogonal to each other. Hence, $V_1 \subset V_0 \oplus^{\perp} W_0$.

Now define W_n to be the closed subspace generated by $\{\psi_{nj} : j \in \mathbb{Z}\}$. With simple changes of variables, we obtain $V_{n+1} = V_n \oplus^{\perp} W_n$. The orthogonality relation $\psi_{nk} \perp \psi_{mj}$ when $m \neq n$ follows from these observations:

(24)   If $n < m$ then $\psi_{nk} \in W_n \subset V_{n+1} \subset V_m$

(25) $\psi_{mj} \in W_m \perp V_m$ ∎

LEMMA 5. *Let X and Y be closed subspaces in a Hilbert space. If $X \perp Y$, then $X + Y$ is closed.*

Proof. Let $z_n \in X + Y$ and $z_n \rightarrow z$. We want to prove that $z \in X + Y$. Write $z_n = x_n + y_n$, where $x_n \in X$ and $y_n \in Y$. We have

$$\|z_n - z_m\|^2 = \|x_n - x_m + y_n - y_m\|^2 = \|x_n - x_m\|^2 + \|y_n - y_m\|^2$$

Since $[z_n]$ is a Cauchy sequence, so are $[x_n]$ and $[y_n]$. Put $x = \lim x_n$ and $y = \lim y_n$. Then $x \in X$ and $y \in Y$. Hence

$$z = \lim (x_n + y_n) = x + y \in X + Y$$ ∎

Straightforward algebraic manipulations lead to the following equations, starting from Equations (16) and (18):

(26) $\phi_{n-1, k} = 2^{-1/2} \sum_j a_j \phi_{n, 2k+j} = 2^{-1/2} \sum_j a_{j-2k} \phi_{nj}$

(27) $\psi_{n-1, k} = 2^{-1/2} \sum_j (-1)^j a_{1-j} \phi_{n, 2k+j} = 2^{-1/2} \sum_j (-1)^j a_{2k+1-j} \phi_{nj}$

THEOREM 3. *Let ϕ and ψ be as in Theorem 2. If $\bigcup V_n$ is dense in $L^2(\mathbb{R})$ and $\bigcap V_n = 0$, then $\{\psi_{nk} : n, k \in \mathbb{Z}\}$ is an orthonormal base for $L^2(\mathbb{R})$.*

Proof. Assume all the hypotheses. By Hilbert space theory, we can prove that $\{\psi_{nk}\}$ is a base by showing that if $f \in L^2(\mathbb{R})$ and $f \perp \psi_{nk}$ for all n and k, then $f = 0$. Adopting these hypotheses on f, we note that $f \perp W_n$ for all n. Let g_n denote the orthogonal projection of f in V_n. Thus $g_n \in V_n$ and $f - g_n \perp V_n$. Since $V_n = V_{n-1} \oplus^\perp W_{n-1}$ (by Theorem 2) we have $f - g_n \perp V_{n-1}$ and $f - g_n \perp W_{n-1}$. Since $f \perp W_{n-1}$ we infer that $g_n \perp W_{n-1}$ and $g_n \in V_{n-1}$. The two properties $g_n \in V_{n-1}$ and $f - g_n \perp V_{n-1}$ imply that $g_n = g_{n-1}$. By the density of $\bigcup V_n$ and the property $V_n \subset V_{n+1}$ for all n, it follows that $g_n \to f$. Since the sequence $[g_n]$ is stationary, $g_n = f$ for all n. Hence $f \in \bigcap V_n$ and $f = 0$. ∎

Example 2. A pair of functions illustrating Theorem 2 is

(28) $$\psi = \mathfrak{h} \text{ (the Haar wavelet)}, \qquad \phi = \mathcal{X}_{[0,1)}$$

To verify this relation, first note that ϕ obeys a simple two-scale equation

(29) $$\phi(t) = \phi(2t) + \phi(2t - 1)$$

Thus the coefficients are $a_0 = a_1 = 1$, and all other a_k are zero. According to Theorem 2, ψ should be given by

(30) $$\psi(t) = \phi(2t) - \phi(2t - 1)$$

and this in fact is correct for the Haar wavelet.

We conclude this chapter with some further properties of the Haar wavelet ψ and its accompanying function ϕ, as in Equation (28).

THEOREM 4. *For the function ϕ in Equation (28), we have $\bigcap \{V_n : n \in \mathbb{Z}\} = 0$.*

Proof. Let $f \in \bigcap V_n$. To prove that $f = 0$, it suffices to show that $\int_I |f(t)| \, dt = 0$ for any interval of the form $I = [-a, a]$.

Since $f \in V_n$, we can write $f = \sum_j c_{nj} \phi_{nj}$. This is an orthonormal expansion, and consequently $\|f\|_2 = \|c_n\|_2$, where c_n denotes the sequence $[c_{nj} : j \in \mathbb{Z}]$. By the definition of ϕ_{nj}, we have, for almost all t,

$$f(t) = \sum_j c_{nj} 2^{n/2} \phi(2^n t - j)$$

Putting $x = 2^n t$, we have, for almost all x,

$$f(2^{-n}x) = \sum_j c_{nj} 2^{n/2} \phi(x - j)$$

From this it follows that, for almost all t,

$$|f(t)| = |f(2^{-n}x)| \le 2^{n/2} \sum_j |c_{nj}| \phi(x - j) \le 2^{n/2} \|c_n\|_\infty$$

$$\le 2^{n/2} \|c_n\|_2 = 2^{n/2} \|f\|_2$$

Here we capitalized on the special form of ϕ, which yields $\sum_j \phi(x-j)=1$. Now we conclude that $\int_I |f(t)|\, dt \le 2^{n/2}\|f\|_2\, \mu(I)$, where μ is Lebesgue measure. Let $n \to -\infty$ to get $\int_I |f(t)|\, dt = 0$.   ∎

In the next proof, we use P_n to denote the orthogonal projection of $L^2(\mathbb{R})$ onto V_n.

THEOREM 5. *For the function ϕ in Equation (28), $\bigcup V_n$ is dense in $L^2(\mathbb{R})$.*

Proof. The space $C_c(\mathbb{R})$ of continuous functions having compact support is dense in $L^2(\mathbb{R})$. Our theorem can therefore be proved by showing that $\bigcup V_n$ is L^2-dense in $C_c(\mathbb{R})$. Accordingly, let $f \in C_c(\mathbb{R})$ and let I be an interval of the form $\bigcup_{j=-N}^{N} I_{mj}$ that contains the support of f. Here I_{mj} is the interval defined by $2^{-m}j \le x < 2^{-m}(j+1)$. Let $n \ge m$ in the following discussion. An interval I_{nk} is contained either in I or $\mathbb{R}\setminus I$. (Cf. Problem 6.) If $x \in I_{nk}$, then

$$
(P_n f)(x) = \sum \langle f, \phi_{nj}\rangle \phi_{nj}(x) = \langle f, \phi_{nk}\rangle \phi_{nk}(x) = 2^{n/2}\int_{-\infty}^{\infty} f(y)2^{n/2}\mathcal{X}_{I_{nk}}(y)\, dy
$$
$$
(31) \qquad\qquad\qquad = 2^n \int_{I_{nk}} f(y)\, dy = \frac{1}{|I_{nk}|}\int_{I_{nk}} f(y)\, dy
$$

From Equation (31) we see that if $x \in I_{nk} \subset \mathbb{R}\setminus I$, then $f(x)=0$ and $(P_n f)(x)=0$. On the other hand, if $x \in I_{nk} \subset I$ then

$$
|f(x)-(P_n f)(x)| = \left| \frac{1}{|I_{nk}|}\int_{I_{nk}} [f(x)-f(y)]\, dy \right|
$$
$$
\le \omega(f; 2^{-n})
$$

where ω is the modulus of continuity of f. We conclude therefore that as $n \to \infty$,

$$
\int |f(x)-(P_n f)(x)|^2\, dx \le \int_I \omega(f; 2^{-n})^2\, dx = |I|\omega(f; 2^{-n})^2 \longrightarrow 0   \qquad ∎
$$

COROLLARY. *For the Haar wavelet $\mathfrak{h}$, $\{\mathfrak{h}_{nk} : n, k \in \mathbb{Z}\}$ is an orthonormal base for $L^2(\mathbb{R})$.*

Proof. By Theorem 1, $\mathfrak{h}$ is an orthonormal wavelet, and hence $\{\mathfrak{h}_{nk} : n, k \in \mathbb{Z}\}$ is an orthonormal system. We know also that the scaling function ϕ associated with $\mathfrak{h}$ as in Equation (28) has these two properties

$$
\bigcap_{n\in\mathbb{Z}} V_n = 0 \qquad\qquad \text{closure}\left(\bigcup_{n\in\mathbb{Z}} V_n\right) = L^2(\mathbb{R})
$$

(See Theorems 4 and 5.) By Theorem 3, $\{\mathfrak{h}_{nk}\}$ is an orthonormal base for $L^2(\mathbb{R})$.   ∎

Problems

1. Prove that if $f \in L^2(\mathbb{R})$ and $|f'(x)| \le M$, then $f(x) \to 0$ as $|x| \to \infty$.

2. Is there a function f in $L^2(\mathbb{R})$ such that the functions $x \mapsto f(\lambda x)$, $(\lambda \in \mathbb{R})$, form a fundamental set in $L^2(\mathbb{R})$?

3. Let ψ be the Haar wavelet, as in Equation (11). Prove that for integers $n < m$, $\int_n^m \psi(x)\, dx = 0$.

4. Let ψ be the Haar wavelet. Prove that $\psi_{kj}(x) > 0$ if and only if $j \le x < j + 2^{-k-1}$. Prove that $\psi_{kj}(x) < 0$ if and only if $j + 2^{-k-1} \le x < j + 2^{-k}$.

5. Fix a pair of real numbers (α, β) with $\alpha > 0$, and define an operator T on $L^2(\mathbb{R})$ by writing
$$(Tf)(x) = \alpha^{1/2} f(\alpha x + \beta)$$
Prove that T is unitary; i.e., $\langle Tf, Tg \rangle = \langle f, g \rangle$.

6. A dyadic interval is defined (here only) as an interval $[n2^m, (n+1)2^m)$ where $n, m \in \mathbb{Z}$. Prove that if two dyadic intervals intersect, then one contains the other.

7. Prove Equations (26) and (27).

8. Let $[u_n]$ be an orthonormal sequence in a Hilbert space. Let α and β be members of ℓ^2 such that $\alpha \perp \beta$. Define $w = \sum_k \alpha_k u_k$ and $v = \sum_k \beta_k u_k$. Prove that $w \perp v$. Is the converse true?

9. Let $[u_n]$ be an orthonormal sequence in a Hilbert space. Let $\alpha \in \ell^2$, and $w = \sum_k \alpha_k u_k$, $v = \sum (-1)^k \alpha_{p-k} u_k$, where p is an odd integer. Prove that $w \perp v$.

10. Let ϕ be an element of $L^2(\mathbb{R})$ such that $\{\phi_{0n} : n \in \mathbb{Z}\}$ is orthonormal. Prove that for each $k \in \mathbb{Z}$, $\{\phi_{kn} : n \in \mathbb{Z}\}$ is orthonormal.

11. Prove that V_k consists of all $\sum_j c_j \phi_{kj}$, where $c \in \ell^2(\mathbb{Z})$.

12. Which of these two subspaces contains the other?
$$Y_1 = \text{span}\{T_n S_2 \phi : n \in \mathbb{Z}\}$$
$$Y_2 = \text{span}\{S_2 T_n \phi : n \in \mathbb{Z}\}$$

13. Prove in full that for all n, $V_{n+1} = V_n \oplus^\perp W_n$, under the hypotheses of Theorem 2.

14. Extend Equation (19) to this more general result:
$$2^{1/2} \phi_{kn} = \sum_j \left[a_{n-2j} \phi_{k-1,j} + (-1)^n a_{1-n+2j} \psi_{k-1,j} \right]$$

15. Describe the subspace closure$(\text{span}\{\mathfrak{h}_{0n} : n \in \mathbb{Z}\})$, where $\mathfrak{h}$ is the Haar wavelet. Explain why $\{\mathfrak{h}_{0n}\}$ is not an orthonormal base for $L^2(\mathbb{R})$.

16. Find a necessary and sufficient condition on $a \in \ell^2$ in order that a solution of the equation $\phi(t) = \sum a_j \phi(2t - j)$ will have the property that $\{\phi_{0n} : n \in \mathbb{Z}\}$ is orthonormal.

17. Prove that for any $\phi \in L^2(\mathbb{R})$, the subspace V_0 is shift-invariant. This term means that $f(\cdot - n)$ is in V_0 if f is in V_0.

18. For what values of n do we have $\sum_i (-1)^i c_j c_{n-j} = 0$ for all $c \in \ell^2(\mathbb{Z})$?

19. For what values of p do we have $\langle \phi_{kj}, \phi_{ni} \rangle = \langle \phi, \phi_{n-k,p} \rangle$ for all $\phi \in L^2(\mathbb{R})$?

20. Prove that if the support of ϕ is contained in the interval $[0, 1]$, then (for each k) $\{\phi_{ki} \ : \ i \in \mathbb{Z}\}$ is orthogonal.

21. Fix ψ in $L^2(\mathbb{R})$, n in $\mathbb{Z}$, and i in $\mathbb{Z}$. Define $\phi = \psi_{ni}$. Show that

$$\mathrm{span}\{\phi_{kj} \ : \ k, j \in \mathbb{Z}\} = \mathrm{span}\{\psi_{kj} \ : \ k, j \in \mathbb{Z}\}$$

22. Generalize the result of Problem 21.

23. Let ϕ and V_n be as described just after Theorem 1, and suppose that $\phi \in V_1$. Show that the coefficients in Equation (16) are given by $a_j = 2 \int \phi(t) \phi(2t - j) \, dt$.

24. Let ϕ and V_n be as in Problem 23. Prove that for all n, $V_n \subset V_{n+1}$.

25. Let f, $g \in L^2(\mathbb{R})$, and suppose that $f_{0j} \perp g_{0i}$ for all i and j. Prove that $f_{nj} \perp g_{ni}$ for all n, i and j.

26. Find the values of k for which this is true: If $c \in \ell^2(\mathbb{Z})$ and $\{u_n\}$ is orthonormal in a Hilbert space, then $\sum c_j u_j \perp \sum c_{j+k}(-1)^j u_j$.

27. Use the proof of Theorem 4 as a model to prove the following stronger result. *If ϕ is a bounded, measurable, and compactly supported function on $\mathbb{R}$, and if $\{\phi_{0n} \ : \ n \in \mathbb{Z}\}$ is orthonormal, then $\bigcap V_n = 0$.*

References

[Chu2] Chui, Charles K. *An Introduction to Wavelets.* Academic Press, New York, 1992.

[CMP] Chui, C. K., L. Montefusco, and L. Puccio (eds.). *Wavelets: Theory, Algorithms, and Applications.* Academic Press, New York, 1994.

[Dau] Daubechies, I. *Ten Lectures on Wavelets.* SIAM, Philadelphia, 1992.

[DMZ] Davis, G., S. Mallat, and Zhifeng Zhang. "Adaptive time-frequency approximations with matching pursuits." In *Wavelets: Theory, Algorithms, and Applications,* ed. by C. K. Chui, L. Montefusco, and L. Puccio. Academic Press, New York, 1994, pp. 271–293.

[JS] Jawerth, B., and W. Sweldens. "An overview of wavelet based multiresolution analyses." *SIAM Review 36* (1994), 377–412.

[Mal] Mallat, S. G. "Multiresolution approximations and wavelet orthonormal bases of $L^2(R)$." *Trans. Amer. Math. Soc. 315* (1989), 69–87.

[MalZ] Mallat, S., and Zhifeng Zhang. "Matching pursuits with time-frequency dictionaries." *IEEE Trans. on Signal Processing 41* (1993), 3397–3415.

[Mey1] Meyer, Yves. *Wavelets: Algorithms and Applications.* SIAM, Philadelphia, 1993.

[SchW] Schumaker, L. L., and G. Webb (eds.). *Recent Advances in Wavelet Analysis.* Academic Press, New York, 1994.

[Walt] Walter, G. G. *Wavelets and Other Orthogonal Systems with Applications.* CRC Press, Boca Raton, FL, 1994.

[Woj2] Wojtaszczyk, P. *A Mathematical Introduction to Wavelets.* Cambridge Univ. Press, Cambridge, UK, 1997.

35
Wavelets, II

This chapter addresses three main issues. First, recall that Theorem 2 in Chapter 34 concerns any function ϕ in $L^2(\mathbb{R})$ having these two properties:

(1) $$\{\phi(\cdot - n) \ : \ n \in \mathbb{Z}\} \text{ is orthonormal}$$

(2) $$\phi(t) = \sum_n a_n \phi(2t - n) \qquad (\text{for some } a = [a_n] \text{ in } \ell^2(\mathbb{Z}))$$

From the function ϕ, an orthonormal wavelet ψ can be defined in such a way that the spaces V_n and W_n generated by integer translates of $\phi(2^n t)$ and $\psi(2^n t)$ will have the property

(3) $$V_{n+1} = V_n \oplus^{\perp} W_n \qquad (n \in \mathbb{Z})$$

We now intend to investigate the situation in which $\{\phi_{0n} \ : \ n \in \mathbb{Z}\}$ is not orthonormal, yet has a related property strong enough to lead to wavelets.

Second, we discuss the decomposition and reconstruction algorithms using wavelets.

Third, we construct the compactly supported wavelets of Daubechies.

We begin with a theorem that helps us to understand the property in (1).

THEOREM 1. *For a function ϕ in $L^2(\mathbb{R})$, the following three properties are equivalent:*

a. $$\{\phi(\cdot - n) : n \in \mathbb{Z}\} \text{ is orthonormal}$$

b. $$\int_{-\infty}^{\infty} |\hat{\phi}(t)|^2 e^{-2\pi i n t} \, dt = \delta_{0n} \qquad (n \in \mathbb{Z})$$

c. $$\sum_{n=-\infty}^{\infty} |\hat{\phi}(t + n)|^2 = 1 \text{ for almost all } t \text{ in } \mathbb{R}$$

Proof. Let F be the function defined by the left-hand side of the equality in **c**. Then F is nonnegative and integer-periodic. Furthermore, $F|[0,1] \in L^1[0, 1]$, since (and here we use Plancherel's Theorem),

$$\int_0^1 |F(t)| \, dt = \sum_{n=-\infty}^{\infty} \int_0^1 |\hat{\phi}(t + n)|^2 \, dt$$

$$= \sum_{n=-\infty}^{\infty} \int_n^{n+1} |\hat{\phi}(t)|^2 \, dt = \int_{-\infty}^{\infty} |\hat{\phi}(t)|^2 \, dt = \|\hat{\phi}\|^2 = \|\phi\|^2 < \infty$$

The same manipulations lead to

$$\int_0^1 F(t) e^{-2\pi i k t} \, dt = \int_{-\infty}^{\infty} |\hat{\phi}(t)|^2 e^{-2\pi i k t} \, dt$$

If $F(t) = 1$ almost everywhere, then this last equation yields **b**. Hence **c** implies **b**. If **b** holds, then the above equations show that the Fourier coefficients of F are $\langle F, e_k \rangle = \delta_{0k}$. (We use $e_k(t) = e^{2\pi i k t}$.) To see that $F(t) = 1$ almost everywhere, define a continuous linear functional L on $C[0, 1]$ by writing

$$L(g) = \int_0^1 g(x)[F(x) - 1] \, dx \qquad (g \in C[0, 1])$$

A routine calculation shows that $L(e_k) = 0$ for all k. Since $\{e_k : k \in \mathbb{Z}\}$ is fundamental in $C[0, 1]$, we conclude that $L = 0$ and that $F(x) - 1 = 0$ almost everywhere. Hence **b** implies **c**.

   In order to complete the proof, we will show that **a** is equivalent to **b**. Let $\langle f, g \rangle = \int_{-\infty}^{\infty} f(t)\overline{g(t)} \, dt$ and $(T_k f)(t) = f(t - k)$. Then the Parseval equation for Fourier transforms gives us

$$\langle T_j \phi, T_k \phi \rangle = \langle \phi, T_{-j} T_k \phi \rangle = \langle \phi, T_{k-j} \phi \rangle$$

$$= \langle \hat{\phi}, \widehat{T_{k-j} \phi} \rangle = \langle \hat{\phi}, e_{k-j} \hat{\phi} \rangle$$

$$= \int_{-\infty}^{\infty} |\hat{\phi}(t)|^2 e^{-2\pi i (k-j) t} \, dt$$

Thus, **a** is equivalent to the equation $\langle T_j \phi, T_k \phi \rangle = \delta_{jk} = \delta_{0, k-j}$, and this in turn is equivalent to **b**.   ∎

   For a given function ϕ in $L^2(\mathbb{R})$, the orthonormality of $\{\phi_{0n}\}$ is equivalent to the equation

(4) $$\sum_n |\hat{\phi}(t + n)|^2 = 1 \text{ almost everywhere.}$$

If $\{\phi_{0n}\}$ is not orthonormal, it seems natural to define Φ by requiring its Fourier transform to be

(5) $$\hat{\Phi}(t) = \hat{\phi}(t) \Big/ \left(\sum_n |\hat{\phi}(t + n)|^2 \right)^{1/2}$$

Then the property expressed in (1) will be true for Φ. Theorem 2 below justifies this construction.

Definition 1. A sequence $[u_n]$ in a Hilbert space is said to be **stable** if there are positive constants A and B such that

$$(6) \qquad A\|c\| \leq \left\| \sum_n c_n u_n \right\| \leq B\|c\| \qquad (c \in \ell^2)$$

LEMMA 1. *Let $[u_k]$ be a stable sequence in a Hilbert space. Then the closed linear subspace generated by $[u_k]$ is*

$$V = \left\{ \sum_k c_k u_k \; : \; \sum_k c_k^2 < \infty \right\}$$

Proof. Let H be the Hilbert space under consideration. First, we note that if $\sum_k c_k^2 < \infty$, then the series $\sum_k c_k u_k$ converges in the Hilbert space. Indeed, if we examine the sequence of partial sums, $x_n = c_1 u_1 + \cdots + c_n u_n$, we find that it has the Cauchy property:

$$\|x_m - x_n\| = \left\| \sum_{k=n+1}^m c_k u_k \right\| \leq B \left[\sum_{k=n+1}^m c_k^2 \right]^{1/2}$$

Secondly, we note that elements of V are limits of linear combinations of the u_k. Hence V is contained in the closed linear subspace generated by $[u_k]$. Thirdly, it is clear that V contains the linear span of the set $[u_k]$. If we can prove that V is closed, V will of necessity contain the closure of the span of $[u_k]$. Let $[y_n]$ be a sequence in V that converges to a point $y \in H$. Define $T : \ell^2 \to H$ by $Tc = \sum_k c_k u_k$, where $c = [c_k]$. The definition of a stable sequence gives us $A\|c\| \leq \|Tc\| \leq B\|c\|$. It follows that T is continuous. Let $y_n = Tc^{(n)}$. Since $[y_n]$ is Cauchy in H, $[c^{(n)}]$ is Cauchy in ℓ^2. Hence $c^{(n)} \to c^* \in \ell^2$, and $y_n = Tc^{(n)} \to Tc^* \in V$. $\blacksquare$

THEOREM 2. *Let ϕ be an element of $L^2(\mathbb{R})$, and let $0 < A \leq B < \infty$. The following assertions are equivalent:*

$$(7) \qquad A\|c\| \leq \left\| \sum_n c_n \phi_{0n} \right\| \leq B\|c\| \qquad \text{(for all c in $\ell^2(\mathbb{Z})$)}$$

$$(8) \qquad A^2 \leq \sum_n |\hat{\phi}(t + n)|^2 \leq B^2 \qquad \text{(for almost all t)}$$

Proof. The Plancherel Theorem gives us, for any $f \in L^2(\mathbb{R})$,

$$\|f\|^2 = \|\hat{f}\|^2 = \int_{-\infty}^{\infty} |\hat{f}(t)|^2 \, dt = \sum_{k=-\infty}^{\infty} \int_0^1 |\hat{f}(t + k)|^2 \, dt$$

Applying this to $f = \sum_n c_n \phi_{0n}$, and using $e_n(t) = e^{2\pi i n t}$, we obtain

$$
\left\| \sum_{n=-\infty}^{\infty} c_n \phi_{0n} \right\|^2 = \sum_{k=-\infty}^{\infty} \int_0^1 \left| \sum_{n=-\infty}^{\infty} c_n \, \widehat{\phi_{0n}}(t+k) \right|^2 dt
$$

(9)

$$
= \int_0^1 \sum_{k=-\infty}^{\infty} \left| \sum_{n=-\infty}^{\infty} c_n e^{-2\pi i n(t+k)} \hat{\phi}(t+k) \right|^2 dt
$$

$$
= \int_0^1 \sum_{k=-\infty}^{\infty} |\hat{\phi}(t+k)|^2 \left| \sum_{n=-\infty}^{\infty} c_n e^{-2\pi i n t} \right|^2 dt
$$

$$
= \int_0^1 |g_c(t)|^2 \sum_{k=-\infty}^{\infty} |\hat{\phi}(t+k)|^2 \, dt
$$

where $g_c(t) = \sum_n c_n e^{-2\pi i n t} = \sum_n c_n e_{-n}(t)$. Now assume that Inequality (8) holds. Then Equation (9) gives us

$$
A^2 \int_0^1 |g_c(t)|^2 \, dt \le \left\| \sum_n c_n \phi_{0n} \right\|^2 \le B^2 \int_0^1 |g_c(t)|^2 \, dt
$$

But

$$
\int_0^1 |g_c(t)|^2 \, dt = \left\| \sum_n c_n e_{-n} \right\|^2 = \sum_n |c_n|^2
$$

and this establishes (7).

For the other implication, assume that (7) is true. Find a sequence of positive, continuous, integer-periodic functions $[h_k]$ such that $\int_0^1 h_k^2(t) \, dt = 1$ for all $k \in \mathbb{N}$, and such that $\int_\varepsilon^{1-\varepsilon} h_k^2(t) \, dt \to 0$ for all ε in $(0, 1/2)$. By Theorem 10 at the end of this chapter, the relation

(10)
$$
\lim_{k \to \infty} \int_0^1 h_k^2(t-s) f(s) \, ds = f(t)
$$

is valid for all $f \in L^1[0, 1]$ if the limit is understood to be in the L^1-metric. From Hypothesis (7) and Equation (9), we infer that

(11) $A^2 \|c\|^2 \le \int_0^1 |g_c(s)|^2 \sum_n |\hat{\phi}(s+n)|^2 \, ds \le B^2 \|c\|^2$ $(c \in \ell^2)$

Since h_k belongs to $L^2[0, 1]$, it has a Fourier expansion:

$$
h_k = \sum_n \gamma_{kn} e_{-n}
$$

Now,

$$
h_k(t-s) = \sum_n \gamma_{kn} e_{-n}(t-s) = \sum_n \gamma_{kn} e_{-n}(t) e_n(s)
$$

By Parseval's identity and the periodicity of h_k,

$$\sum_n |\gamma_{kn} e_{-n}(t)|^2 = \int_0^1 h_k^2(t - s)\, ds = \int_0^1 h_k^2(s)\, ds = 1$$

On substituting $g_c(s) = h_k(t - s)$ in Inequality (11), we obtain

$$A^2 = A^2 \sum_n |\gamma_{kn} e_{-n}(t)|^2 \le \int_0^1 h_k^2(t - s) \sum_n |\hat{\phi}(s + n)|^2\, ds$$

$$\le B^2 \sum_n |\gamma_{kn} e_{-n}(t)|^2 = B^2$$

As in the proof of Theorem 1, $\sum_n |\hat{\phi}(\cdot + n)|^2$ lies in $L^1[0, 1]$. Letting $k \to \infty$ in the preceding inequality, and using Equation (10), we obtain the required result. ∎

THEOREM 3. *Let ϕ be an element of $L^2(\mathbb{R})$ such that $\{\phi_{0n} : n \in \mathbb{Z}\}$ is stable. Define Φ by*

$$\hat{\Phi} = \left(\sum_n |\hat{\phi}_{0n}|^2 \right)^{-1/2} \hat{\phi}$$

Then $\{\Phi_{0n}\}$ is an orthonormal base for the space V_0 generated by $\{\phi_{0n}\}$.

Proof. Let F be defined by $F(t) = \sum_n |\hat{\phi}_{0n}(t)|^2$, $t \in [0, 1]$. By the preceding theorem $A^2 \le F(t) \le B^2$, for $t \in [0, 1]$. Therefore $F^{1/2}$ and $F^{-1/2}$ are in $L^1[0, 1]$. Since F is integer-periodic,

$$\sum_n |\hat{\Phi}(\cdot + n)|^2 = \sum_n \{ |\hat{\phi}(\cdot + n)|^2 / F(\cdot + n) \}$$

$$= \sum_n |\hat{\phi}(\cdot + n)|^2 / F = F / F = 1$$

By Theorem 1, $\{\Phi_{0n}\}$ is orthonormal.

To establish that V_0 is contained in the closure of the span of $[\Phi_{0n}]$ it suffices to show that $\phi = \sum_n c_n \Phi_{0n}$ for some $c \in \ell^2(\mathbb{Z})$. Let $[c_n]$ be the Fourier coefficients of $F^{1/2}$. Then,

$$\hat{\phi} = \hat{\Phi} F^{1/2} = \sum_n c_n e_{-n} \hat{\Phi} = \left[\sum_n c_n \Phi(\cdot - n) \right]^{\wedge}$$

Taking the inverse Fourier transform gives us $\phi = \sum_n c_n \Phi_{0n}$.

To establish that the closure of the span of $\{\Phi_{0n}\}$ is contained in V_0, we use a similar argument, taking γ_n as the Fourier coefficients of $F^{-1/2}$. Then,

$$\hat{\Phi} = \hat{\phi} F^{-1/2} = \sum_n \gamma_n e_{-n} \hat{\phi} = \left[\sum_n \gamma_n \phi_{0n} \right]^{\wedge}$$

The inverse transform yields $\Phi = \sum_n \gamma_n \phi_{0n}$. ∎

THEOREM 4. *Let ϕ be an element of $L^2(\mathbb{R})$ such that $\{\phi_{0n}\}$ is stable and $\phi \in V_1$. Then Φ (as defined in the preceding theorem) satisfies a two-scale relation,*

$$\Phi(t) = \sum_n a_n \Phi(2t - n) \qquad (a = [a_n] \in \ell^2(\mathbb{Z}))$$

The equation $\psi(t) = \sum_n (-1)^n a_{1-n} \Phi(2t - n)$ defines an orthonormal wavelet.

Proof. Theorem 3 asserts that $\{\Phi_{0n}\}$ is an orthonormal base for V_0. It follows that for each k, $\{\Phi_{kn}\}$ is an orthonormal base for V_k. Since $\phi \in V_1$, it follows that $V_0 \subset V_1$. Since $\Phi \in V_0$, it follows that $\Phi \in V_1$. Hence, for some a_n, $\Phi(t) = \sum_n a_n \Phi(2t - n)$. An application of Theorem 2 in Chapter 34 (page 278) completes the proof. ∎

We turn now to algorithmic considerations. Let us assume that an element ϕ in $L^2(\mathbb{R})$ is given satisfying hypotheses (1) and (2) at the beginning of this chapter. Then the equation

$$(12) \qquad \psi(t) = \sum_j (-1)^j a_{1-j} \phi(2t - j)$$

defines a wavelet. This is the content of Theorem 2 in the preceding chapter. The following are also true, for all $n \in \mathbb{Z}$:

$$(13) \qquad V_n = V_{n-1} \oplus^\perp W_{n-1}$$

$$(14) \qquad \{\phi_{nk} : k \in \mathbb{Z}\} \text{ is an orthonormal base for } V_n$$

$$(15) \qquad \{\psi_{nk} : k \in \mathbb{Z}\} \text{ is an orthonormal base for } W_n$$

Let f be a function in the closure of $\left(\bigcup_{n \in \mathbb{Z}} V_n\right)$. In many cases, this closed subspace will be identical with $L^2(\mathbb{R})$, but we need not make that assumption. Let a positive ε be given. Since $V_n \subset V_{n+1}$ for all n, there exists an integer N such that $\text{dist}(f, V_N) < \varepsilon$. Next, one can write

$$(16) \qquad \begin{aligned} V_N &= V_{N-1} \oplus W_{N-1} \\ &= V_{N-2} \oplus W_{N-2} \oplus W_{N-1} \\ &\quad\vdots \\ &= V_m \oplus W_m \oplus W_{m+1} \oplus \cdots \oplus W_{N-1} \end{aligned}$$

There are orthogonal projections involved in these decompositions. Since V_n has the orthonormal base $\{\phi_{nj} : j \in \mathbb{Z}\}$, the orthogonal projection P_n of $L^2(\mathbb{R})$ onto V_n is given by the formula

$$P_n F = \sum_j \langle F, \phi_{nj} \rangle \phi_{nj} \qquad (F \in L^2(\mathbb{R}))$$

Likewise, the orthogonal projection Q_n of $L^2(\mathbb{R})$ onto W_n is given by

$$Q_n F = \sum_j \langle F, \psi_{nj} \rangle \psi_{nj} \qquad (F \in L^2(\mathbb{R}))$$

Now, start with an element $v_N \in V_N$. (It could be $P_N f$.) Use Equation (16) to write

$$
\begin{aligned}
v_N &= v_{N-1} + w_{N-1} \\
&= v_{N-2} + w_{N-2} + w_{N-1} \\
&\vdots \\
&= v_m + w_m + w_{m+1} + \cdots + w_{N-1}
\end{aligned}
$$

In these equations, $v_n = P_n v_{n+1}$ and $w_n = Q_n v_{n+1}$ for all n. If the coefficients $\langle v_N , \phi_{Nk} \rangle$ are known (for all $k \in \mathbb{Z}$), then the coefficients $\langle v_n , \phi_{nk} \rangle$ and $\langle v_n , \psi_{nk} \rangle$ can be determined recursively from the formulas in the next lemma.

LEMMA 2. *If one adopts the notation*

(17)
$$
v_n = \sum_k c_k^{(n)} \phi_{nk} \qquad\qquad w_n = \sum_k d_k^{(n)} \psi_{nk}
$$

then these recurrence relations are valid:

(18)
$$
c_k^{(n-1)} = 2^{-1/2} \sum_j a_{j-2k} c_j^{(n)}
$$

(19)
$$
d_k^{(n-1)} = 2^{-1/2} \sum_j (-1)^j a_{2k+1-j} c_j^{(n)}
$$

Proof. From previous equations and Equation (26) in Chapter 34 (page 280), we have

$$
v_{n-1} = P_{n-1} v_n = \sum_k \langle v_n , \phi_{n-1,k} \rangle \phi_{n-1,k}
$$

Hence,

$$
\begin{aligned}
c_k^{(n-1)} = \langle v_n , \phi_{n-1,k} \rangle &= \left\langle v_n , 2^{-1/2} \sum_j a_{j-2k} \phi_{nj} \right\rangle \\
&= 2^{-1/2} \sum_j a_{j-2k} c_j^{(n)}
\end{aligned}
$$

Similarly, starting with Equation (27) in Chapter 34 (page 280), one obtains

$$
w_{n-1} = Q_{n-1} v_n = \sum_k \langle v_n , \psi_{n-1,k} \rangle \psi_{n-1,k}
$$

Hence,

$$
\begin{aligned}
d_k^{(n-1)} = \langle v_n , \psi_{n-1,k} \rangle &= \left\langle v_n , 2^{-1/2} \sum_j (-1)^j a_{2k+1-j} \phi_{nj} \right\rangle \\
&= 2^{-1/2} \sum_j (-1)^j a_{2k+1-j} c_j^{(n)} \qquad\qquad \blacksquare
\end{aligned}
$$

What we have just described is the *decomposition algorithm*. It starts with a function v_N in V_N and decomposes it (or *analyzes* it) according to the following scheme:

$$v_N \xrightarrow{\nearrow} \begin{array}{c} w_{N-1} \\ v_{N-1} \end{array} \xrightarrow{\nearrow} \begin{array}{c} w_{N-2} \\ v_{N-2} \end{array} \longrightarrow \cdots \xrightarrow{\nearrow} \begin{array}{c} w_m \\ v_m \end{array}$$

In the numerical computation, one computes the corresponding coefficient sequences $c^{(n)} = [c_j^{(n)}]$ and $d^{(n)} = [d_j^{(n)}]$ in the same order, using the formulas in Lemma 2:

$$c^{(N)} \xrightarrow{\nearrow} \begin{array}{c} d^{(N-1)} \\ c^{(N-1)} \end{array} \xrightarrow{\nearrow} \begin{array}{c} d^{(N-2)} \\ c^{(N-2)} \end{array} \longrightarrow \cdots \xrightarrow{\nearrow} \begin{array}{c} d^{(m)} \\ c^{(m)} \end{array}$$

There is also a *reconstruction algorithm* by which we recover (or *synthesize*) v_N from the components w_n ($m \le n \le N - 1$) and from v_m. Schematically, it looks like this:

$$\begin{array}{c} d^{(m)} \\ c^{(m)} \end{array} \xrightarrow{\searrow} \begin{array}{c} d^{(m+1)} \\ c^{(m+1)} \end{array} \xrightarrow{\searrow} \begin{array}{c} d^{(m+2)} \\ c^{(m+2)} \end{array} \longrightarrow \cdots \longrightarrow \begin{array}{c} d^{(N-1)} \\ c^{(N-1)} \end{array} \xrightarrow{\searrow} c^{(N)}$$

The formulas for this numerical process are given in the next lemma.

LEMMA 3. *Using the notation in Equation* (17), *we have*

$$(20) \qquad c_k^{(n+1)} = 2^{-1/2} \sum_j \left[c_j^{(n)} a_{k-2j} + (-1)^k a_{2j+1-k} d_j^{(n)} \right]$$

Proof. Use Equations (26) and (27) of Chapter 34, page 280, to arrive at

$$\begin{aligned} v_{n+1} &= v_n + w_n \\ &= \sum_j c_j^{(n)} \phi_{nj} + \sum_j d_j^{(n)} \psi_{nj} \\ &= \sum_j c_j^{(n)} \sum_k 2^{-1/2} a_{k-2j} \, \phi_{n+1,k} + \sum_j d_j^{(n)} \sum_k 2^{-1/2} (-1)^k a_{2j+1-k} \phi_{n+1,k} \\ &= \sum_k 2^{-1/2} \sum_j \left[c_j^{(n)} a_{k-2j} + (-1)^k d_j^{(n)} a_{2j+1-k} \right] \phi_{n+1,k} \end{aligned}$$

From this last expression one reads off the coefficients $c_k^{(n+1)}$. ∎

If the above algorithm is to be effective, we need certain assumptions on the subspaces $\{V_n : n \in \mathbb{Z}\}$. These assumptions are given in the following definition.

Definition 2. A **multiresolution** of $L^2(\mathbb{R})$ is a sequence of closed linear subspaces $\{V_n : n \in \mathbb{Z}\}$ having the following four properties:

    a. $V_n \subset V_{n+1}$ *for all* $n \in \mathbb{Z}$

    b. $\bigcup_{n \in \mathbb{Z}} V_n$ *is dense in* $L^2(\mathbb{R})$

    c. $\bigcap_{n \in \mathbb{Z}} V_n = 0$ *(the zero subspace)*

    d. $V_{n+1} = \{x \mapsto f(2x) : f \in V_n\}$ *for all* n.

If $\{V_n\}$ is a multiresolution of $L^2(\mathbb{R})$ and if V_0 is the closed subspace generated by the integer translates of a single function ϕ, then we say that "ϕ generates the multiresolution."

Note that it is not necessary to have all the conditions of the multiresolution analysis in order to be able to implement the decomposition and reconstruction algorithms. For example, we did not assume previously that $\bigcup_{n\in\mathbb{Z}}V_n$ is dense in $L^2(\mathbb{R})$ when discussing those algorithms. The condition that the union of the V_n be dense is present to ensure that given any f in $L^2(\mathbb{R})$ and a tolerance $\varepsilon > 0$, there is an N sufficiently large so that $\text{dist}(f, V_N)$ is less than ε. We can then approximate f by an element of V_N and begin our decomposition from this point without incurring an error larger than the prescribed tolerance. Neither did we assume that the intersection of all V_n was the trivial subspace. If this is not the case, then there are elements in $L^2(\mathbb{R})$ that cannot be decomposed via the wavelet decomposition algorithm—these are precisely the elements in the intersection of all the spaces V_n. Thus if we wish our decomposition to be able to continue down to an "arbitrarily fine" level, we must assume that the intersection of all the V_n is $\{0\}$. This motivates conditions **b** and **c** in Definition 2. The preceding discussion is actually a reiteration in nonmathematical terms of the proof of Theorem 3 in Chapter 34.

In Chapter 34 we investigated the Haar wavelet in some detail. In this case, Theorems 3 and 4 of that chapter assert that conditions **b** and **c** hold in the case of the multiresolution generated by the Haar wavelet. Thus we have the following result.

THEOREM 5. *Let* $\phi = \chi_{[0,1)}$. *Then* ϕ *generates the multiresolution associated with the Haar wavelet.*

We want now to construct other examples of wavelets from the associated multiresolution of $L^2(\mathbb{R})$. The Haar wavelet and the function $\phi = \chi_{[0,1]}$ have the very favorable feature that the sequence $[a_n]$ appearing in the two-scale equation for ϕ has only finitely many nonzero terms. Thus the sums appearing in Equations (18), (19), and (20) of our decomposition and reconstruction algorithms are in fact finite. If $[a_n]$ has infinitely many nonzero terms, then the formulas appearing in these equations will have to be truncated, and errors will result. Special importance therefore attaches to the case when $[a_k]$ is finitely supported; that is, it has only a finite number of nonzero terms. The Haar wavelet has the drawback that it is discontinuous. Thus our discussion will focus on producing wavelets that are continuous. We have to investigate the two-scale equation

$$(21) \qquad \phi(t) = \sum_{k=-\infty}^{\infty} a_k \phi(2t - k)$$

in more detail. To make progress, we make some assumptions on ϕ and on the sequence $[a_n]$. If $\{\phi(\cdot - n) : n \in \mathbb{Z}\}$ forms a stable basis for V_0, then it follows (since $\phi \in V_1$), that $\sum_{n=-\infty}^{\infty} a_n^2 < \infty$. However, it will often help to assume a little more, namely $\sum_{n=-\infty}^{\infty} |a_n| < \infty$. Taking Fourier transforms of both sides of the two-scale equation gives

$$(22) \qquad \hat{\phi}(t) = \frac{1}{2} \sum_{k=-\infty}^{\infty} a_k e^{-\pi i k t} \hat{\phi}\left(\frac{t}{2}\right)$$

Define P by

(23)
$$P(z) = \frac{1}{2} \sum_{k=-\infty}^{\infty} a_k z^k \qquad (z \in \mathbb{C}, |z| = 1)$$

Then Equation (22) becomes

(24)
$$\hat{\phi}(t) = \hat{\phi}\left(\frac{t}{2}\right) Q\left(\frac{t}{2}\right) \qquad \left(Q(t) = P(e^{-2\pi i t})\right)$$

The assumption $\sum_{n=-\infty}^{\infty} |a_n| < \infty$ guarantees that the series defining P converges absolutely and uniformly on the unit circle in $\mathbb{C}$. Hence, P is continuous on $|z| = 1$.

Definition 3. Let ϕ be an element of $L^2(\mathbb{R})$ satisfying the two-scale relation in Equation (21), where $\sum_{n=-\infty}^{\infty} |a_n| < \infty$. The **symbol** for ϕ is defined to be the Laurent series $P(z) = \frac{1}{2} \sum_{n=-\infty}^{\infty} a_n z^n$, where $z \in \mathbb{C}$ and $|z| = 1$.

In Theorem 1, page 285, we gave conditions for the orthonormality of the set $\{\phi(\cdot - n) : n \in \mathbb{Z}\}$. This was done without assuming that ϕ satisfied a two-scale equation. For our wavelet construction as given in Theorem 2 of Chapter 34 to be successful, we must require the orthonormality of the family of translates of ϕ as well as the two-scale equation. It is quite easy to get necessary conditions that follow from the orthonormality of $\{\phi(\cdot - n) : n \in \mathbb{Z}\}$ given that ϕ satisfies a two-scale equation. Useful sufficient conditions are somewhat more elusive. The next two results address the issue of necessary conditions.

LEMMA 4. *Let ϕ be an element of $L^2(\mathbb{R})$ that satisfies the two-scale relation in Equation (2). Assume that $\sum_{n=-\infty}^{\infty} |a_n| < \infty$ and that P is as defined in Definition 3. Then,*

$$\sum_{n=-\infty}^{\infty} |\hat{\phi}(2t + n)|^2 = |P(e^{-2i\pi t})|^2 \sum_{n=-\infty}^{\infty} |\hat{\phi}(t + n)|^2$$

$$+ |P(-e^{-2i\pi t})|^2 \sum_{n=-\infty}^{\infty} \left|\hat{\phi}\left(t + n + \frac{1}{2}\right)\right|^2$$

Proof. In the two-scale equation we take the Fourier transform, obtaining

$$\hat{\phi}(t) = \frac{1}{2} \sum_{k=-\infty}^{\infty} a_k e^{-i\pi k t} \hat{\phi}\left(\frac{t}{2}\right)$$

Define F by $F(t) = \sum_{n=-\infty}^{\infty} |\hat{\phi}(t + n)|^2$. Then, as was noticed in the proof of Theorem 1, F is an integer-periodic function such that $F|[0, 1] \in L^1[0, 1]$. Thus

$$F(2t) = \sum_{n=-\infty}^{\infty} |\hat{\phi}(2t + n)|^2 = \sum_{n=-\infty}^{\infty} \left|\frac{1}{2} \sum_{k=-\infty}^{\infty} a_k e^{-i\pi k(2t+n)} \hat{\phi}\left(t + \frac{n}{2}\right)\right|^2$$

$$= \sum_{n=-\infty}^{\infty} \left|\hat{\phi}\left(t + \frac{n}{2}\right) \frac{1}{2} \sum_{k=-\infty}^{\infty} (-1)^{nk} a_k e^{-2\pi i k t}\right|^2$$

The outer sum in this last line is split into the even and odd values of n as follows:

$$F(2t) = \sum_{n=-\infty}^{\infty} \left| \frac{1}{2}\hat{\phi}\left(t + \frac{2n}{2}\right) \sum_{k=-\infty}^{\infty} (-1)^{2nk} a_k e^{-2\pi i k t} \right|^2$$

$$+ \sum_{n=-\infty}^{\infty} \left| \frac{1}{2}\hat{\phi}\left(t + \frac{2n+1}{2}\right) \sum_{k=-\infty}^{\infty} (-1)^{(2n+1)k} a_k e^{-2\pi i k t} \right|^2$$

A direct comparison with Definition 3 yields

$$F(2t) = \sum_{n=-\infty}^{\infty} |\hat{\phi}(t+n)P(e^{-2i\pi t})|^2 + \sum_{n=-\infty}^{\infty} \left| \hat{\phi}\left(t + n + \frac{1}{2}\right)P(-e^{-2\pi i t}) \right|^2$$

$$= |P(e^{-2\pi i t})|^2 F(t) + |P(-e^{-2\pi i t})|^2 F\left(t + \frac{1}{2}\right) \qquad \blacksquare$$

THEOREM 6. *Let ϕ be an element of $L^2(\mathbb{R})$ satisfying the two-scale relation in Equation (21). If $\sum_{n=-\infty}^{\infty} |a_n| < \infty$ and $\{\phi(\cdot - n) : n \in \mathbb{Z}\}$ is orthonormal, then the symbol P in Definition 3 satisfies $|P(z)|^2 + |P(-z)|^2 = 1$ when $|z| = 1$. Furthermore, $\sum_{n=-\infty}^{\infty} a_n a_{n-2v}$ is 2 if $v = 0$ and is zero if $v \neq 0$ and $v \in \mathbb{Z}$.*

Proof. Since the sequence $[\phi(\cdot - n) : n \in \mathbb{Z}]$ is orthonormal, Theorem 1 shows that $\sum_{n=-\infty}^{\infty} |\hat{\phi}(t+n)|^2 = 1$. Now from Lemma 4,

$$1 = |P(e^{-i\pi t})|^2 \sum_{n=-\infty}^{\infty} \left| \hat{\phi}\left(\frac{t}{2} + n\right) \right|^2 + |P(-e^{-i\pi t})|^2 \sum_{n=-\infty}^{\infty} \left| \hat{\phi}\left(t + \frac{1}{2} + n\right) \right|^2$$

Another application of Theorem 1, together with the substitution $z = e^{-i\pi t}$, produces

$$1 = |P(z)|^2 + |P(-z)|^2 \qquad (|z| = 1)$$

By the definition of P and the assumption that $|z| = 1$, we can write

$$4 = 4|P(z)|^2 + 4|P(-z)|^2 = 4P(z)\overline{P(z)} + 4P(-z)\overline{P(-z)}$$

$$= \left(\sum_{n=-\infty}^{\infty} a_n z^n \right) \left(\overline{\sum_{k=-\infty}^{\infty} a_k z^k} \right) + \left(\sum_{n=-\infty}^{\infty} a_n (-z)^n \right) \left(\overline{\sum_{k=-\infty}^{\infty} a_k (-z)^k} \right)$$

$$= \left(\sum_{n=-\infty}^{\infty} a_n z^n \right) \left(\sum_{k=-\infty}^{\infty} a_k z^{-k} \right) + \left(\sum_{n=-\infty}^{\infty} (-1)^n a_n z^n \right) \left(\sum_{k=-\infty}^{\infty} (-1)^k a_k z^{-k} \right)$$

$$= \sum_{n,k=-\infty}^{\infty} (1 + (-1)^{n+k}) a_n a_k z^{n-k}$$

The terms in this last summation corresponding to odd values of $n + k$ are all zero. Hence, the substitution $n - k = 2v$ allows us to conclude that

$$2 = \sum_{v=-\infty}^{\infty} z^{2v} \sum_{n=-\infty}^{\infty} a_n a_{n-2v} \qquad (z \in \mathbb{C}, |z| = 1)$$

Thus $\sum_{n=-\infty}^{\infty} a_n a_{n-2v}$ must be 2 when $v = 0$ and must otherwise be zero.   ∎

As we have just seen, establishing necessary conditions that follow from the ortho-normality of $[\phi(\cdot - n) : n \in \mathbb{Z}]$ and the two-scale equation is not difficult. We turn next to sufficient conditions. These are rather harder to prove, unless we make further assumptions. We have already assumed that the sequence $[a_n]$ appearing in the two-scale equation satisfies $\sum_{n=-\infty}^{\infty} |a_n| < \infty$. If we assume in addition that $\phi \in L^1(\mathbb{R})$, then $\hat{\phi} \in C(\mathbb{R})$. This has the consequence that in the equation

$$\hat{\phi}(t) = \frac{1}{2} \sum_{n=-\infty}^{\infty} a_n e^{-i\pi nt} \hat{\phi}\left(\frac{t}{2}\right) = P(e^{-i\pi t})\hat{\phi}\left(\frac{t}{2}\right)$$

both $\hat{\phi}$ and P are continuous, so that the above equation holds for all $t \in \mathbb{R}$.

LEMMA 5. *Let ϕ be a function in $L^2(\mathbb{R})$ that vanishes outside the interval $[-M/2, M/2]$. Define $F(t) = \sum_{n=-\infty}^{\infty} |\hat{\phi}(t+n)|^2$. Then there exist coefficients c_v such that $F(t) = \sum_{v=-M}^{M} c_v e^{-2i\pi vt}$.*

Proof. From the proof of Theorem 1, we see that F is integer-periodic and that the restriction of F to $[0, 1]$ is in $L^1[0, 1]$. Also, for any integer v, the corresponding Fourier coefficient of F is

$$c_v := \int_0^1 F(t)e^{-2\pi ivt}\, dt = \sum_{n=-\infty}^{\infty} \int_0^1 |\hat{\phi}(t+n)|^2 e^{-2\pi ivt}\, dt$$

$$= \sum_{n=-\infty}^{\infty} \int_n^{n+1} |\hat{\phi}(u)|^2 e^{-2\pi iv(u-n)}\, du$$

$$= \sum_{n=-\infty}^{\infty} e^{2\pi ivn} \int_n^{n+1} |\hat{\phi}(u)|^2 e^{-2\pi ivu}\, du$$

$$= \int_{-\infty}^{\infty} |\hat{\phi}(u)|^2 e^{-2\pi ivu}\, du$$

Now recall the computational rule that relates the translation operator and the Fourier transform: $(T_v \phi)^\wedge = e_v \hat{\phi}$, where $(T_v \phi)(t) = \phi(t - v)$ and $e_v(t) = e^{-2\pi ivt}$. We also require the Plancherel-Parseval identity: $\langle \phi, \psi \rangle = \langle \hat{\phi}, \hat{\psi} \rangle$. Continuing our calculation, we get

$$c_v = \int_{-\infty}^{\infty} \overline{\hat{\phi}(u)}\hat{\phi}(u)e_v(u)\, du = \int_{-\infty}^{\infty} \overline{\hat{\phi}(u)}(T_v\phi)^\wedge(u)\, du = \int_{-\infty}^{\infty} \overline{\phi(u)}(T_v\phi)(u)\, du$$

Since the support of ϕ is contained in $[-M/2, M/2]$, it is clear that the last integral above is zero if $v \geq M$ or $v \leq -M$. To see that $F(t) = \sum_{v=-M}^{M} c_v e^{-2\pi ivt}$, we define the continuous linear functional L on $C[0, 1]$ by

$$L(g) = \int_0^1 g(x)\left(F(x) - \sum_{v=-M}^{M} c_v e^{-2\pi ivx}\right) dx \qquad (g \in C[0, 1])$$

A simple calculation shows that $L(e_k) = 0$ for all $k \in \mathbb{Z}$. Since $\{e_k : k \in \mathbb{Z}\}$ is fundamental in $C[0, 1]$, we conclude that $L = 0$. Hence $F(x) = \sum_{v=-M}^{M} c_v e^{-2\pi i v x}$ for $x \in [0, 1]$. The integer-periodicity of F now shows that $F(x) = \sum_{v=-M}^{M} c_v e^{-2\pi i v x}$ for all $x \in \mathbb{R}$. ∎

> **LEMMA 6.** *Let ϕ be a function in $L^1(\mathbb{R})$ that satisfies the two-scale relation $\phi(t) = \sum_{n=-\infty}^{\infty} a_n \phi(2t - n)$ with $\sum_{n=-\infty}^{\infty} |a_n| < \infty$.*
>
> > **a.** *If the associated symbol P satisfies $|P(z)|^2 + |P(-z)|^2 = 1$ for all z on the unit circle, and if $\hat{\phi}(0) \neq 0$, then $P(1) = 1$ and $P(-1) = 0$.*
> >
> > **b.** *If $P(-1) = 0$, then $\hat{\phi}(n) = 0$ for all nonzero integers.*

Proof. Part **a**. We previously observed that under the hypotheses of the lemma, the equation

(25)
$$\hat{\phi}(t) = P(e^{-i\pi t}) \hat{\phi}\left(\frac{t}{2}\right)$$

holds for all $t \in \mathbb{R}$. Putting $t = 0$, we see that, since $\hat{\phi}(0) \neq 0$, $P(1) = 1$. From the equation $|P(1)|^2 + |P(-1)|^2 = 1$, it follows that $P(-1) = 0$.

Part **b**. Let n be a nonzero integer. Let p be the nonnegative integer such that $n/2^p \in \mathbb{Z}$ and $n/2^{p+1} \notin \mathbb{Z}$. Then $n = 2^p(n/2^p)$, where $n/2^p$ is an odd integer. Applying Equation (25) repeatedly, we see that

$$\hat{\phi}(n) = \hat{\phi}\left(\frac{n}{2^{p+1}}\right) \prod_{j=0}^{p} P(e^{-i\pi n/2^j})$$

The product term contains $P(e^{-i\pi n/2^p}) = P(e^{-i\pi v})$, where $v = n/2^p$ is an odd integer. Thus $P(e^{-i\pi n/2^p}) = P(-1) = 0$, and hence $\hat{\phi}(n) = 0$. ∎

We now come to the sufficient condition for the orthonormality of the family $\{\phi(\cdot - n) : n \in \mathbb{Z}\}$.

> **THEOREM 7.** *Let ϕ be a function in $L^2(\mathbb{R})$ that is compactly supported and satisfies a finite two-scale equation $\phi(t) = \sum_{n=0}^{N} a_n \phi(2t - n)$. If ϕ and its associated symbol P satisfy*
>
> > **a.** $|P(z)|^2 + |P(-z)|^2 = 1$ *when* $|z| = 1$
> >
> > **b.** $P(1) = 1$
> >
> > **c.** $P(e^{i\pi t}) \neq 0$ *for all* $t \in [-1/2, 1/2]$
> >
> > **d.** $\hat{\phi}(0) = 1$
>
> *then $\{\phi(\cdot - n) : n \in \mathbb{Z}\}$ is orthonormal.*

Proof. Set $F(t) = \sum_{n=-\infty}^{\infty} |\hat{\phi}(t + n)|^2$ for $t \in \mathbb{R}$. We have already observed that F is an integer-periodic function. From Lemma 5, we deduce that $F|[-1/2, 1/2]$ belongs to the class $C[-1/2, 1/2]$. From Lemma 4,

$$F(2t) = |P(e^{-2\pi i t})|^2 F(t) + |P(-e^{-2\pi i t})|^2 F(t + 1/2)$$

Let m be the minimum of $F(t)$ on the interval $[-1/2, 1/2]$, and let t_0 be a point in that interval where $F(t_0) = m$. Then

$$m = F(t_0) = |P(e^{-i\pi t_0})|^2 F(t_0/2) + |P(-e^{-i\pi t_0})|^2 F((t_0 + 1)/2)$$
$$\geq m|P(e^{-i\pi t_0})|^2 + m|P(-e^{-i\pi t_0})|^2 = m$$

In this equation, since $t_0 \in [1/2, -1/2]$, $P(e^{-i\pi t_0}) \neq 0$. It then follows that $F(t_0/2) = m$. Repeating this argument, we obtain $F(2^{-j}t_0) = m$, for $j = 0, 1, \ldots$. Since F is continuous at zero, we conclude that $m = \lim_{j \to \infty} F(2^{-j}t_0) = F(0)$. Lemma 6 now shows that

$$m = F(0) = \sum_{n=-\infty}^{\infty} |\hat{\phi}(n)|^2 = |\hat{\phi}(0)|^2 = 1$$

Now let M be the maximum of $F(t)$ in the interval $[-1/2, 1/2]$, and let t_1 be a point in that interval where $F(t_1) = M$. Then

$$M = F(t_1) = |P(e^{-i\pi t_1})|^2 F\left(\frac{t_1}{2}\right) + |P(-e^{-i\pi t_1})|^2 F\left(\frac{t_1 + 1}{2}\right)$$
$$\leq M|P(e^{-i\pi t_1})|^2 + M|P(-e^{-i\pi t_1})|^2 = M$$

As before, since $t_1 \in [-1/2, 1/2]$, $P(e^{-i\pi t_1}) \neq 0$, and so we conclude that $F(t_1/2) = M$. As with the previous argument, $M = \lim_{j \to \infty} F(2^{-j}t_1) = F(0) = 1$. Hence, $F(t) = 1$ for all $t \in [-1/2, 1/2]$. The integer-periodicity of F shows that $F(t) = 1$ for all $t \in \mathbb{R}$, and then Theorem 1 shows that $\{\phi(\cdot - n)\}$ is orthonormal. ∎

A related question concerns how to solve the two-scale equation, interpreting it as a fixed–point problem. This brings us to the cascade algorithm, which we now describe. Let ϕ be an element of $L^2(\mathbb{R})$ that satisfies, for some $a \in \ell^1$,

$$\phi(t) = \sum_n a_n \phi(2t - n)$$

Starting with any nonzero element ϕ_0 in $L^2(\mathbb{R})$, define the sequence $[\phi_m]_{m=0}^{\infty}$ recursively by

(26) $$\phi_m(t) = \sum_n a_n \phi_{m-1}(2t - n) \qquad (m = 1, 2, \ldots)$$

We naturally hope that the sequence $[\phi_m]$ converges to a solution of the two-scale equation. Problem 18 indicates that the question of convergence is rather delicate. The iteration described in Equation (26) is known as the **cascade algorithm,** and the next few results concern its convergence. Because we shall eventually want to construct a function ϕ in $L^2(\mathbb{R})$ such that $\{\phi(\cdot - n) : n \in \mathbb{Z}\}$ is orthonormal, we freely assume in the following discussion that the necessary condition $P(1) = 1$ (Lemma 6) is satisfied. This is equivalent to $\sum_n a_n = 2$, by virtue of Definition 3 on page 294. If we assume also that $P(-1) = 0$, then $\sum_n (-1)^n a_n = 0$ by Lemma 6. Thus

$$\sum_{\substack{n \text{ odd}}} a_n = \sum_{\substack{n \text{ even}}} a_n = 1$$

Throughout our discussion of the cascade algorithm, convergence of the algorithm will be understood in the L^2-sense.

LEMMA 7. *Let* $[\phi_m : m = 0, 1, 2, \dots]$ *be constructed as described above. Set*

$$c_{ij}(k) = \int_{-\infty}^{\infty} \phi_i(t)\phi_j(t + k)\, dt \qquad (k \in \mathbb{Z},\, i, j \in \mathbb{Z}_+)$$

Then

$$c_{ij}(k) = \sum_m \left(\sum_n \frac{1}{2} a_n a_{n+m-2k} \right) c_{i-1, j-1}(m) \qquad (k \in \mathbb{Z},\, i, j \in \mathbb{N})$$

Proof. In the following calculation, the steps are justified by the absolute convergence of the series involved.

$$c_{ij}(k) = \int_{-\infty}^{\infty} \phi_i(t)\phi_j(t + k)\, dt$$

$$= \int_{-\infty}^{\infty} \left(\sum_n a_n \phi_{i-1}(2t - n) \right) \left(\sum_m a_m \phi_{j-1}(2t + 2k - m) \right) dt$$

$$= \sum_n \sum_m a_n a_m \int_{-\infty}^{\infty} \phi_{i-1}(2t - n)\phi_{j-1}(2t + 2k - m)\, dt$$

$$= \sum_n \sum_m a_n a_m \int_{-\infty}^{\infty} \phi_{i-1}(u)\phi_{j-1}(u + n + 2k - m)\, \frac{du}{2}$$

$$= \frac{1}{2} \sum_n \sum_m a_n a_m c_{i-1, j-1}(n + 2k - m)$$

We now make the substitution $n = v - 2k + m$, to get

$$c_{ij}(k) = \frac{1}{2} \sum_v \sum_m a_{v-2k+m} a_m c_{i-1, j-1}(v) \qquad \blacksquare$$

LEMMA 8. *If* $a \in \ell^1(\mathbb{Z})$, *then the same is true of the sequence* b *defined by* $b(n) = \sum_k a_k a_{k+n}$.

Proof. We note that

$$\sum_k \sum_n |a_k a_{n+k}| = \sum_k |a_k| \sum_n |a_{n+k}| = \|a\|_1^2 < \infty$$

Hence, by the Fubini Theorem (applied to $\mathbb{Z}$ with counting measure) the order of summation can be reversed to obtain

$$\infty > \sum_n \sum_k |a_k a_{k+n}| \geq \sum_n \left| \sum_k a_k a_{k+n} \right| = \sum_n |b_n| \qquad \blacksquare$$

Now, define a doubly infinite matrix, T, whose elements are

$$t_{km} = \frac{1}{2} \sum_n a_n a_{n+m-2k}$$

By the preceding lemma, we have

$$\sum_k |t_{km}| < \infty \qquad \sum_m |t_{km}| < \infty$$

Therefore, T defines continuous linear operators on $\ell^\infty(\mathbb{Z})$ in two ways: $x \mapsto y$ by $y = Tx$ or $y = xT$. That is,

$$y_k = \sum_m t_{km} x_m \qquad \text{or} \qquad y_m = \sum_k t_{km} x_k$$

In particular, we have, by Lemma 7, $c_{ij} = T c_{i-1, j-1}$.

LEMMA 9. *If* $\sum a_{2n} = \sum a_{2n+1} = 1$ *and if* $e_n = 1$ *for all* n *in* $\mathbb{Z}$, *then* $eT = e$ *and* $Te = 2e$.

Proof. We use Lemma 8 in the calculation:

$$(eT)_m = \sum_k e_k t_{km} = \sum_k t_{km} = \sum_k \frac{1}{2} \sum_n a_n a_{n+m-2k}$$

$$= \frac{1}{2} \sum_n a_n \sum_k a_{n+m-2k}$$

Now the sum $\sum_k a_{n+m-2k}$ equals $\sum a_{2k}$ if $n + m$ is even, and it equals $\sum a_{2k+1}$ if $n + m$ is odd. In either case the sum is 1 by observations made following Equation (26). Hence the sum being computed reduces to $(1/2) \sum_n a_n = 1$. The sum needed to establish $Te = 2e$ is similar:

$$(Te)_k = \sum_m t_{km} = \sum_m \frac{1}{2} \sum_n a_n a_{n+m-2k}$$

$$= \frac{1}{2} \sum_n \sum_m a_n a_{n+m-2k} = \frac{1}{2} \sum_n a_n \sum_m a_{n-2k+m}$$

$$= \frac{1}{2} \sum_n a_n \sum_m a_m = \frac{1}{2}(2)(2) = 2 \qquad \blacksquare$$

LEMMA 10. *If the coefficient sequence* $[a_n]$ *in the two-scale equation is supported on the set* $\{0, 1, \ldots, N\}$ *and satisfies* $\sum a_{2n} = \sum a_{2n+1} = 1$, *then*

$$\sum_{k=1-N}^{N-1} t_{km} = 1 \qquad (|m| \le N - 1)$$

Proof. By Lemma 9 we know that $\sum_k t_{km} = 1$ for all m. Thus in the present circumstances, we only have to show that $t_{km} = 0$ when $|m| \le N - 1$ and $|k| \ge N$. Now,

$$t_{km} = \frac{1}{2} \sum_{n=0}^{N} a_n a_{n+m-2k}$$

By our assumptions,

$$|m - 2k| \geq 2|k| - |m| \geq 2N - (N - 1) = N + 1$$

Hence, if $m - 2k$ is positive, we have $n + m - 2k \geq 0 + N + 1$ and $a_{n+m-2k} = 0$. If $m - 2k$ is negative, we have

$$n + m - 2k \leq N + m - 2k = N - |m - 2k| \leq N - (N + 1) = -1$$

whence again $a_{n+m-2k} = 0$. ∎

LEMMA 11. *Let ϕ be a function in $L^2(\mathbb{R})$ whose support lies in $[0, N]$. Assume that ϕ satisfies the two-scale relation in which the coefficient sequence is supported on $\{0, 1, \ldots, N\}$. For $k \in \mathbb{Z}$, define*

$$v_k = \int_{-\infty}^{\infty} \phi(t)\phi(t + k)\, dt \qquad (k \in \mathbb{Z})$$

Then $v_k = 0$ when $|k| \geq N$, and

$$\sum_{j=-\infty}^{\infty} t_{kj} v_j = \sum_{|j| < N} t_{kj} v_j = v_k \qquad \text{for all } k \in \mathbb{Z}$$

Proof. Since the support of ϕ lies in $[0, N]$, we obviously have

$$v_k = \int_0^N \phi(t)\phi(t + k)\, dt$$

If $k \geq N$, then in the above integral $t + k \geq N$ and $\phi(t + k) = 0$. Hence $v_k = 0$. Similarly, if $k \leq -N$, then $v_k = 0$. Now, just as in the proof of Lemma 7, we can establish that

$$v_k = \sum_m \sum_n \frac{1}{2} a_n a_{n+m-2k} v_m = \sum_m t_{km} v_m \qquad (k \in \mathbb{Z})$$

Since $v_m = 0$ when $|m| \geq N$, we can write

$$v_k = \sum_{|m| < N} t_{km} v_m \qquad (|k| < N) \qquad ∎$$

From this point on, let T be the square matrix whose elements are t_{km} for $|k| < N$ and $|m| < N$. The crux of our investigation of the convergence of the cascade algorithm is that 1 should be a simple eigenvalue of T and that all other eigenvalues of T should be less than 1 in modulus. Such a matrix is said to satisfy *Condition E.* If T satisfies this condition, then the power method for computing eigenvectors simply consists of forming the sequence $\{T^n b\}_{n=0}^{\infty}$, where b is a fixed vector in $\mathbb{R}^{2N-1}$. Problem 19 shows that this sequence converges either to zero or to an eigenvector of T corresponding to the eigenvalue 1, provided that T has a complete set of eigenvectors. In fact, this result continues to hold if T fails to have a complete set of eigenvectors. We give a proof of this fact on the following page, in Lemma 12.

For a function ϕ in $L^2(\mathbb{R})$, the support of ϕ is taken to be a set on the complement of which $\phi(x) = 0$. Thus, the equivalence class of functions equal to ϕ almost everywhere should contain a member having that property.

LEMMA 12. *Let A be an $n \times n$ matrix satisfying Condition E. Let b and d be points in $\mathbb{R}^n$ such that $Ad = d$, $b^T A = b^T$ and $b^T d = 1$. Then for each $x \in \mathbb{R}^n$, $A^k x \longrightarrow db^T x$. Furthermore, this convergence is uniform on bounded subsets of $\mathbb{R}^n$.*

Proof. By the Jordan decomposition theorem, there exist matrices S and J such that $A = SJS^{-1}$, and J is upper triangular. From this equation, it follows that $A^k = SJ^k S^{-1}$. Since A satisfies Condition E, we can find matrices B, C, D such that

$$A = (d \ \ C)\begin{pmatrix} 1 & 0 \\ 0 & B \end{pmatrix}\begin{pmatrix} b^T \\ D \end{pmatrix}$$

Here, d is $n \times 1$, C is $n \times (n-1)$, B is $(n-1) \times (n-1)$, b^T is $1 \times n$, and D is $(n-1) \times n$. Because A satisfies Condition E, the matrix B consists of Jordan blocks corresponding to the eigenvalues that are less than 1 in modulus. Hence, $B^k \longrightarrow 0$ as $k \longrightarrow \infty$. (See Problems 22 and 23.) Now,

$$A^k x = (d \ \ C)\begin{pmatrix} 1 & 0 \\ 0 & B^k \end{pmatrix}\begin{pmatrix} b^T \\ D \end{pmatrix} x = db^T x + CB^k Dx$$

It is now clear that $A^k x \longrightarrow db^T x$. Moreover, if $\|x\| \leq M$ then

$$\|db^T x - A^k x\| = \|CB^k Dx\| \leq \|C\|\|B^k\|\|D\|M \longrightarrow 0$$

as $k \longrightarrow \infty$. This establishes the required uniform convergence. ∎

Let us now review our progress in understanding the cascade algorithm. Suppose that $[\phi_m]$ is the sequence in $L^2(\mathbb{R})$ generated by this algorithm, as described in Equation (26). Suppose also that the initial function ϕ_0 has support in $[0, N]$. Our aim will be to show that $[\phi_m]$ is a Cauchy sequence in $L^2(\mathbb{R})$. From Lemma 7, we see that, for $m > n$,

$$\|\phi_m - \phi_n\|_2^2 = \|\phi_m\|_2^2 - 2\langle \phi_m, \phi_n \rangle + \|\phi_n\|_2^2 = c_{mm}(0) - 2c_{mn}(0) + c_{nn}(0)$$

$$= (T^m c_{00})(0) - 2(T^n c_{m-n,0})(0) + (T^n c_{00})(0)$$

If T satisfies condition E, then $T^m c_{00} \longrightarrow c$ as $m \longrightarrow \infty$, where c is a right eigenvector for T corresponding to eigenvalue 1. To make the quantity $\|\phi_m - \phi_n\|_2$ small as m and n become large, it must be that $T^n c_{m-n,0}$ converges uniformly to this same vector c. Now, the power method applied to T will always produce a sequence $[T^m v_0]$ that converges to αc for some $\alpha \in \mathbb{R}$, since T satisfies Condition E. The number α depends on the choice of starting vector v_0, as we saw in Lemma 11. We must therefore control the vectors c_{i0} in some suitable fashion so as to ensure that $T^n c_{m-n,0} \longrightarrow c$ as $m, n \longrightarrow \infty$. It is not surprising that this control is achieved by making a suitable choice for ϕ_0, as in Theorem 8, below.

LEMMA 13. *In the cascade algorithm, let the initial function ϕ_0 be supported on $[0, N]$, and let the coefficient sequence $[a_n]$ be supported on $\{0, 1, \ldots, N\}$. Then each ϕ_m is supported on $[0, N]$.*

Proof. Use induction on m. The algorithm gives us

$$\phi_m(t) = \sum_n a_n \phi_{m-1}(2t - n) = \sum_{n=0}^{N} a_n \phi_{m-1}(2t - n)$$

Assume that $\phi_{m-1}(x) = 0$ when $x \notin [0, N]$. If $t > N$, then in the equation above $2t - n > 2N - N = N$ and $\phi_{m-1}(2t - n) = 0$. If $t < 0$, then $2t - n < 0$, and again $\phi_{m-1}(2t - n) = 0$. ∎

THEOREM 8. *In the cascade algorithm, assume that*

    a. $\operatorname{supp}(\phi_0) \subset [0, N]$
    b. $\operatorname{supp}(a) \subset \{0, 1, \ldots, N\}$
    c. $\sum a_{2n} = \sum a_{2n+1} = 1.$

If the sequence $[\phi_m]$ generated by the algorithm converges in $L^2(\mathbb{R})$, then $\sum_n \phi_0(t - n)$ is constant.

Proof. Let $\phi_m \to \phi$. By Lemma 13, $\operatorname{supp}(\phi_m) \subset [0, N]$ for all m. It then follows that $\operatorname{supp}(\phi) \subset [0, N]$. Indeed, let $E = \mathbb{R} \setminus [0, N]$, so that $\phi_m \mid E = 0$. Then $\phi \mid E = 0$, since $\| \phi \mid E \| = \|(\phi - \phi_m) \mid E\| \leq \|\phi - \phi_m\| \to 0$. Define

$$f(t) = \sum_n \phi(t - n) \qquad \text{and} \qquad f_m(t) = \sum_n \phi_m(t - n) \qquad (m = 0, 1, \ldots)$$

These functions are integer-periodic, and we interpret them as functions in $L^2[0, 1]$. (In this connection, see Problem 17.) With that understood, the sums defining f and f_m are finite, and consequently, $f_m \to f$. Now we wish to prove that

(27) $$f_m(t) = f_{m-1}(2t) = f_{m-2}(4t) = \cdots = f_0(2^m t)$$

The computation accomplishing this is as follows:

$$f_m(t) = \sum_n \phi_m(t - n) = \sum_n \sum_j a_j \phi_{m-1}(2t - 2n - j)$$

$$= \sum_j a_{2j} \sum_n \phi_{m-1}(2t - 2n - 2j) + \sum_j a_{2j+1} \sum_n \phi_{m-1}(2t - 2n - 2j - 1)$$

$$= \sum_j a_{2j} \sum_n \phi_{m-1}(2t - 2n) + \sum_j a_{2j+1} \sum_n \phi_{m-1}(2t - 2n - 1)$$

$$= \sum_n \phi_{m-1}(2t - 2n) + \sum_n \phi_{m-1}(2t - 2n - 1)$$

$$= \sum_n \phi_{m-1}(2t - n)$$

$$= f_{m-1}(2t)$$

Next we compute the k-th Fourier coefficient A_k of f_0.

$$A_k = \int_0^1 f_0(t)e^{2\pi ikt}\, dt = \frac{1}{2^m}\int_0^{2^m} f_0(t)e^{2\pi ikt}\, dt$$

(28)
$$= \frac{1}{2^m}\int_0^{2^m} f_m(t/2^m)e^{2\pi ikt}\, dt$$

$$= \int_0^1 f_m(u)e^{2^{m+1}\pi iku}\, du$$

By the Cauchy-Schwarz inequality, we have, for all $m \in \mathbb{N}$,

$$\left|\int_0^1 f_0(t)e^{2\pi ikt}\, dt\right| \leq \left|\int_0^1 (f_m(u)-f(u))e^{2^{m+1}\pi iku}\, du\right| + \left|\int_0^1 f(u)e^{2^{m+1}\pi iku}\, du\right|$$

$$\leq \|f_m - f\| + |A_{k2^m}|$$

We have already established that $\|f_m - f\| \to 0$ as $m \to \infty$. The Fourier coefficient A_{k2^m} also converges to 0 when $k \neq 0$. We therefore conclude that all the Fourier coefficients of f_0 are zero except the zeroth coefficient. It follows that f_0 is constant almost everywhere on $[0, 1]$. The integer periodicity of f_0 allows us to conclude that f_0 is constant almost everywhere on $\mathbb{R}$.   ∎

LEMMA 14. *Assume the hypotheses of Theorem 8. If the sequence $[\phi_m]$ converges in $L^2(\mathbb{R})$ to ϕ, then $\sum_n \phi_0(t - n) = \hat{\phi}(0)$.*

Proof. We continue to employ the notation in the proof of Theorem 8. By Equation (28), with $k = 0$, we have

$$\int_0^1 f_0(t)\, dt = \int_0^1 f_m(t)\, dt$$

Recall that the Fourier transform, considered as an operator from $(L^1(\mathbb{R}), \|\cdot\|_1)$ to $(C_0(\mathbb{R}), \|\cdot\|_\infty)$, is continuous. (The space C_0 contains all continuous functions that vanish at infinity.) Since $\|\phi_m - \phi\|_2 \to 0$, and since these functions have compact support, we have $\|\phi_m - \phi\|_1 \to 0$, by Problem 17. Hence $\hat{\phi}_m(0) \to \hat{\phi}(0)$, by Problem 27. The following calculation can now be made.

$$\hat{\phi}(0) = \lim_m \hat{\phi}_m(0) = \lim_m \int_{\mathbb{R}} \phi_m(t)\, dt = \lim_m \int_0^N \phi_m(t)\, dt$$

$$= \lim_m \sum_{n=0}^{N-1} \int_n^{n+1} \phi_m(t)\, dt = \lim_m \sum_{n=1-N}^0 \int_0^1 \phi_m(t - n)\, dt$$

$$= \lim_m \int_0^1 \sum_n \phi_m(t - n)\, dt = \lim_m \int_0^1 f_m(t)\, dt$$

$$= \int_0^1 f_0(t)\, dt = \int_0^1 \sum_n \phi_0(t - n)\, dt$$

Since $\sum \phi_0(t - n)$ is a constant (by Theorem 8), the calculation shows that the constant must be $\hat{\phi}(0)$.   ∎

Lemma 14 is useful because Theorem 7 (page 297) shows that if we are to manufacture a compactly supported ϕ with orthonormal translates, then a convenient normalization to adopt is $\hat{\phi}(0) = 1$. To obtain such a function as a limit of the cascade algorithm, we must start with ϕ_0 that satisfies $\sum_{n=-\infty}^{\infty} \phi_0(t - n) = 1$ for almost all $t \in \mathbb{R}$. Now let us return to the question of the convergence of the cascade algorithm. The discussion prior to Theorem 8 emphasized the importance of establishing that $T^n c_{m-n,0}$ converges to a vector c independent of m. This is the substance of the next two results.

LEMMA 15. *In the cascade algorithm, assume that the coefficient sequence $[a_n]$ and the initial function ϕ_0 are compactly supported. Then for all m, $\int \phi_m(t)\, dt = c \int \phi_{m-1}(t)\, dt$, where $c = \sum_n a_n / 2$.*

Proof. From the definition of the cascade algorithm in Equation (26) on page 298 we have

$$\int \phi_m(t)\, dt = \int \sum_n a_n \phi_{m-1}(2t - n)\, dt = \sum_n a_n \int \phi_{m-1}(2t - n)\, dt$$

$$= \sum_n a_n \int \phi_{m-1}(2t)\, dt = \sum_n a_n \int \phi_{m-1}(u)\, du/2$$

$$= c \int \phi_{m-1}(u)\, du \qquad \blacksquare$$

LEMMA 16. *Assume the hypotheses of Theorem 8. If T satisfies condition E and if $\sum_n \phi_0(t + n) = 1$ almost everywhere, then $\lim_k T^k c_{i0}$ exists and is independent of i.*

Proof. The definition of c_{i0} in Lemma 7 (page 299) is

$$c_{i0} = \int \phi_i(t)\phi_0(t + n)\, dt$$

From this it is easily deduced that $c_{i0}(n) = 0$ when $|n| \geq N$. By Lemma 15,

$$\sum_n c_{i0}(n) = \sum_n \int \phi_i(t)\phi_0(t + n)\, dt = \int \phi_i(t) \sum_n \phi_0(t + n)\, dt$$

$$= \int \phi_i(t)\, dt = \int \phi_{i-1}(t)\, dt = \cdots = \int \phi_0(t)\, dt$$

$$= \sum_n \int_0^1 \phi_0(t + n)\, dt = \int_0^1 \sum_n \phi_0(t + n)\, dt$$

$$= \int_0^1 1\, dt = 1$$

In Lemma 12 (page 302), let $b = (1, \ldots, 1)^T$ and let d be a vector such that $b^T d = 1$ and $Td = d$. We know by Lemma 10 (page 300) that $b^T T = b^T$. By Lemma 12, $T^k c_{i0} \rightarrow d b^T c_{i0} = d$. (The calculation at the beginning of the proof shows that $b^T c_{i0} = 1$.) $\qquad \blacksquare$

THEOREM 9. *Let the coefficient sequence* $[a_n]$ *be supported on the set* $\{0, 1, \dots, N\}$, *and assume that* $\sum_n a_{2n} = \sum_n a_{2n+1} = 1$. *Let the matrix* T *given by*

$$T = (t_{km}) = \left(\frac{1}{2} \sum_{n=0}^{N} a_n a_{n+m-2k} \right) \qquad (|k|, |m| \le N - 1)$$

have 1 *as a simple eigenvalue and all other eigenvalues less than* 1 *in modulus. Let* $\phi_0 \in L^2(\mathbb{R})$ *have support in* $[0, N]$, *and satisfy* $\sum_{n=-\infty}^{\infty} \phi_0(t - n) = 1$ *for almost all* $t \in \mathbb{R}$. *Define* $[\phi_m]$ *in* $L^2(\mathbb{R})$ *by*

$$\phi_m(t) = \sum_{n=0}^{N} a_n \phi_{m-1}(2t - n) \qquad (t \in \mathbb{R}, m = 1, 2, \dots)$$

Then the sequence ϕ_m *converges in* $L^2(\mathbb{R})$ *to a function* ϕ *whose support is contained in* $[0, N]$ *and satisfies* $\hat{\phi}(0) = 1$.

Proof. We have already computed that for $m > n$,

$$\|\phi_m - \phi_n\|_2^2 = (T^m c_{00})(0) - 2(T^n c_{m-n,0})(0) + (T^m c_{00})(0)$$

Let $d \in \mathbb{R}^{2N-1}$ be as specified in the proof of Lemma 16. Then

$$\|\phi_m - \phi_n\|_2^2 \le |(T^m c_{00})(0) - d(0)| + 2|(T^n c_{m-n,0})(0) - d(0)|$$
$$+ |(T^m c_{00})(0) - d(0)|$$
$$\le \|T^m c_{00} - d\|_\infty + 2\|T^n c_{m-n,0} - d\|_\infty + \|T^m c_{00} - d\|_\infty$$

Lemma 12 enables us to see that the three quantities on the right-hand side of the above inequality can be made as small as we wish, simply by choosing m and n sufficiently large. Hence, $[\phi_m]$ is a Cauchy sequence. Since $L^2(\mathbb{R})$ is complete, there is a $\phi \in L^2(\mathbb{R})$ such that $\|\phi - \phi_m\|_2 \to 0$ as $m \to \infty$. The fact that ϕ has compact support follows now from the first part of the proof of Theorem 8. Lemma 14 (page 304) implies that $\hat{\phi}(0) = 1$. ∎

Next on the agenda is to apply the knowledge developed so far in this chapter to the construction of one of the compactly supported orthonormal wavelets first discovered by Daubechies. We suppose that the scaling function ϕ satisfies

$$\phi(t) = \sum_{n=0}^{3} a_n \phi(2t - n)$$

for almost all $t \in \mathbb{R}$. We want $\{\phi(\cdot - n) : n \in \mathbb{Z}\}$ to be an orthonormal set, and we therefore impose the necessary conditions on the symbol (Theorem 6, page 295):

$$|P(z)|^2 + |P(-z)|^2 = 1 \qquad (z \in \mathbb{C}, |z| = 1)$$

We will also assume $\hat{\phi}(0) = 1$. Then $P(1) = 1$ and $P(-1) = 0$ by Lemma 6, page 297. Thus P contains $(1 + z)$ as a factor. Since P is a cubic polynomial, we will try to construct P with the form

$$P(z) = \left(\frac{1 + z}{2} \right)^2 S(z) \qquad (z \in \mathbb{C}, |z| = 1)$$

This choice guarantees that $P(-1) = 0$. We see also that the equation $P(1) = 1$ will hold if and only if $S(1) = 1$. In fact, the assumption on P is a special case of a general procedure, in which one would assume

$$P(z) = \left(\frac{1+z}{2}\right)^N S(z)$$

where N is at the user's disposal. Writing

$$P(z) = \left(\frac{1+z}{2}\right)^2 (p_0 + p_1 z)$$

and using the equation $P(1) = 1$ gives

(29) $$p_0 + p_1 = 1$$

The equation $|P(i)|^2 + |P(-i)|^2 = 1$ leads to

$$1 = |P(i)|^2 + |P(-i)|^2 = \left|\left(\frac{1+i}{2}\right)^2 (p_0 + ip_1)\right|^2 + \left|\left(\frac{1-i}{2}\right)^2 (p_0 - ip_1)\right|^2$$

(30) $$= \left|\frac{i}{2}(p_0 + ip_1)\right|^2 + \left|-\frac{i}{2}(p_0 - ip_1)\right|^2$$

$$= \frac{1}{2}(p_0^2 + p_1^2)$$

Solving Equations (29) and (30) gives either

$$p_0 = \frac{1 + \sqrt{3}}{2} \qquad \text{and} \qquad p_1 = \frac{1 - \sqrt{3}}{2}$$

or vice versa. This leads to

$$P(z) = \frac{p_0}{2} + \frac{p_0 + 1}{2} z + \frac{2 - p_0}{2} z^2 + \frac{1 - p_0}{2} z^3$$

Hence, the two-scale equation is

$$\phi(t) = \frac{p_0}{2}\phi(2t) + \frac{p_0 + 1}{2}\phi(2t - 1) + \frac{2 - p_0}{2}\phi(2t - 2) + \frac{1 - p_0}{2}\phi(2t - 3)$$

It is now a tedious computation (or a task for a computer algebra package) to check that $|P(z)|^2 + |P(-z)|^2 = 1$ for all $z \in \mathbb{C}$ with $|z| = 1$.

At this stage, we can determine whether the two-scale equation has a solution with compact support. To do this, we have to compute the matrix T as described in Theorem 9, page 306. This produces

$$T = \begin{pmatrix} 0 & -1/16 & 0 & 0 & 0 \\ 1 & 9/16 & 0 & -1/16 & 0 \\ 0 & 9/16 & 1 & 9/16 & 0 \\ 0 & -1/16 & 0 & 9/16 & 1 \\ 0 & 0 & 0 & -1/16 & 0 \end{pmatrix}$$

A simple computation reveals that the eigenvalues of T are 1, 0.5, 0.25, 0.25 and 0.125. Thus T satisfies Condition E, and so by Theorem 9, the two-scale equation has a solution with support in $[0, 3]$. Note that we do not know whether the two-scale equation has other solutions. It certainly has the trivial solution, $\phi = 0$, and any solution is only well-determined up to a constant (although in our case the constant is determined by the assumption $\hat{\phi}(0) = 1$). We assume therefore from now on that the solution to be used has support in $[0, 3]$.

We now need to consider whether the translates of this solution of the two-scale equation form an orthonormal set. It is elementary to see that the trigonometric polynomial $t \mapsto P(e^{-i\pi t})$ has no zeros in $[-1/2, 1/2]$. It therefore follows from Theorem 7 that $\{\phi(\cdot - n) : n \in \mathbb{Z}\}$ is an orthonormal set. An application of Theorem 2 in Chapter 34 (page 278) then leads us to the Daubechies wavelet:

$$\psi(t) = \frac{1 - \sqrt{3}}{4}\phi(2t + 2) - \frac{3 - \sqrt{3}}{4}\phi(2t + 1) + \frac{3 + \sqrt{3}}{4}\phi(2t) - \frac{1 + \sqrt{3}}{4}\phi(2t - 1)$$

A routine calculation shows that the support of ψ is contained in the interval $[-1, 2]$.

In our discussion, we did not treat the smoothness properties of ϕ. Note that because ψ is a finite combination of translates of ϕ, ψ inherits the smoothness of ϕ. In fact, $\phi \in \text{Lip}^{0.55}(\mathbb{R})$, which means that an inequality of the following form is valid:

$$|\phi(t + h) - \phi(t)| \le |h|^{0.55}$$

We now turn to the question of how to evaluate the function ϕ that has just been constructed. The following lemma is helpful.

LEMMA 17. *Let ϕ be a function in $L^2(\mathbb{R})$ that satisfies the usual two-scale relation (Equation 20 on page 278). Assume that ϕ and the sequence $[a_n]$ are compactly supported, and that $\sum_n a_{2n} = \sum_n a_{2n+1} = 1$. Then $\sum_n \phi(t + n)$ is constant and equal to $\int \phi(u)\,du$.*

Proof. Apply Theorem 8 (page 303), letting $\phi_0 = \phi$. The cascade algorithm is then stationary: $\phi_m = \phi$ for all m. By Theorem 8, $\sum_n \phi(t - n)$ is constant. By Lemma 14 (page 304), the value of this constant is $\hat{\phi}(0)$. ∎

Now let us return to the explicit ϕ constructed prior to Lemma 17. Since we ensured that $P(1) = 1$, it follows that $\sum_{k=0}^{3} a_k = 2$. Clearly, the other conditions of Lemma 17 are satisfied. The fact that $\hat{\phi}(0) = 1$ allows us to conclude that

$$\phi(1) + \phi(2) = \int_{-\infty}^{\infty} \phi(t)\,dt = \hat{\phi}(0) = 1$$

By substituting $t = 1$ in the two-scale equation, we obtain

$$\phi(1) = \frac{1 + \sqrt{3}}{4}\phi(2) + \frac{3 + \sqrt{3}}{4}\phi(1).$$

These two equations give $\phi(1) = 1 + \sqrt{3}$ and $\phi(2) = 1 - \sqrt{3}$. With these values in hand, the two-scale equation can be used repeatedly to compute values of ϕ at any dyadic rational point. For example,

$$\phi\left(\frac{1}{2}\right) = a_0\phi(1) + a_1\phi(0) + a_2\phi(-1) + a_3\phi(-2) = a_0\phi(1) = 1 + \frac{1}{2}\sqrt{3}$$

The following theorem is adapted from Katznelson's book [Ka, page 11]. It was required in the proof of Theorem 2.

THEOREM 10. *Let* $[h_n]$ *be a sequence of continuous, nonnegative, integer-periodic functions such that* $\int_0^1 h_n(t)\, dt = 1$ *for all n and* $\lim_{n\to\infty} \int_\varepsilon^{1-\varepsilon} h_n(t)\, dt = 0$ *for each ε in* $(0, 1/2)$. *Let* $f \in L^1[0, 1]$ *and be extended to be integer-periodic. Define* $f_n(t) = \int_0^1 h_n(t - s)f(s)\, ds$. *Then* $\|f_n - f\|_1 \to 0$.

Proof. In the integral defining f_n, let $\sigma = t - s$. Then use the periodicity of h_n and f to obtain

$$f_n(t) = \int_t^{t-1} h_n(\sigma)f(t - \sigma)(-d\sigma) = \int_{-1/2}^{1/2} h_n(\sigma)f(t - \sigma)\, d\sigma$$

Since $\int_{-1/2}^{1/2} h_n(\sigma)\, d\sigma = 1$, we have

$$f_n(t) - f(t) = \int_{-1/2}^{1/2} h_n(\sigma)[f(t - \sigma) - f(t)]\, d\sigma$$

Therefore

$$\|f_n - f\| = \int_{-1/2}^{1/2} |f_n(t) - f(t)|\, dt \le \int_{-1/2}^{1/2} \int_{-1/2}^{1/2} h_n(\sigma)|f(t - \sigma) - f(t)|\, d\sigma\, dt$$

Now use the notation $(T_\sigma f)(t) = f(t - \sigma)$ and recall Lemma 2 in Chapter 20 (page 155), according to which $\|T_\sigma f - f\|_1 \to 0$ as $\sigma \to 0$. By the Fubini Theorem,

$$\|f_n - f\| \le \int_{-1/2}^{1/2} h_n(\sigma) \int_{-1/2}^{1/2} |f(t - \sigma) - f(t)|\, dt\, d\sigma$$

$$= \int_{-1/2}^{1/2} h_n(\sigma)\|T_\sigma f - f\|_1\, d\sigma$$

To complete the proof, divide the integral into two parts, the first on $I = \{\sigma : |\sigma| < \delta\}$ and the second on $J = \{\sigma : \delta \le |\sigma| \le 1/2\}$. If $\varepsilon > 0$, we can select δ so that $\|T_\sigma f - f\|_1 < \varepsilon$ when $|\sigma| < \delta$. We can select m so that $\int_J h_n(\sigma)\, d\sigma < \varepsilon$ when $n \ge m$. Then if $n \ge m$,

$$\|f_n - f\| \le \int_I h_n(\sigma)\varepsilon\, d\sigma + \int_J h_n(\sigma)2\|f\|_1\, d\sigma$$

$$\le \varepsilon + 2\|f\|_1\varepsilon \qquad \blacksquare$$

Problems

1. Prove that if $\phi \in L^2(\mathbb{R})$, then any finite linear combination of its integer shifts will satisfy the equation

$$\left\| \sum_{n\in\mathbb{Z}} c_n\phi_{0n} \right\| = \left\| \sum_{n\in\mathbb{Z}} c_n\phi_{kn} \right\| \qquad (k \in \mathbb{Z})$$

   Hence deduce that $\{\phi_{kn} : n \in \mathbb{Z}\}$ is stable if and only if $\{\phi_{0n} : n \in \mathbb{Z}\}$ is stable.

2. Give an example of a function ϕ in $L^2(\mathbb{R})$ such that no inequality of the form $\|\sum c_n\phi_{0n}\| \le B\|c\|$ is valid for all $c \in \ell^2(\mathbb{Z})$.

3. Establish the equation

$$2^{1/2} \phi_{kn} = \sum_j \left[a_{n-2j} \phi_{k-1,j} + (-1)^n a_{1-n+2j} \psi_{k-1,j} \right]$$

starting with Equation (21) in Chapter 34.

4. Prove that if ϕ is an element of $L^2(\mathbb{R})$ that obeys the two-scale relation in Equation (2) and if P is its "symbol," then

$$\hat{\phi}(t) = \hat{\phi}\left(\frac{t}{2}\right) P(e^{-\pi i t})$$

5. Prove directly that if ϕ is the characteristic function of the interval $[0, 1)$, then its "symbol" satisfies $|P(z)|^2 + |P(-z)|^2 = 1$.

6. In the two-scale relation, assume that $a_j \neq 0$ only if $0 \leq j \leq m$. Assume the support of ϕ is $[\alpha, \beta]$. What can be said of the support of ψ?

7. Give a new version of Theorem 3 in which Φ is defined by $\Phi = \sum_n \gamma_n \phi_{0n}$, where $\gamma_n = \int_0^1 F^{-1/2}(t) e^{2\pi i n t} \, dt$.

8. Prove that $\bigoplus_{n \in \mathbb{Z}} W_n = \bigcup_{n \in \mathbb{Z}} V_n$ and that for all $m \in \mathbb{Z}$, $\bigcup_{n \in \mathbb{Z}} V_n = \bigcup_{n > m} V_n$.

9. What is the difference between $(\phi_{0n})^\wedge$ and $(\hat{\phi})_{0n}$? In particular, when are they identical?

10. Prove that the coefficients a_n in Theorem 4 are given by $a_n = \sqrt{2} \langle \Phi, \Phi_{1n} \rangle$.

11. Is Theorem 2 true without the hypotheses $A > 0$ and $B < \infty$?

12. Prove that the support of the Daubechies wavelet is $[-1, 2]$.

13. Let ϕ be a nontrivial function in $L^2(\mathbb{R})$ having compact support. Does it follow that $\{\phi_{0n} : n \in \mathbb{Z}\}$ is stable?

14. Let $\{u_n : n \in \mathbb{Z}\}$ be an orthonormal sequence, and define $v_n = \alpha u_n + \beta u_{n+1}$. Find the necessary and sufficient conditions on α and β in order that $\{v_n\}$ be stable.

15. Let ϕ be a function in $L^2(\mathbb{R})$ such that $\langle \phi_{0n}, \phi_{0m} \rangle = c_n \delta_{nm}$. Find the exact conditions on the sequence c_n so that $\{\phi_{0n}\}$ is stable.

16. Consider the two-scale equation that arose in defining the Daubechies wavelet. Is there a solution supported on $[0, 4]$ for which $\phi(1)$, $\phi(2)$, and $\phi(3)$ are nonzero?

17. Prove that $L^2[0, 1] \subset L^1[0, 1]$ by establishing the inequality $\|f\|_1 \leq \|f\|_2$ if $f \in L^2[0, 1]$. What is the corresponding inequality for an arbitrary compact interval?

18. The question of the convergence of the cascade algorithm is rather delicate, as the following example shows. Consider the two-scale equation $\phi(t) = \phi(2t) + \phi(2t - 1)$, and the associated cascade

$$\phi_m(t) = \phi_{m-1}(2t) + \phi_{m-1}(2t - 1) \qquad (t \in \mathbb{R}, m \in \mathbb{N})$$

Suppose ϕ_0 is the piecewise linear "hat" function with support on $[0, 1]$ that is linear in $[0, 1/2]$ and $[1/2, 1]$ and has value 2 at $t = 1/2$. Describe the sequence $[\phi_m]_0^\infty$ exactly. This sequence does not converge in $L^2(\mathbb{R})$, although it does converge weakly to the function ϕ that is the scaling function for the Haar wavelet (see Example 2 of Chapter 34, page 281).

19. Let A be an $n \times n$ matrix that satisfies Condition E. Suppose A has a complete set of eigenvectors. Let c be an eigenvector of A corresponding to the eigenvalue 1. Let b be any vector in $\mathbb{R}^n$. Show that $A^m b \to \alpha c$ as $m \to \infty$, $\alpha \in \mathbb{R}$.

20. Let $[f_m]$ be a sequence of functions in $L^2[0, 1]$ such that $\|f_m - f\|_2 \to 0$ as $m \to \infty$. Show that $\int_0^1 f_m(t)\, dt \to \int_0^1 f(t)\, dt$ as $m \to \infty$.

21. Let $[\phi_m]$ be the sequence defined by the cascade algorithm in Equation (26), where ϕ_0 has support in $[0, N]$. Show that $\|\phi_m\|_2 = \|\phi_0\|_2$ for all $m \in \mathbb{N}$.

22. Suppose that the $n \times n$ matrix B consists of a single Jordan block. Assume that this Jordan block has all its diagonal entries equal to λ and all superdiagonal entries equal to 1. Show that if $|\lambda| < 1$, then $B^m \to 0$ as $m \to \infty$. (*Hint:* B satisfies the equation $(B - \lambda I)^n = 0$.)

23. Use Problem 22 to show that if J is a collection of Jordan blocks, all corresponding to eigenvalues that are less than 1 in modulus, then $J^m \to 0$ as $m \to \infty$.

24. Prove that if $\sum_n |a_n| < \sqrt{2}$, then the two-scale equation has only the 0 solution.

25. Carry out the same analysis as was done for the Daubechies wavelet, but beginning with the two-scale equation $\phi(t) = a_0 \phi(2t) + a_1 \phi(2t - 1)$. Determine the values of a_0 and a_1. Verify that the matrix T satisfies condition E. Prove that $\{\phi(\cdot - n) : n \in \mathbb{Z}\}$ is an orthonormal set of functions. Construct the wavelet.

26. Generalize Lemma 15 to the case when $a \in \ell^1(\mathbb{Z})$ and $\phi_0 \in L^1(\mathbb{R}) \cap L^2(\mathbb{R})$. Indicate where the proof may fail if it be assumed only that $\phi_0 \in L^2(\mathbb{R})$.

References

[Chu2] Chui, C. K. *An Introduction to Wavelets*. Academic Press, New York, 1992.

[Dau] Daubechies, I. *Ten Lectures on Wavelets*. SIAM, Philadelphia, 1992.

[DLu] DeVore, R. A., and B. J. Lucier. "Wavelets." *Annals of Numerical Mathematics* (1991), 1–56.

[Ka] Katznelson, Y. *Harmonic Analysis*. Wiley, New York, 1968. Reprint, Dover, New York, 1976.

[Mey1] Meyer, Y. *Wavelets: Algorithms and Applications*. SIAM, Philadelphia, 1993.

[Ru0] Rudin, W. *Real and Complex Analysis*. 2nd ed., McGraw-Hill, New York, 1974.

[Stra2] Strang, G. "Wavelets and dilation equations." *SIAM Review 31* (1989), 613–627.

[Stra3] Strang, G. "Eigenvalues and the convergence of the cascade algorithm." *IEEE Trans. Signal Proc. 44* (1996), 233–238.

[StN] Strang, G., and T. Nguyen. *Wavelets and Filter Banks*. Wellesley-Cambridge Press, Wellesley, MA, 1996.

[Walt] Walter, G. G. *Wavelets and Other Orthogonal Systems with Applications*. CRP Press, Boca Raton, FL, 1994.

36
Quasi-Interpolation

Chapter 20 was devoted to approximation by convolution. The fundamental idea there was to approximate a function f by taking its convolution with the dilates of a suitable kernel, H. Thus, we selected $H \in L^1(\mathbb{R}^s)$ and defined $H_k(x) = k^s H(kx)$, where $k \in \mathbb{N}$ and $x \in \mathbb{R}^s$. Theorem 2 of Chapter 20 established that the sequence $H_k * f$ converges to f if $f \in C(\mathbb{R}^s)$ and $\int H(x)\,dx = 1$. The topology for this convergence is that of uniform convergence on compact sets. The convolution is defined by

$$(H_k * f)(x) = \int_{\mathbb{R}^s} H_k(y) f(x - y)\,dy$$

For such an approximation to have a practical value, some numerical integration process must be applied to this integral. An alternative strategy, which may be more efficient, is to use a discrete convolution in the first place. This chapter will develop this idea.

A **discrete convolution** will be denoted by $\star$. Its definition is

$$(1) \qquad (f \star g)(x) = \sum_{v \in \mathbb{Z}^s} f(x - v) g(v) \qquad (x \in \mathbb{R}^s)$$

The definition makes sense formally if g is defined on $\mathbb{Z}^s$ (or on any larger subset of $\mathbb{R}^s$). Our first goal is to establish conditions under which the series in Equation (1) converges.

Definition. Having fixed the dimension s, we define for each $\lambda \in \mathbb{R}$ a set of functions:

$$(2) \qquad E_\lambda = \left\{ f \in C(\mathbb{R}^s) \ : \ \sup_{x \in \mathbb{R}^s} |f(x)| (1 + \|x\|_\infty)^\lambda < \infty \right\}$$

A number of pleasant properties of the space E_λ are collected in Problem 1.

> **LEMMA 1.** *If $f \in E_\lambda$ for some λ greater than s, then the series $\sum_{v \in \mathbb{Z}^s} f(x + v)$ is absolutely convergent (and uniformly so in x).*

312

Proof. Assume that $f \in E_\lambda$ so that the quantity

$$M = \sup_{x \in \mathbb{R}^s} |f(x)|(1 + \|x\|)^\lambda$$

is finite. If the series in question converges, it defines an integer-periodic function. Thus, it suffices to consider an x in the unit "cube," $[0, 1)^s$. With the help of Problems 2 and 3, we can write

$$\sum_v |f(x + v)| \leq \sum_{k=0}^\infty \sum_{\|v\|_\infty = k} M(1 + \|x + v\|_\infty)^{-\lambda}$$

$$\leq \sum_{\|v\|_\infty \leq 1} M + \sum_{k=2}^\infty \sum_{\|v\|_\infty = k} M(1 + \|v\|_\infty - \|x\|_\infty)^{-\lambda}$$

$$\leq 3^s M + \sum_{k=2}^\infty \sum_{\|v\|_\infty = k} M(1 + k - 1)^{-\lambda}$$

$$\leq 3^s M + \sum_{k=2}^\infty M k^{-\lambda} 4^s k^{s-1}$$

$$\leq 4^s M \sum_{k=1}^\infty k^{-1-\lambda+s} < \infty \qquad ∎$$

The basic theorem about convergence in Equation (1) can now be stated.

THEOREM 1. *If $f \in E_\lambda$, $g \in E_\mu$, and $\lambda + \mu > s$, then $|f| \star |g|$ exists at each point of $\mathbb{R}^s$.*

Proof. We begin by contemplating the definition:

$$(|f| \star |g|)(x) = \sum_{v \in \mathbb{Z}^s} |f(x - v)|\,|g(v)|$$

This last sum can be interpreted as

(3) $$\sum_{v \in \mathbb{Z}^s} |F(v)g(v)|$$

by defining $F(y) = f(x - y)$. (We regard x as fixed.) By Problem 1, we have $F \in E_\lambda$ and $Fg \in E_{\lambda+\mu}$. Since $\lambda + \mu > s$, the series in (3) converges, by Lemma 1. $\qquad ∎$

It facilitates our work to introduce **normalized monomials:**

(4) $$V_\alpha(x) = \frac{x^\alpha}{\alpha!} \qquad (x \in \mathbb{R}^s, \alpha \in \mathbb{Z}_+^s)$$

For example, the Binomial Theorem takes this form:

(5) $$V_\alpha(x + y) = \sum_{0 \leq \beta \leq \alpha} V_\beta(x) V_{\alpha-\beta}(y)$$

Furthermore, we have this simple rule for differentiation:

$$(6) \qquad\qquad D^\beta V_\alpha = V_{\alpha - \beta} \qquad (\alpha \ge \beta)$$

LEMMA 2. *In order that the discrete convolutions* $|f| \star |V_\alpha|$ *and* $|V_\alpha| \star |f|$ *exist pointwise, it is sufficient that* $f \in E_\lambda$ *for some* λ *greater than* $s + |\alpha|$.

Proof. We have $|V_\alpha(x)| \le \|x\|_\infty^{|\alpha|}$ by Problem 4. Hence $V_\alpha \in E_{-|\alpha|}$. If $\lambda > s + |\alpha|$ and $f \in E_\lambda$, then $-|\alpha| + \lambda > s$, and Theorem 1 applies. ∎

Most of the ensuing analysis will take place in $C(\mathbb{R}^s)$. Since our approximations will be discrete convolutions, they can be drawn from a space of the following type. Let $\phi \in E_\lambda$, where $\lambda > s$. Define

$$(7) \qquad\qquad \mathcal{A}(\phi) = \{\phi \star c : c \in \ell_\infty(\mathbb{Z}^s)\}$$

The functions in $\mathcal{A}(\phi)$ are limits of linear combinations of integer shifts of the function ϕ.

Two useful operators S_h and T_w are defined as follows:

$$(8) \qquad\qquad (S_h f)(x) = f(hx) \qquad (x \in \mathbb{R}^s, h > 0)$$

$$(9) \qquad\qquad (T_w f)(x) = f(x - w) \qquad (x \in \mathbb{R}^s, w \in \mathbb{R}^s)$$

These are called the **scaling** operator and the **translation** operator, respectively. If $v \in \mathbb{Z}^s$, T_v is called a **shift** operator. Notice that if $f \in \mathcal{A}(\phi)$, then $T_v f \in \mathcal{A}(\phi)$, since we can change the summation variable like this:

$$[T_v(\phi \star c)](x) = (\phi \star c)(x - v) = \sum_\mu \phi(x - v - \mu)c(\mu)$$

$$= \sum_{\mu'} \phi(x - \mu')c(\mu' - v) = (\phi \star T_v c)(x)$$

We say therefore that $\mathcal{A}(\phi)$ is **shift-invariant.**

It is clear that the discrete convolution $\phi \star f$ cannot be expected to produce a good approximation to f because the discrete convolution has a very crude "sampling rate" for the function f. Indeed, only the values of f at the multi-integers are used in the calculation of $\phi \star f$. For this reason, a scaling or dilation is used to increase the sampling rate. Let L be the operator corresponding to discrete convolution with a fixed kernel ϕ:

$$(10) \qquad\qquad (Lf)(x) = \sum_{v \in \mathbb{Z}^s} \phi(x - v)f(v) \qquad (x \in \mathbb{R}^s)$$

The **dilations** of L are denoted by L_h and defined by

$$(11) \qquad\qquad L_h = S_{1/h} L S_h \qquad (h > 0)$$

In greater detail, we can write

$$(12) \qquad\qquad (L_h f)(x) = \sum_{v \in \mathbb{Z}^s} \phi\left(\frac{x}{h} - v\right) f(hv) \qquad (x \in \mathbb{R}^s, h > 0)$$

This operator samples f at the points hv, and as $h \to 0$, one might expect that $L_h f \to f$ for all f in some suitable class of functions (about which we are being deliberately vague). Our goal, in fact, is to establish error estimates of the type $\| f - L_h f \|_\infty = O(h^k)$.

Since the analysis makes use of the Fourier transform, we review the pertinent properties of this important operator. (See also Chapter 15.) The definition is

$$(13) \qquad \hat{f}(x) = \int_{\mathbb{R}^s} f(y) e^{-2\pi i x y} \, dy \qquad (x \in \mathbb{R}^s)$$

(Recall that xy is the inner product or "dot" product of two points in $\mathbb{R}^s$.) A useful function in this context is

$$(14) \qquad e_x(y) = e^{2\pi i x y}$$

With this notation, we have

$$(15) \qquad \hat{f}(x) = \langle f, e_x \rangle = \int_{\mathbb{R}^s} f(y) \overline{e_x(y)} \, dy$$

Computational rules, which the reader should verify, are as follows:

$$(16) \qquad (T_x f)^\wedge = e_{-x} \hat{f}$$

$$(17) \qquad (e_x f)^\wedge = T_x \hat{f}$$

$$(18) \qquad (f * g)^\wedge = \hat{f} \hat{g}$$

$$(19) \qquad (S_h f)^\wedge = h^{-s} S_{1/h} \hat{f}$$

As is customary, we set

$$(20) \qquad D = \left(\frac{\partial}{\partial \xi_1}, \frac{\partial}{\partial \xi_2}, \ldots, \frac{\partial}{\partial \xi_s} \right)$$

where $x \in \mathbb{R}^s$ and $x = (\xi_1, \xi_2, \ldots, \xi_s)$. For any multi-index α in $\mathbb{Z}_+^s$, we write

$$(21) \qquad D^\alpha = \frac{\partial^{\alpha_1}}{\partial \xi_1^{\alpha_1}} \frac{\partial^{\alpha_2}}{\partial \xi_2^{\alpha_2}} \cdots \frac{\partial^{\alpha_s}}{\partial \xi_s^{\alpha_s}}$$

For any polynomial p, $p(D)$ is a differential operator. Its interaction with the Fourier transform obeys these rules:

$$(22) \qquad p\!\left(\frac{iD}{2\pi}\right) \hat{f} = (pf)^\wedge$$

$$(23) \qquad [p(D)f]^\wedge = p(2\pi i \cdot) \hat{f}$$

In many discussions of this subject, the **Poisson summation formula** is invoked. Its use here is avoided by substituting the following more elementary lemma.

LEMMA 3. *Let $f \in E_\lambda$, where $\lambda > s$. If $\hat{f}(v) = 0$ for all v in $\mathbb{Z}^s$ except for $v = 0$, then $\hat{f}(0) = \sum_{v \in \mathbb{Z}^s} f(v)$.*

Proof. Define F by the equation

$$(24) \qquad F(x) = \sum_{v \in \mathbb{Z}^s} f(x + v)$$

The definition is a proper one, because the series is absolutely convergent (by Lemma 1). The function F is integer-periodic: $F(x + \mu) = F(x)$ for all $x \in \mathbb{R}^s$ and all $\mu \in \mathbb{Z}^s$. It is continuous, since the series is uniformly convergent (also by Lemma 1). Let Q be the "cube" in $\mathbb{R}^s$ consisting of points x for which $x \geq 0$ and $\|x\|_\infty < 1$. Then $\mathbb{R}^s = \bigcup_{v \in \mathbb{Z}^s} (v + Q)$.

The Fourier coefficients of the continuous, periodic function F are given by

$$c_\mu = \int_Q F(x) e^{-2\pi i x \mu}\, dx = \int_Q \sum_{v \in \mathbb{Z}^s} f(x + v) e^{-2\pi i x \mu}\, dx$$

$$= \sum_{v \in \mathbb{Z}^s} \int_Q f(x - v) e^{-2\pi i x \mu}\, dx = \sum_{v \in \mathbb{Z}^s} \int_{v+Q} f(y) e^{-2\pi i (y-v) \mu}\, dy$$

$$= \int_{\mathbb{R}^s} f(y) e^{-2\pi i y \mu}\, dy = \hat{f}(\mu) \qquad (\mu \in \mathbb{Z}^s)$$

By our hypothesis, $\hat{f}(\mu) = 0$ for all μ in $\mathbb{Z}^s$ with the sole exception of $\mu = 0$. It follows that $F - c_0$ is a continuous periodic function all of whose Fourier coefficients are zero. Hence $F - c_0 = 0$, and in particular $F(0) = c_0 = \hat{f}(0)$. ∎

LEMMA 4. *Let $k \in \mathbb{N}$, $\lambda \in \mathbb{R}$, and $\lambda \geq k + s$. Let ϕ be an element of E_λ such that $\hat{\phi}(0) = 1$ and such that $(D^\alpha \hat{\phi})(v) = 0$ when $v \in \mathbb{Z}^s \backslash 0$ and $|\alpha| < k$. Then $V_\alpha - \phi \star V_\alpha \in \Pi_{|\alpha|-1}(\mathbb{R}^s)$ when $|\alpha| < k$.*

Proof. It is convenient to define the "backward" operator B by $(Bf)(x) = f(-x)$. Then we can write

$$(25) \qquad (\phi \star V_\alpha)(x) = \sum_v \phi(x - v) V_\alpha(v) = \sum_v (V_\alpha T_x B\phi)(v)$$

Now assume the hypotheses, including the inequality $|\alpha| < k$. Since $\lambda \geq k + s > |\alpha| + s$, Lemma 2 guarantees the existence of the convolution in Equation (25). By Problem 1, part (f), we have $V_\alpha T_x B\phi \in E_{\lambda - |\alpha|}$. Lemma 3 now applies, if λ in Lemma 3 is taken to be $\lambda - |\alpha|$ in the present argument. We verify the crucial hypothesis in Lemma 3, using Equations (22), (16), and the Leibniz rule.

$$(V_\alpha T_x B\phi)^\wedge = V_\alpha \left(\frac{i}{2\pi} D \right) (T_x B\phi)^\wedge$$

$$= V_\alpha \left(\frac{i}{2\pi} D \right) [e_{-x} (B\phi)^\wedge]$$

$$= \left(\frac{i}{2\pi} \right)^{|\alpha|} V_\alpha(D) [e_{-x} B\hat{\phi}]$$

$$= \left(\frac{i}{2\pi} \right)^{|\alpha|} \sum_{0 \leq \beta \leq \alpha} [V_\beta(D) e_{-x}][V_{\alpha-\beta}(D) B\hat{\phi}]$$

In the sum, we have $|\alpha - \beta| \leq |\alpha| < k$. Hence, by hypothesis, $[V_{\alpha-\beta}(D)B\hat{\phi}](v) = 0$ for $v \in \mathbb{Z}^s \backslash 0$. From this we conclude that $(V_\alpha T_x B\phi)^\wedge(v) = 0$ when $v \neq 0$. Lemma 3 then implies that

$$\sum_v (V_\alpha T_x B\phi)(v) = (V_\alpha T_x B\phi)^\wedge(0)$$

$$= \left(\frac{i}{2\pi}\right)^{|\alpha|} \sum_{0 \leq \beta \leq \alpha} V_\beta(-2\pi i x)[V_{\alpha-\beta}(D)B\hat{\phi}](0)$$

This final expression is a polynomial in x, and in it the term corresponding to $\beta = \alpha$ is

$$\left(\frac{i}{2\pi}\right)^{|\alpha|} V_\alpha(-2\pi i x)(B\hat{\phi})(0) \doteq V_\alpha(x)$$

Hence $V_\alpha - \phi \star V_\alpha$ is a polynomial in $\Pi_{|\alpha|-1}(\mathbb{R}^s)$. ∎

LEMMA 5. *If the formula $Lp = \phi \star p$ produces a polynomial whenever $p \in \Pi_k(\mathbb{R}^s)$, then $T_x Lp = LT_x p$ for all x and all $p \in \Pi_k(\mathbb{R}^s)$.*

Proof. It suffices to prove that if $x \in \mathbb{R}^s$, $y \in \mathbb{R}^s$, and $|\alpha| \leq k$, then

(26) $$(T_x LV_\alpha)(y) = (LT_x V_\alpha)(y)$$

Let y be fixed. Each side of this equation is then a polynomial in x. To verify this assertion for the function on the right, use the Binomial Theorem:

$$(LT_x V_\alpha)(y) = \sum_v \phi(y - v)V_\alpha(v - x)$$

$$= \sum_v \phi(y - v) \sum_{0 \leq \beta \leq \alpha} V_\beta(v)V_{\alpha-\beta}(-x)$$

To prove that the polynomials in x in Equation (26) are the same, it suffices to show that they agree on the integer grid. Therefore, let $x = \mu$, where $\mu \in \mathbb{Z}^s$. We have

$$(T_\mu LV_\alpha)(y) = (LV_\alpha)(y - \mu) = \sum_v \phi(y - \mu - v)V_\alpha(v)$$

$$= \sum_v \phi(y - v)V_\alpha(v - \mu)$$

$$= (LT_\mu V_\alpha)(y)$$ ∎

LEMMA 6. *Fix s and k. Let $\phi \in E_\lambda$, where $\lambda > s + k$. If $\phi \star V_\alpha - V_\alpha \in \Pi_{|\alpha|-1}$ whenever $|\alpha| \leq k$, then there exists a mapping $b : \mathbb{Z}^s \to \mathbb{R}$ having finite support such that*

$$\phi \star b \star p = p \qquad (p \in \Pi_k(\mathbb{R}^s))$$

Proof. Define L on $\Pi_k(\mathbb{R}^s)$ by writing $Lp = \phi \star p$. Since ϕ is in E_λ and $\lambda > s + k$, Lemma 2 ensures that Lp is well-defined. Our hypothesis on ϕ implies that

$L(\Pi_k(\mathbb{R}^s)) \subset \Pi_k(\mathbb{R}^s)$. It also implies that $I - L$ is degree reducing on $\Pi_k(\mathbb{R}^s)$. The operator L is injective, because if $Lp = 0$ then

$$\text{degree } (p) = \text{degree } \left[(I - L)p\right] < \text{degree } (p)$$

and this can occur only if $p = 0$. For each $v \in \mathbb{Z}^s$ define the point-evaluation functional v^* by writing $v^*(f) = f(v)$. The set $\{v^* : v \in \mathbb{Z}^s_+, |v| \le k\}$ is a basis for $[\Pi_k(\mathbb{R}^s)]^*$, by the corollary to Theorem 4 in Chapter 10, page 65. Consequently there exist coefficients a_v such that

$$(L^{-1}p)(0) = \sum_{0 \le v, |v| \le k} a_v p(v) \qquad (p \in \Pi_k(\mathbb{R}^s))$$

Define $b(v) = a_{-v}$. Then we have, for any x in $\mathbb{R}^s$,

$$\left[(\phi \star b) \star p\right](x) = \sum_\mu (\phi \star b)(x - \mu)p(\mu) = \sum_\mu \sum_v \phi(x - \mu - v)b(v)p(\mu)$$

$$= \sum_\mu \sum_v \phi(x - \mu + v)b(-v)p(\mu) = \sum_v a_v(\phi \star p)(x + v)$$

$$= \sum_v a_v(Lp)(x + v) = \sum_v a_v(T_{-x}Lp)(v)$$

$$= (L^{-1}T_{-x}Lp)(0) = (T_{-x}L^{-1}Lp)(0) = (T_{-x}p)(0) = p(x) \qquad \blacksquare$$

Define, as usual,

(27) $$\|f\|_\infty = \sup_{x \in \mathbb{R}^s} |f(x)|$$

We shall need also the following seminorms:

(28) $$|f|_k = \max_{|\alpha|=k} \|D^\alpha f\|_\infty \qquad (k = 0, 1, 2, \dots)$$

Functions f for which the derivatives $D^\alpha f$ are bounded and continuous for $|\alpha| \le k$ will be of special interest in what follows.

THEOREM 2. *Fix s and k, and let $\phi \in E_\lambda$, where $\lambda > s + k$. Suppose that $\phi \star p = p$ for all $p \in \Pi_k(\mathbb{R}^s)$. Then there is a constant C such that*

$$\|f - \phi \star f\|_\infty \le C|f|_k$$

for functions f all of whose derivatives $D^\alpha f$, $(|\alpha| \le k)$, are bounded and continuous.

Proof. Fix an $x \in \mathbb{R}^s$, and let p be the Taylor polynomial of degree $k - 1$ for f at x. If r is the remainder in Taylor's Theorem, then $f = p + r$ and

$$r(y) = \sum_{|\alpha|=k} (D^\alpha f)(\xi_{\alpha y})V_\alpha(y - x)$$

Hence,

$$\left| (f - \phi \star f)(x) \right| = \left| p(x) - (\phi \star f)(x) \right| = \left| (\phi \star p)(x) - (\phi \star f)(x) \right|$$

$$= \left| (\phi \star r)(x) \right| = \left| \sum_{v \in \mathbb{R}^s} \phi(x - v) r(v) \right|$$

$$\leq \sum_{v \in \mathbb{R}^s} \left| \phi(x - v) \right| \sum_{|\alpha| = k} \left| (D^\alpha f)(\xi_{\alpha v}) \right| \left| V_\alpha(v - x) \right|$$

$$\leq \max_{|\alpha| \leq k} \| D^\alpha f \|_\infty \sum_{|\alpha| = k} \sum_{v \in \mathbb{Z}^s} \left| \phi(x - v) V_\alpha(x - v) \right|$$

Since $\phi \in E_\lambda$, we have (by Problem 1)

$$\phi V_\alpha \in E_{\lambda - |\alpha|} \subset E_{\lambda - k}$$

Since $\lambda > s + k$ by hypothesis, we have $\lambda - k > s$, and Lemma 1 applies to yield

$$\sum_{v \in \mathbb{Z}^s} \left| \phi(x - v) V_\alpha(x - v) \right| \leq M \qquad (x \in \mathbb{R}^s)$$

Hence,

$$\sum_{|\alpha| = k} \sum_{v \in \mathbb{Z}^s} \left| \phi(x - v) V_\alpha(x - v) \right| \leq C$$

for some C. ∎

THEOREM 3. *Fix s and k, and let $\lambda > s + k$. Let ϕ be an element of E_λ such that $\hat{\phi}(0) = 1$ and $(D^\alpha \hat{\phi})(v) = 0$ when $v \in \mathbb{Z}^s \backslash 0$ and $|\alpha| < k$. Then there exist a finitely supported function b and a constant C such that*

$$\| f - S_{1/h}(\phi \star b \star S_h f) \|_\infty \leq C h^k |f|_k \qquad (h > 0)$$

whenever f has bounded continuous derivatives $D^\alpha f$ for $|\alpha| \leq k$.

Proof. By Lemma 4, whenever $|\alpha| < k$ we have

$$V_\alpha - \phi \star V_\alpha \in \prod_{|\alpha| - 1} (\mathbb{R}^s)$$

By Lemma 6, there is a finitely supported function b such that

$$\phi \star b \star p = p \qquad \left(p \in \prod_{k-1} (\mathbb{R}^s) \right)$$

Since S_h is an isometry, and since Theorem 2 is applicable, we have

$$\| f - S_{1/h}(\phi \star b \star S_h f) \|_\infty = \| S_{1/h}(S_h f - \phi \star b \star S_h f) \|_\infty$$

$$= \| S_h f - \phi \star b \star S_h f \|_\infty \leq C |S_h f|_k$$

To complete the proof, we note that

$$|S_h f|_k = \max_{|\alpha|=k} \|D^\alpha(S_h f)\|_\infty = h^k \max_{|\alpha|=k} \|S_h D^\alpha f\|_\infty$$

$$= h^k \max_{|\alpha|=k} \|D^\alpha f\|_\infty = h^k |f|_k \qquad \blacksquare$$

Theorem 3 provides a lower bound on the "degree of approximation" of a smooth function f by $S_{1/h}\phi \star S_h f$ as h approaches 0. Given the conditions imposed on ϕ, is this the best degree of approximation? In other words, are there examples of smooth f for which no better rate of convergence occurs? In fact, we can establish the converse of Theorem 3, and this will show that the estimate in Theorem 3 is best possible. The next four lemmas lead up to the desired result in Theorem 4.

LEMMA 7. *Let* $\lambda > s$ *and* $\phi \in E_\lambda$. *If* f *is continuous and if* $f(x)$ *vanishes when* $\|x\|_\infty > \rho$, *then*

$$[S_{1/h}(\phi \star S_h f)](x) \le c h^{\lambda-s}\|x\|_\infty^{-\lambda}\|f\|_\infty \qquad (\|x\|_\infty \ge 2\rho)$$

Proof. Abbreviate $S_{1/h}(\phi \star S_h f)$ by f_h. For all x we have

$$|f_h(x)| = \left|(\phi \star S_h f)\left(\frac{x}{h}\right)\right| \le \sum_v \left|\phi\left(\frac{x}{h}-v\right)f(hv)\right|$$

$$\le \sum_{\|hv\|_\infty \le \rho} \left|\phi\left(\frac{x}{h}-v\right)\right|\|f\|_\infty$$

$$\le c_1\|f\|_\infty \sum_{\|hv\|_\infty \le \rho} \left(1 + \left\|\frac{x}{h}-v\right\|_\infty\right)^{-\lambda}$$

$$= c_1 h^\lambda\|f\|_\infty \sum_{\|hv\|_\infty \le \rho} (h + \|x-hv\|_\infty)^{-\lambda}$$

For $\|x\|_\infty \ge 2\rho$, we have in this sum

$$h + \|x-hv\|_\infty \ge \|x\|_\infty - \|hv\|_\infty \ge \|x\|_\infty - \rho \ge \|x\|_\infty/2$$

Hence, for $\|x\|_\infty \ge 2\rho$,

$$|f_h(x)| \le c_1 h^\lambda\|f\|_\infty \sum_{\|hv\|_\infty \le \rho} (\|x\|_\infty/2)^{-\lambda}$$

$$\le c_1 h^\lambda\|x\|_\infty^{-\lambda}\|f\|_\infty \sum_{\|v\|_\infty \le \rho/h} 2^\lambda$$

$$\le c_2 h^\lambda\|x\|_\infty^{-\lambda}\|f\|_\infty 2^\lambda (\rho/h)^s$$

$$= c_3 h^{\lambda-s}\|x\|_\infty^{-\lambda}\|f\|_\infty$$

We note that c_3 depends on ϕ, s, and ρ. $\qquad \blacksquare$

LEMMA 8. *Let* $\lambda > k + s$ *and* $\phi \in E_\lambda$. *Put* $f_h = S_{1/h}(\phi \star S_h f)$. *If* $\|f - f_h\|_\infty = O(h^k)$ *as* $h \to 0$, *then for* $|\alpha| < k$, $\|D^\alpha(\hat{f} - \hat{f}_h)\|_\infty = O(h^k)$.

Proof. From Equation (22) describing the interaction between the Fourier transform and differentiation, we have

$$\left| \left[V_\alpha\left(\frac{iD}{2\pi}\right)(\hat{f} - \hat{f}_h) \right](x) \right| = |[V_\alpha(f - f_h)]^\wedge(x)|$$

$$= \left| \int_{\mathbb{R}^s} e^{-2\pi i x y} V_\alpha(y)[f(y) - f_h(y)] \, dy \right|$$

$$\leq \int_{\mathbb{R}^s} |V_\alpha(y)| \, |f(y) - f_h(y)| \, dy$$

By Lemma 7, we have $|f_h(y)| \leq c h^{\lambda - s} \|y\|_\infty^{-\lambda}$ when $\|y\|_\infty \geq 2\rho$. We assume, without loss of generality, that $2\rho > 1$. Using Problems 4 and 7, we have

$$\int_{\|y\|_\infty \geq 2\rho} |V_\alpha(y)| \, |f(y) - f_h(y)| \, dy = \int_{\|y\|_\infty \geq 2\rho} |V_\alpha(y)| \, |f_h(y)| \, dy$$

$$\leq c h^{\lambda - s} \int_{\|y\|_\infty \geq 2\rho} \|y\|_\infty^{|\alpha|} \|y\|_\infty^{-\lambda} \, dy$$

$$\leq c h^{\lambda - s} \int_{\|y\|_\infty \geq 2\rho} \|y\|_\infty^{k - 1 - \lambda} \, dy$$

$$\leq c_1 h^{\lambda - s} = O(h^k)$$

For the other part of the integral, we have

$$\int_{\|y\|_\infty < 2\rho} |V_\alpha(y)| \, |f(y) - f_h(y)| \, dy \leq \|f - f_h\|_\infty \int_{\|y\|_\infty \leq 2\rho} \|y\|_\infty^{|\alpha|} \, dy = O(h^k) \quad \blacksquare$$

LEMMA 9. *Let* $\lambda > s$, $\phi \in E_\lambda$, *and* $f_h = S_{1/h}(\phi \star S_h f)$. *Then*

$$\hat{f}_h = h^s (S_h \hat{\phi}) \sum_{v \in \mathbb{Z}^s} f(hv) e_{-hv}$$

Proof. By the definition of f_h,

$$f_h(x) = \sum_{v \in \mathbb{Z}^s} \phi\left(\frac{x}{h} - v\right) f(hv) = \sum_v (S_{1/h} T_v \phi)(x) f(hv)$$

Both series converge because f has compact support. Now

$$\hat{f}_h(x) \sum_v (S_{1/h} T_v \phi)^\wedge(x) f(hv) = \sum_v h^s [S_h(T_v \phi)^\wedge](x) f(hv)$$

$$= \sum_v h^s S_h(e_{-v} \hat{\phi})(x) f(hv)$$

$$= \sum_v h^s e_{-hv}(x)(S_h \hat{\phi})(x) f(hv) \quad \blacksquare$$

LEMMA 10. *Let* $\lambda > s + k$, $\phi \in E_\lambda$, *and* $f_h = S_{1/h}(\phi \star S_h f)$. *Then, for* $|\alpha| \leq k$,

$$V_\alpha(D)\hat{f}_h = h^{s+|\alpha|}(-i)^{|\alpha|} \sum_{0 \leq \beta \leq \alpha} [S_h V_\beta(iD)\hat{\phi}] \sum_v f(hv)V_{\alpha-\beta}(2\pi v)e_{-hv}$$

Proof. Use the formula for f_h in the preceding lemma and Leibniz's rule to obtain

$$V_\alpha(D)\hat{f}_h = V_\alpha(D)\left[h^s(S_h\hat{\phi}) \sum_{v \in \mathbb{Z}^s} f(hv)e_{-hv} \right]$$

$$= h^s \sum_v f(hv)V_\alpha(D)[e_{-hv}S_h\hat{\phi}]$$

$$= h^s \sum_v f(hv) \sum_{0 \leq \beta \leq \alpha} [V_\beta(D)S_h\hat{\phi}][V_{\alpha-\beta}(D)e_{-hv}]$$

$$= h^s \sum_v f(hv) \sum_\beta [h^{|\beta|}S_h V_\beta(D)\hat{\phi}][V_{\alpha-\beta}(2\pi ihv)e_{-hv}]$$

$$= h^{s+|\alpha|}(-i)^{|\alpha|} \sum_\beta [S_h V_\beta(iD)\hat{\phi}] \sum_v f(hv)V_{\alpha-\beta}(2\pi v)e_{-hv} \quad\blacksquare$$

Let B^k denote the univariate B-spline of degree k having support $\left[-\frac{k+1}{2}, \frac{k+1}{2}\right]$. For $x = (\xi_1, \xi_2, \ldots, \xi_s) \in \mathbb{R}^s$, we let

$$(29) \qquad u(x) = \prod_{i=1}^s B^k(\xi_i)$$

This is the multivariable tensor-product B-spline. From Chapter 29, we have

$$(30) \qquad \hat{u}(x) = \prod_{i=1}^s \left[\frac{\sin(\pi\xi_i)}{\pi\xi_i} \right]^{k+1}$$

and

$$(31) \qquad (D^\alpha\hat{u})(x/h) = o(h^k) \qquad (x \neq 0, \quad h \to 0, \quad |\alpha| < k)$$

LEMMA 11. *Let* u *be defined as in Equation* (29). *Let* $\lambda > s + k$ *and* $\phi \in E_\lambda$. *Define* $u_h = S_{1/h}(\phi * S_h u)$. *If* $\|u - u_h\|_\infty = O(h^k)$ *as* $h \to 0$, *then for* $|\alpha| < k$ *and* $m \in \mathbb{Z}^s\backslash 0$,

$$\sum_{0 \leq \beta \leq \alpha} [V_\beta(iD)\hat{\phi}](m) \sum_{v \in \mathbb{Z}^s} u(hv)V_{\alpha-\beta}(2\pi v) = O(h^{1-s})$$

Proof. Apply Lemma 10 to conclude that

$$V_\alpha(D)\hat{u}_h = h^{s+|\alpha|}(-i)^{|\alpha|} \sum_{0 \leq \beta \leq \alpha} [S_h V_\beta(iD)\hat{\phi}] \sum_v u(hv)V_{\alpha-\beta}(2\pi v)e_{-hv}$$

Evaluating at m/h yields

$$(32) \quad [V_\alpha(D)\hat{u}_h]\left(\frac{m}{h}\right) = h^{s+|\alpha|}(-i)^{|\alpha|} \sum_{0 \le \beta \le \alpha} [V_\beta(iD)\hat{\phi}](m) \sum_v u(hv) V_{\alpha-\beta}(2\pi v)$$

From Equation (31), we have

$$(33) \quad (V_\alpha(D)\hat{u})\left(\frac{m}{h}\right) = o(h^k) \qquad (h \to 0, \, |\alpha| < k)$$

From the hypothesis $\|u - u_h\|_\infty = O(h^k)$ and from Lemma 8, we get

$$(34) \quad \|V_\alpha(D)(\hat{u} - \hat{u}_h)\|_\infty = O(h^k) \qquad (|\alpha| < k)$$

Thus, from Equations (33) and (34), we conclude that

$$[V_\alpha(D)\hat{u}_h]\left(\frac{m}{h}\right) = O(h^k)$$

This term is the left-hand side of Equation (32). Therefore the right-hand side of Equation (32) is also $O(h^k)$. The desired result follows, since $k - |\alpha| \ge 1$. ∎

LEMMA 12. *Let u be as in Equation (29), and let $|\alpha| < k$. Then*

$$\lim_{h \to 0, \, h^{-1} \in \mathbb{Z}} h^s \sum_{v \in \mathbb{Z}^s} V_\alpha(v) u(hv) = \delta_{\alpha 0}$$

Proof. By using Equations (22) and (19), we obtain

$$(35) \quad \begin{aligned} (V_\alpha S_h u)^\wedge(v) &= V_\alpha\left(\frac{iD}{2\pi}\right)(S_h u)^\wedge(v) = V_\alpha\left(\frac{iD}{2\pi}\right)(h^{-s} S_{1/h} \hat{u})(v) \\ &= h^{-s-|\alpha|}\left[V_\alpha\left(\frac{iD}{2\pi}\right)\hat{u}\right]\left(\frac{v}{h}\right) \end{aligned}$$

Inspection of Equation (30) reveals that $\hat{u}$ has a zero of multiplicity $k + 1$ at each point $v \in \mathbb{Z}^s \backslash 0$. Therefore, if $h^{-1} \in \mathbb{Z}$, then $\left[V_\alpha\left(\frac{iD}{2\pi}\right)\hat{u}\right](v/h) = 0$ when $|\alpha| \le k$ and $v \in \mathbb{Z}^s \backslash 0$. It follows from Equation (35) that $(V_\alpha S_h u)^\wedge$ vanishes at all $v \in \mathbb{Z}^s$ except $v = 0$. By Lemma 3 and Equation (35),

$$(36) \quad h^s \sum_v V_\alpha(v) u(hv) = h^s (V_\alpha S_h u)^\wedge(0) = h^{-|\alpha|}\left[V_\alpha\left(\frac{iD}{2\pi}\right)\hat{u}\right](0)$$

If $\alpha = 0$, this last expression reduces to $\hat{u}(0)$, which is 1. If $\alpha \ne 0$, then by Equation (31),

$$\lim_{h \to 0} h^{-|\alpha|}\left[V_\alpha\left(\frac{iD}{2\pi}\right)\hat{u}\right](0) = 0$$

By Equation (36) we arrive at the assertion in the lemma. ∎

THEOREM 4. *Fix s and k. Let $\lambda > s + k$ and $\phi \in E_\lambda$. Let u and u_h be as in Equation (29) and Lemma 11. If $\|u - u_h\|_\infty = O(h^k)$ as $h \to 0$, then $\hat{\phi}(0) = 1$ and $(D^\alpha \hat{\phi})(v) = 0$ when $|\alpha| < k$ and $v \in \mathbb{Z}^s \backslash 0$.*

Proof. From Lemma 12 and the observation that $V_\alpha(x) = (2\pi)^{-|\alpha|} V_\alpha(2\pi x)$, we have

$$(37) \qquad \lim_{h \to 0, h^{-1} \in \mathbb{Z}} h^s \sum_{v \in \mathbb{Z}^s} V_\alpha(2\pi v) u(hv) = \delta_{\alpha 0}$$

If $|\alpha| < k$ and $m \in \mathbb{Z}^s \backslash 0$, then by Lemma 11,

$$\begin{aligned} 0 &= \lim_{h \to 0, h^{-1} \in \mathbb{Z}} h^s \sum_{0 \leq \beta \leq \alpha} [V_\beta(iD)\hat{\phi}](m) \sum_{v \in \mathbb{Z}^s} u(hv) V_{\alpha - \beta}(2\pi v) \\ &= \sum_{0 \leq \beta \leq \alpha} [V_\beta(iD)\hat{\phi}](m) \lim_h h^s \sum_v u(hv) V_{\alpha - \beta}(2\pi v) \\ &= \sum_{0 \leq \beta \leq \alpha} [V_\beta(iD)\hat{\phi}](m) \delta_{\alpha\beta} = [V_\beta(iD)\hat{\phi}](m) \end{aligned}$$

To prove that $\hat{\phi}(0) = 1$, start by applying Lemma 9 to u, obtaining

$$\hat{u}_h = h^s (S_h \hat{\phi}) \sum_{v \in \mathbb{Z}^s} u(hv) e_{-hv}$$

Evaluating at 0 gives us

$$(38) \qquad \hat{u}_h(0) = h^s \hat{\phi}(0) \sum_v u(hv)$$

From Lemma 12, with $\alpha = 0$, we have

$$(39) \qquad \lim_{h \to 0, h^{-1} \in \mathbb{Z}} h^s \sum_v u(hv) = 1$$

Combining Equations (38) and (39) leads to

$$(40) \qquad \lim_{h \to 0, h^{-1} \in \mathbb{Z}} \hat{u}_h(0) = \hat{\phi}(0)$$

Now $\|u - u_h\|_\infty = O(h^k)$ by hypothesis. Hence by Lemma 8, we have $\|D^\alpha(\hat{u} - \hat{u}_h)\|_\infty = O(h^k)$, for $|\alpha| < k$. In particular, we have $\hat{u}_h(0) \to \hat{u}(0)$, the latter being equal to 1 by Equation (30). Equation (40) now shows that $\hat{\phi}(0) = 1$. ∎

Problems

1. Let $f \in E_\lambda$ and $g \in E_\mu$. Prove that
   a. $fg \in E_{\lambda + \mu}$
   b. $|f| \in E_\lambda$
   c. $T_w f \in E_\lambda$ (T_w is the translation operator.)

    d. $S_h f \in E_\lambda$ (S_h is the scaling operator.)

    e. $Bf \in E_\lambda$ $((Bf)(x) = f(-x))$

    f. $V_\alpha f \in E_{\lambda - |\alpha|}$ $(V_\alpha(x) = x^\alpha/\alpha!)$

2. Prove that the number of multi-integers v in $\mathbb{Z}^s$ that satisfy $\|v\|_\infty \le k$ (where $k \in \mathbb{N}$) is $(2k+1)^s$.

3. Prove that the number of v in $\mathbb{Z}^s$ that satisfy $\|v\|_\infty = k$ ($k \in \mathbb{N}$) is at most $4^s k^{s-1}$.

4. Prove that $|V_\alpha(x)| \le \|x\|_\infty^{|\alpha|}$.

5. Prove that if $\phi \in E_\lambda$, $\lambda > s$, and $c \in \ell_\infty(\mathbb{Z}^s)$ then $\phi \star c$ is well-defined.

6. Use the ideas in the proof of Lemma 3 to prove this version of the Poisson summation formula: If f and $\hat{f}$ belong to E_λ, for some λ greater than s, then $\sum f(v) = \sum \hat{f}(v)$, both sums being over $\mathbb{Z}^s$.

7. Prove that if $\lambda > s$ then $\int_A \|x\|_\infty^{-\lambda} dx < \infty$ for any closed set A in $\mathbb{R}^s$ that does not contain 0.

8. Let $\lambda > s$ and $f \in E_\lambda$. Prove that $f \in L^1(\mathbb{R}^s)$. Find an inequality relating $\|f\|_1$ to the quantity

$$M(f) = \sup_{x \in \mathbb{R}^s} |f(x)|(1 + \|x\|_\infty)^\lambda$$

9. Prove the following computational rules for the Fourier transform:

    a. $(T_x f)^\wedge = e_{-x}\hat{f}$

    b. $(pf)^\wedge = p\left(\dfrac{iD}{2\pi}\right)\hat{f}$ (p is any polynomial)

10. If V is a shift-invariant subspace in $\ell^2(\mathbb{Z}^s)$ does it follow that $\{\phi \star c : c \in V\}$ is shift-invariant? (Prove a reasonable theorem.)

11. Verify Equations (4) and (5).

References

[Bo5]   de Boor, C. "Quasi-interpolants and approximation power of multivariate splines." In *Computation of Curves and Surfaces*, ed. by W. Dahmen, M. Gasca, and C. A. Micchelli. Kluwer, Amsterdam, 1990, pp. 313–345.

[Bo7]   de Boor, C. "The quasi-interpolant as a tool in elementary polynomial spline theory." In *Approximation Theory*, ed. by G. G. Lorentz. Academic Press, 1973, 269–276.

[BJ]   de Boor, C., and R. Q. Jia. "Controlled approximation and a characterization of the local approximation order." *Proc. Amer. Math. Soc. 95* (1985), 547–553.

[Bu8]   Buhmann, M. D. "On quasi-interpolation with radial basis functions." *J. Approx. Theory 72* (1993), 103–130.

[BuDL]   Buhmann, M. D., N. Dyn, and D. Levin. "On quasi-interpolation by radial basis functions with scattered centers." *Constr. Approx. 11* (1995), 239–254.

[BuLe]   Burchard, H. G., and Junjiang Lei. "Coordinate order of approximation by functional-based aproximation operators." *J. Approx. Theory 82* (1995), 240–256.

[ChuDi]   Chui, Charles, and H. Diamond. "A natural formulation of quasi-interpolation by multi-variate splines." *Proc. Amer. Math. Soc. 99* (1987), 643–646.

[CL]   Cheney, E. W., and Junjiang Lei. "Quasi-interpolation on irregular points." In *Approximation and Computation: A Festschrift in Honor of Walter Gautschi,* ed. by R. V. M. Zahar. Birkhäuser, Boston, Series ISNM, Vol. 119, 1994, pp. 121–135.

[CLX]   Cheney, E. W., W. A. Light, and Yuan Xu. "On Kernels and approximation orders." In *Approximation Theory: Proceedings of the Sixth Southeastern International Approximation Conference,* ed. by G. Anastassiou. Marcel Dekker Publisher, New York, 1992, pp. 227–242.

[DM2]   Dahmen, W., and C. Micchelli. "On the approximation order from certain multivariate spline spaces." *J. Australian Math. Soc.,* Ser. B, 26 (1984), 233–246.

[FS]   Fix, G., and G. Strang. "Fourier analysis of the finite element method in Ritz-Galerkin theory." *Stud. Appl. Math. 48* (1969), 265–273.

[LC2]   Light, W. A., and E. W. Cheney. "Quasi-interpolation with translates of a function having non-compact support." *Constr. Approx. 8* (1992), 35–48.

[JL]   Jia, R. Q., and Junjiang Lei. "Approximation by multiinteger translates of functions having global support." *J. Approx. Theory 72* (1993), 2–23.

[JL2]   Jia, R. Q., and Junjiang Lei. "A new version of the Strang-Fix conditions." *J. Approx. Theory 74* (1993), 221–225.

[Ru2]   Rudin, W. *Functional Analysis.* McGraw–Hill, New York, 1973.

[SW]   Stein, E. M., and G. Weiss. *Introduction to Fourier Analysis on Euclidean Spaces.* Princeton University Press, Princeton, NJ, 1971.

Bibliography

Textbooks and Treatises on General Approximation Theory

[Ak2]   Achiezer, N. I. *Theory of Approximation.* Frederick Ungar, New York, 1956. Reprint, Dover, New York.

[Burk] Burkill, J. C. *Lectures on Approximation by Polynomials.* Tata Institute, Bombay, 1959.

[C1]   Cheney, E. W. *Introduction to Approximation Theory.* McGraw-Hill, New York, 1966. 2nd ed., Chelsea, New York, 1982. Amer. Math. Soc., Providence, RI, 1998. Japanese edition, Kyoritsu Shuppan, Tokyo, 1978. Chinese edition, Beijing, 1981.

[Da2]   Davis, P. J. *Interpolation and Approximation.* Blaisdell, New York, 1963. Reprint, Dover, New York.

[DL]   DeVore, R. A., and G. G. Lorentz. *Constructive Approximation.* Springer-Verlag, New York, 1993.

[FN]   Feinerman, R. P., and D. J. Newman. *Polynomial Approximation.* Williams & Wilkins, Baltimore, 1974.

[Gol2] Golomb, M. *Lectures on Theory of Approximation.* Argonne National Laboratory, Argonne, IL, 1962.

[GWR]   Gustafson, S.-A., P.-A. Wedin and A. Ruhe. *Föreläsningar för Kursen Approximation.* Department of Information Processing, University of Umeå, Sweden, 1972.

[HS]   Holland, A. S. B., and B. N. Sahney. *The General Problem of Approximation and Spline Functions.* Kreiger, Huntington, New York, 1979.

[Hol]   Holmes, R. B. *A Course on Optimization and Best Approximation.* Lecture Notes in Math., vol. 257, Springer-Verlag, Berlin, 1972.

[J]   Jackson, D. *The Theory of Approximation.* Amer. Math, Soc., Providence, RI, 1930.

[LS]   Lancaster, P., and K. Salkauskas. *Curve and Surface Fitting.* Academic Press, New York, 1986.

[Lau]   Laurent, P.-J. *Approximation et Optimisation.* Hermann, Paris, 1972.

[Lor]   Lorentz, G. G. *Approximation of Functions.* Holt, Rinehart and Winston, New York, 1966.

[LGM]   Lorentz, G. G., M. von Golitschek, and Y. Makovoz. *Constructive Approximation: Advanced Problems.* Springer-Verlag, Berlin, 1996.

[Meina] Meinardus, G. *Approximation of Functions: Theory and Numerical Methods.* Springer-Verlag, Berlin, 1967.

[Nata]   Natanson, I. P. *Constructive Theory of Functions.* U.S. Atomic Energy Commission, Office of Technical Information, AEC-tr-4503, 1961. Original Russian text, Moscow, 1949. German translation, *Konstructive Funktionentheorie,* Akademie-Verlag, Berlin, 1955.

[P5]   Powell, M. J. D. *Approximation Theory and Methods.* Cambridge University Press, Cambridge, UK, 1981.

[Rei3]   Reimer, M. *Constructive Theory of Multivariate Functions.* BI-Wissenschaftsverlag, Mannheim, 1990.

[Rice]   Rice, J. R. *The Approximation of Functions* (2 vols.). Addison-Wesley, London, 1969.

[Riv3]   Rivlin, T. J. *An Introduction to the Approximation of Functions.* Dover, New York, 1981.

[Scho]   Schönhage, A. *Approximationstheorie.* De Gruyter, Berlin, 1971.

[Shap]   Shapiro, H. S. *Topics in Approximation Theory.* Lecture Notes in Math., vol. 187. Springer-Verlag, Berlin, 1971.

[Tim]   Timan, A. F. *Theory of Approximation of Functions of a Real Variable.* Macmillan, New York, 1963. Reprint, Dover, New York.

[Wat]   Watson, G. A. *Approximation Theory and Numerical Methods.* Wiley, New York, 1980.

Papers, Monographs, and Proceedings

[AS]   Abramowitz, M., and I. A. Stegun. *Handbook of Mathematical Functions.* National Bureau of Standards, Washington, 1964. Reprint, Dover, New York.

[Ad]   Adams, R. A. *Sobolev Spaces.* Academic Press, Boston, 1978.

[AW]   Agarwal, R. P., and P. J. Y. Wong. *Error Inequalities in Polynomial Interpolation and Their Applications.* Kluwer, Dordrecht, 1993.

[Ak1]   Akhiezer, N. I. *The Classical Moment Problem and Some Related Questions in Analysis.* Oliver and Boyd, London, 1965. Also Hafner, New York, 1965.

[Aki]   Akima, H. "A method of bivariate interpolation and smooth surface fitting based on local procedures." *Comm. Assoc. Comput. Mach. 17* (1974), 18–20. Algorithm, pp. 26–31.

[ANS]   Alfeld, P., M. Neamtu, and L. L. Schumaker. "Scattered data interpolation on the sphere using spherical Bernstein-Bézier polynomials." Forthcoming.

[All1]   Allasia, G. "A class of interpolating positive linear operators: Theoretical and computational aspects." In *Approximation Theory, Wavelets and Applications,* ed. by S. P. Singh, A. Carbone, and B. Watson. Kluwer, Dordrecht, 1995, pp. 1–36.

[All2]   Allasia, G., R. Besenghi, and V. Demichelis. "Weighted arithmetic means possessing the interpolation property." *Calcolo 25* (1988), 203–217.

[All3]   Allasia, G. "Some physical and mathematical properties of inverse distance weighted methods for scattered data interpolation." *Calcolo 29* (1992), 97–109.

[All4]   Allasia, G., R. Basenghi, and V. Demichelis. "Multivariate interpolation by weighted arithmetic means at arbitrary points." *Calcolo 29* (1992), 301–311.

[AMSa]   Amir, D., J. Mach, and K. Saatkamp. "Existence of Chebyshev centers." *Trans. Amer. Math. Soc. 271* (1982), 513–524.

[Al]   Alexsandrov, P. S. *Combinatorial Topology.* Graylock Press, Rochester, NY, 1956.

[AZ]   Amir, D., and Z. Ziegler. "Construction of elements of the relative Chebyshev center." In *Approximation Theory and Applications,* ed. by Z. Ziegler. Academic Press, New York, 1981. pp. 1–12.

[AKHL]   Angelos, J. R., E. H. Kaufman, Jr., M. S. Henry, and T. D. Lenker. "Optimal nodes for polynomial interpolation." In *Approximation Theory VI,* ed. by C. K. Chui, L. L. Schumaker, and J. D. Ward. Academic Press, New York, 1989, vol. 1, pp. 17–20.

[AG]   Arcangeli, R., and J. L. Gout. "Sur l'interpolation de Lagrange dans R^n." *Comptes Rendues Acad. Sci. Paris 281,* Series A (1975), 357–359.

[Arm]   Armitage, D. H. "A non-constant continuous function on the plane whose integral on every line is zero." *Amer. Math. Monthly 101* (1994), 892–894.

[Aron]   Aronszajn, N. "Theory of reproducing kernels." *Trans. Amer. Math. Soc. 68* (1950), 337–404.

[As]   Askey, R. *Orthogonal Polynomials and Special Functions.* Regional Conference Series in Applied Mathematics, vol. 21. SIAM, Philadelphia, 1975.

[AsH]   Askey, R., and I. I. Hirschman. "Mean summability for ultraspherical polynomials." *Math. Scand. 12* (1963), 167–177.

[Att]   Atteia, M. *Hilbertian Kernels and Spline Functions.* North-Holland, Amsterdam, 1992.

[Ba2]   Ball, K. "Eigenvalues of Euclidean distance matrices." *J. Approx. Theory 68* (1992), 74–82.

[BlSW]   Ball, K., N. Sivakumar, and J. W. Ward. "On the sensitivity of radial basis interpolation to minimal data separation distance." *Constr. Approx. 8* (1992), 401–426.

[Ban]   Banach, S. *Théorie des Opérationes Linéaires.* Warsaw, 1932. Reprint, Haffner, New York, 1952.

[BBG]   Barnhill, R. E., G. Birkhoff, and W. J. Gordon. "Smooth interpolation in triangles." *J. Approx. Theory 8* (1973), 214–225.

[Barn1]   Barnhill, R. E. "Representation and approximation of surfaces." In *Mathematical Software II,* ed. by J. R. Rice. Academic Press, New York, 1977, pp. 69–120.

[BDL]   Barnhill, R. E., R. P. Dube, and F. F. Little. "Properties of Shepard's surfaces." *Rocky Mt. J. Math. 13* (1983), 365–382.

[Bar]   Barron, A. R. "Universal approximation bounds for superpositions of a sigmoidal function." *IEEE Trans. Inform. Theory 39,* no. 3 (1993), 930–945.

[Bar2]   Barron, A. R. "Approximation bounds for superpositions of a sigmoidal function." *Proc. IEEE Internat. Symp. Inform. Theory.* Budapest, June 1991.

[Bart1]   Bartle, R. G. *Elements of Integration Theory.* Wiley, New York, 1966. Reprinted, enlarged, and retitled: *Elements of Integration and Lebesgue Measure,* 1995.

[BasD]   Baszenski, G., and F.-J. Delvos. "Boolean algebra and multivariate interpolation." In *Approximation and Function Spaces.* Banach Center Publications, vol. 22. PWN-Polish Scientific Publishers, Warsaw, 1989, pp. 25–44.

[Bax2]   Baxter, B. J. C. "Norm estimates for inverses of Toeplitz distance matrices." *J. Approx. Theory 79* (1994), 222–242.

[Bax3]   Baxter, B. J. C. "On the asymptotic cardinal function for the multiquadric." *Computers Math. Applic. 24* (1992), 1–6.

[BaxM]   Baxter, B. J. C., and C. Micchelli. "Norm estimates for the inverses of multivariate Toeplitz matrices." *Numer. Algorithms 1* (1994), 103–117.

[BS]   Baxter, B. J. C., and N. Sivakumar. "On shifted cardinal interpolation by Gaussians and multiquadrics." *J. Approx. Theory 87* (1996), 36–59.

[BxSW]   Baxter, B. J. C., N. Sivakumar, and J. D. Ward. "Regarding the p-norms of radial basis interpolation matrices." *Constr. Approx. 10* (1994), 451–468.

[BL]   Beatson, R. K., and W. A. Light. "Quasi-interpolation in the absence of polynomial reproduction." In *Numerical Methods of Approximation Theory.* Internat. Ser. Numer. Math., vol. 105. Birkhäuser, Basel, 1992, pp. 21–39.

[BN]   Beatson, R. K., and G. N. Newsam. "Fast evaluation of radial basis funcions: Moment-based methods." Society for Industrial and Applied Mathematics (SIAM). *J. Sci. Comput. 19* (1998), 1428–1449.

[BPo]   Beatson, R. K., and M. J. D. Powell. "Univariate interpolation on a regular finite grid by a multiquadric plus a linear ploynomial." *IMA J. Numer. Anal. 12* (1992), 107–133.

[Bel]   Bellman, R. *Introduction to Matrix Analysis.* McGraw-Hill, New York, 1960. 2nd ed., 1970.

[Bere]   Berezanskii, Yu. M. "A generalization of a multidimensional theorem of Bochner." *Sov. Math. Doklady 2* (1961), 143–147.

[BCR]   Berg, C., J. P. R. Christensen, and P. Ressel. *Harmonic Analysis on Semigroups.* Springer-Verlag, Berlin, 1984.

[Berg]   Bergman, S. *The Kernel Function and Conformal Mapping.* Mathematical Surveys, vol. 5. Amer. Math. Soc., Providence, RI, 1950. 2nd (rev.) ed., 1970.

[Be]   Bernstein, S. N. "Sur les fonctions absolument monotones." *Acta Math. 52* (1929), 1–66.

[Be2]   Bernstein, S. N. "Sur la limitation des valeurs d'un polynôme $P_n(x)$ de degré n sur tout un segment par ses valeurs en $n + 1$ pointes du segment." *Bull. Acad. Sci. URSS, Leningrad* (1931), 1025–1050. *Zentralblatt 4* (1932), 6.

[Bi]   Bingham, N. H. "Positive definite functions on spheres." *Proc. Camb. Phil. Soc. 73* (1973), 145–156.

[Bir]   Birkhoff, G. "Interpolation to boundary data in triangles." *J. Math. Anal. Appl. 42* (1973), 199–208.

[Bir2]   Birkhoff, G. "The algebra of multivariate interpolation." In *Constructive Approaches to Mathematical Models,* ed. by C. V. Coffman and G. J. Fix. Academic Press, New York, 1979, pp. 345–363.

[Bla2]   Blatter, J. *Grothendieck Spaces in Approximation Theory.* Memoirs Amer. Math. Soc. vol. 120. Amer. Math. Soc., Providence, RI, 1972.

[Blo2]   Bloom, T. "On the convergence of multivariable Lagrange interpolants." *Constr. Approx. 5* (1989), 415–435.

[Blo1]   Bloom, T. "The Lebesgue constant for Lagrange interpolation in the simplex." *J. Approx. Theory 54* (1988), 338–353.

[Blo3]   Bloom, T. "Lagrange interpolants at equally spaced points in the simplex." In *Approximation Theory V,* ed. by C. K. Chui, L. L. Schumaker, and J. D. Ward. Academic Press, New York, 1986, pp. 267–269.

[Bls]   Blumenson, L. E. "A derivation of n-dimensional spherical coordinates." *Amer. Math. Monthly 67* (1960), 63–66.

[Blu]   Blumenthal, L. M. "Theory and Applications of Distance Geometry," Chelsea, New York, 1970. Amer. Math. Soc., Providence, RI, 1998.

[Boc1] Bochner, S. "A theorem on Fourier-Stieltjes integrals." *Bull. Amer. Math. Soc. 40* (1934), 271–276.

[Boc3] Bochner, S. "Monotone Funktionen, Stieltjes Integrale und harmonische Analyse." *Math. Ann. 108* (1933), 378–410.

[Boc2] Bochner, S. *Vorlesungen über Fouriersche Integrale.* Leipzig, 1932.

[Boc4] Bochner, S. "Hilbert distances and positive definite functions." *Annals of Math. 42* (1941), 647–656.

[Bog]   Bognar, J. *Indefinite Inner Product Spaces.* Ergebnisse Series, vol. 78. Springer-Verlag, Berlin, 1974.

[BHS]   Bojanov, B. D., H. A. Hakopian, and A. A. Sahakian. *Spline Functions and Multivariate Interpolations.* Kluwer Academic, Dordrecht, 1993.

[BJ]   de Boor, C., and R. Q. Jia. "Controlled approximation and a characterization of the local approximation order." *Proc. Amer. Math. Soc. 95* (1985), 547–553.

[BP1]   de Boor, C., and A. Pinkus. "Proof of the conjectures of Bernstein and Erdős concerning the optimal nodes for polynomial interpolation." *J. Approx. Theory 24* (1978), 289–303.

[BR2]   de Boor, C., and A. Ron. "Computational aspects of polynomial interpolation in several variables." *Math. Comp. 58* (1992), 705–727.

[BR1]   de Boor, C., and A. Ron. "On multivariate polynomial interpolation." *Constr. Approx. 6* (1990), 287–302.

[BR4]   de Boor, C., and A. Ron. "Fourier analysis of the approximation power of principal shift-invariant spaces." *Constr. Approx. 8* (1992), 427–462.

[BRi]   de Boor, C., and J. Rice. "An adaptive algorithm for multivariate approximation giving optimal convergence rates." *J. Approx. Theory 25* (1979), 337–359.

[Bo1]   de Boor, C. "Polynomial interpolation in several variables." In *Proceedings of a Conference Honoring S. D. Conte,* ed. by R. DeMillo and J. R. Rice. Plenum Press, New York, 1994.

[Bo3]   de Boor, C. "The polynomials in the linear span of integer translates of a compactly supported function." *Constr. Approx. 3* (1987), 199–208.

[BF]   de Boor, C., and G. Fix. "Spline approximation by quasi-interpolants." *J. Approx. Theory 8* (1973), 19–45.

[BHR]   de Boor, C., K. Höllig, and S. Riemenschneider. *Box Splines.* Springer-Verlag, New York, 1993.

[BDR1] de Boor, C., R. DeVore, and A. Ron. "Approximation from shift-invariant subspaces of $L_2(R^d)$." *Trans. Amer. Math. Soc. 341* (1994), 787–806.

[Bo4]   de Boor, C. *A Practical Guide to Splines.* Springer-Verlag, New York, 1978.

[Bo2]   de Boor, C. "Multivariate approximation." In *State of the Art in Numerical Analysis,* ed. by A. Iserles and M. J. D. Powell. Oxford University Press, Oxford, UK, 1987, pp. 87–109.

[Bo5]   de Boor, C. "Quasiinterpolants and approximation power of multivariate splines." In *Computation of Curves and Surfaces,* ed. by W. Dahmen, M. Gasca, and C. A. Micchelli. Kluwer, Dordrecht, 1990, pp. 313–345.

[Bo7]   de Boor, C. "The quasi-interpolant as a tool in elementary polynomial spline theory." In *Approximation Theory,* ed. by G. G. Lorentz. Academic Press, 1973, pp. 269–276.

[BorE]   Borwein, P., and T. Erdelyi. *Polynomials and Polynomial Inequalities.* Springer-Verlag, Berlin, 1995.

[Bos1]   Bos, L. "A characteristic of points in R^2 having Lebesgue function of polynomial growth." *J. Approx. Theory 56* (1989), 316–329.

[Bos2]   Bos, L. "Some remarks on the Fejér problem for Lagrange interpolation in several variables." *J. Approx. Theory 60* (1990), 133–140.

[Bos3]   Bos, L. "Bounding the Lebesgue function for Lagrange interpolation in a simplex." *J. Approx. Theory 38* (1983), 43–59.

[Bos5]   Bos, L. "On certain configurations of points in R^n which are unisolvent for polynomial interpolation." *J. Approx. Theory 64* (1991), 271–280.

[Bos6]   Bos, L. "On Lagrange interpolation at points in a disk." In *Approximation Theory V,* ed. by C. K. Chui, L. L. Schumaker, and J. D. Ward. Academic Press, Boston, 1986, pp. 275–278.

[BS]   Bos, L. P., and K. Salkauskas. "On the matrix $|x_i - x_j|^3$ and the cubic spline continuity equation." *J. Approx. Theory 51* (1987), 81–88.

[BP]   Braess, D., and A. Pinkus. "Interpolation by ridge functions." *J. Approx. Theory 73* (1993), 218–236.

[BFF]   Brand, R., W. Freeden, and J. Frolich. "An adaptive hierarchical approximation method on the sphere using axisymmetric locally supported basis functions." University of Kaiserslautern, Report No. 30, February 1995.

[Br]   Branham, R. L. *Scientific Data Analysis: An Introduction to Overdetermined Systems.* Springer-Verlag, Berlin, 1990.

[Brez]   Brezinski, C. *Biorthogonality and Its Applications to Numerical Analysis.* Pure and Applied Mathematics vol. 156. Dekker, Basel, 1992.

[Brez2] Brezinski, C. "The generalization of Newton's interpolation formula due to Mühlbach and Andoyer." *Electronic Transactions on Numerical Analysis 2* (1994), 130–137.

[Bro]   Brown, A. L. "Uniform approximation by radial basis functions." In *Advances in Numerical Analysis,* vol. 2, ed. by W. A. Light. Oxford University Press, Oxford, UK, 1992, pp. 203–206.

[Bru]   Brutman, L. "On the Lebesgue function for polynomial interpolation." *SIAM J. Numerical Analysis 15* (1978), 694–704.

[Bru2]   Brutman, L. "Lebesgue functions for polynomial interpolation: A survey." *Ann. Numer. Math. 4* (1997) 111–127.

[BruT]   Brutman, L., and D. Toledano. "On an extremal problem of Erdős in interpolation theory." *Computers Math. Appl. 34* (1997), 37–47.

[Bu2]   Buhmann, M. D. "Multivariate interpolation in odd-dimensional Euclidean spaces using multi-quadrics." *Constr. Approx. 6* (1990), 21–34.

[Bu1]   Buhmann, M. D. "Cardinal interpolation with radial basis functions: An integral transform approach." In *Multivariate Approximation Theory IV,* ed. by W. Schempp and K. Zeller. ISNM, vol. 90. Birkhäuser, Basel, 1989, pp. 41–64.

[Bu7]   Buhmann, M. D. "Multivariate cardinal-interpolation with radial-basis functions." *Constr. Approx. 6* (1990), 225–255.

[Bu8]   Buhmann, M. D. "On quasi-interpolation with radial basis functions." *J. Approx. Theory 72* (1993), 103–130.

[Bu9]   Buhmann, M. D. "New developments in the theory of radial basis function interpolation." In *Multivariate Approximation and Wavelets,* ed. by K. Jetter and F. I. Utreras. World Scientific Publishers, Singapore, 1993, pp. 35–75.

[BuD1]   Buhmann, M. D., and N. Dyn. "Error estimates for multiquadric interpolation." In *Curves and Surfaces,* ed. by P.-J. Laurent, A. LeMéhauté, and L. L. Schumaker. Academic Press, New York, 1991, pp. 51–58.

[BuDL]   Buhmann, M. D., N. Dyn, and D. Levin. "On quasi-interpolation by radial basis functions with scattered centers." *Constr. Approx. 11* (1995), 239–254.

[BuM4]   Buhmann, M. D., and C. A. Micchelli. "Multiquadric interpolation improved." *Computers Math. Appl. 24,* no. 12 (1992), 27–34.

[BuP1]   Buhmann, M. D., and M. J. D. Powell. "Radial basis function interpolation on a infinite regular grid." In *Algorithms for Approximation II,* ed. by J. C. Mason and M. G. Cox. Chapman and Hall, London, 1990, pp. 146–169.

[BuLe]   Burchard, H. G., and Junjiang Lei. "Coordinate order of approximation by functional-based approximation operators." *J. Approx. Theory 82* (1995), 240–256.

[But1]   Butzer, P. L., "Representation and approximation of functions by general singular integrals." *Neder. Akad. Wetensch. Proc. Ser. 63A (Indag. Math.) 22* (1960), 1–24.

[But2]   Butzer, P. L. "A survey of the Whittaker-Shannon sampling theorem and some of its extensions." *J. Math. Res. Expositions 3* (1983), 185–212.

[But3]   Butzer, P. L. "The Shannon sampling theorem and some of its generalizations. An overview." In *Constructive Function Theory '81.* Sofia, 1983, pp. 258–273.

[BO]   Butzer, P. L., and W. Oberdörster. "Linear functionals defined on various spaces of continuous functions on *R*." *J. Approx. Theory 13* (1975), 451–469.

[ButN]   Butzer, P. L., and R. J. Nessel. *Fourier Analysis and Approximation.* Birkhäuser, Basel, 1971.

[BRS]   Butzer, P. L., S. Ries, and R. L. Stens. "Approximation of continuous and discontinuous functions by generalized sampling series." *J. Approx. Theory 50* (1987), 25–39.

[BSS]   Butzer, P. L., W. Splettstösser, and R. L. Stens. "The sampling theorem and linear predictions in signal analysis." *Jahresber. Deutsch. Math.–Verein 90* (1988), 1–60.

[BuyS]   Buyn, D. W., and S. Saitoh. "Approximation by the solutions of the heat equation." *J. Approx. Theory 78* (1994), 226–238.

[CMS]   Cavaretta, A. S., C. A. Micchelli, and A. Sharma. "Multivariate interpolation and the Radon Transform." Part I, *Math. Zeit. 174* (1980), 263–279. Part II, *Qualitative Approximation,* ed. by R. A. DeVore and K. Scherer. Academic, New York, 1980. Part III, *Canadian Math. Soc. Conference Proceedings 3* (1983), 37–50.

[CEH]   Censor, Y., T. Elfving, and G. Herman (eds.) Linear algebra in image reconstruction from projections. *Linear Alg. Appl. 130* (1990), 1–305.

[CMe]   Chalmers, B. L., and F. T. Metcalf. "Determination of a minimal projection from *C*[−1, 1] onto the quadratics." *Numer. Func. Anal. Optim. 11* (1990), 1–10.

[CMe2]   Chalmers, B. L., and F. T. Metcalf. "The determination of minimal projections and extensions in L^1." *Trans. Amer. Math. Soc. 329* (1992), 289–308.

[Cha] Chalmers, B. L. "A natural simple projection . . ." *J. Approx. Theory 32* (1981), 226–232.

[Chand] Chandrasekharan, K. *Classical Fourier Transforms.* Springer-Verlag, New York, 1989.

[Chan2] Chang, Kuei-Fang. "Strictly positive definite functions." *J. Approx. Theory 87* (1996), 148–158. Addendum, *J. Approx. Theory 88* (1997), 384.

[ChMu] Chanillo, S., and B. Muckenhoupt. *Weak Type Estimates for Cesaro Sums of Jacobi Polynomial Series.* Memoirs, Amer. Math. Soc., Amer. Math. Soc., no. 487, Providence, RI, 1993.

[CCL] Chen, T., H. Chen, and R. Liu. "A constructive proof and extension of Cybenko's approximation theorem." In *Computing Science and Statistics: Proceedings of the 22nd Symposium on the Interface.* Springer-Verlag, New York, 1991, pp. 163–168.

[DC1] Chen, Debao. "Degree of approximation by superpositions of a sigmoidal function." *Approx. Theory Appl. 9* (1993), 17–28.

[C3] Cheney, E. W. "Projection operators in approximation theory." In *Studies in Functional Analysis,* ed. by R. G. Bartle. Math. Assoc. of Amer., Washington, DC, 1980, pp. 50–80.

[CL] Cheney, E. W., and Junjiang Lei. "Quasi-interpolation on irregular points." In *Approximation and Computation: A Festschrift in Honor of Walter Gautschi,* ed. by R. V. M. Zahar. ISNM, vol. 119. Birkhäuser, Boston, 1994, pp. 121–135.

[CP] Cheney, E. W., and K. H. Price. "Minimal projections." In *Approximation Theory,* ed. by A. Talbot. Academic Press, New York, 1970, pp. 261–289.

[CX3] Cheney, E. W., and Yuan Xu. "A set of research problems in approximation theory," In *Topics in Polynomials of One and Several Variables and Their Applications,* ed. by T. M. Rassias, H. M. Srivastava, and A. Yanushauskas. World Scientific Publishers, Singapore, 1992, pp. 109–123.

[CM] Cheney, E. W., and P. D. Morris. "On the existence and characterization of minimal projections." *J. Reine Angew. Math. 270* (1974), 215–227.

[CLX] Cheney, E. W., W. A. Light, and Yuan Xu. "On kernels and approximation orders." In *Approximation Theory: Proceedings of the Sixth Southeastern International Approximation Conference,* ed. by G. Anastassiou. Dekker, New York, 1992, pp. 227–242.

[Cho] Choquet, G. *Lectures on Analysis* (3 vols.). Benjamin, New York, 1969.

[ChRe] Christensen, J. P. R., and P. Ressel. "Functions operating on positive definite matrices, and a theorem of Schoenberg." *Trans. Amer. Math. Soc. 243* (1978), 89–95.

[ChRe2] Christensen, J. P. R., and P. Ressel. "Positive definite kernels on the complex Hilbert sphere." *Math. Zeit. 180* (1982), 193–201.

[Chu2] Chui, Charles K. *An Introduction to Wavelets.* Academic Press, New York, 1992.

[ChuDi] Chui, Charles, and H. Diamond. "A natural formulation of quasi-interpolation by multivariate splines." *Proc. Amer. Math. Soc. 99* (1987), 643–646.

[ChuLi] Chui, Charles, and Xin Li. "Approximation by ridge functions and neural networks with one hidden layer." *J. Approx. Theory 70* (1992), 131–141.

[CLM1] Chui, Charles, Xin Li, and H. N. Mhaskar. "Neural networks for localized approximation." *Math. Comp. 63* (1994), 607–623.

[CLM2] Chui, Charles, Xin Li, and H. N. Mhaskar. "Limitations of the approximation capabilities of neural networks with one hidden layer." *Adv. Comput. Math. 5* (1996), 233–243.

[ChuLa] Chui, Charles, and M. J. Lai. "Vandermonde determinant and Lagrange interpolation in R^s." In *Nonlinear and Convex Analysis,* ed. by B. L. Lin and L. Simons. Dekker. New York, 1988, pp. 23–36.

[CMP] Chui, Charles, L. Montefusco, and L. Puccio (eds.). *Wavelets: Theory, algorithms, and Applications.* Academic Press, New York, 1994.

[CSZ] Chui, Charles, X. C. Shen, and L. Zhong. "On Lagrange polynomial quasi-interpolation." In *Topics in Polynomials of One and Several Variables and Their Applications,* ed. by T. H. Rassias et al. World Scientific Publishers, Singapore, 1993, pp. 125–142.

[Chu1] Chui, Charles K. *Multivariate Splines.* SIAM, Philadelphia, 1988.

[CY] Chung, K. C., and T. H. Yao. "On lattices admitting unique Lagrange interpolation." *SIAM J. Numer. Anal. 14* (1977), 735–741.

[CR] Ciarlet, P. G., and P.-A. Raviart. "Interpolation de Lagrange dans R^n." *Comptes Rendus Acad. Sci. Paris 273* (1971), 578–581.

[Cli] Cline, A. K. "Scalar and planar valued curve fitting using splines under tension." *Comm. Assoc. Comput. Mach. 17* (1974), 218–223.

[Cr] Crum, M. M. "On positive definite functions." *Proc. London Math. Soc. 6* (1956), 548–560.

[Cu] Curtis, P. C. "N-Parameter families and best approximation." *Pacific J. Math. 9* (1959), 1013–1027.

[Cy] Cybenko, G. "Approximation by superpositions of a sigmoidal function," *Math. Control Signals Systems 2* (1989), 203–314.

[Dae] Daehlen, M., and M. Floater. "Iterative polynomial interpolation and data compression." *Numer. Algorithms 5* (1993), 165–177.

[DM1] Dahmen, W., and C. Micchelli. "Some remarks on ridge functions." *Approx. Theory Appl. 3* (1987), 139–143.

[DM2] Dahmen, W., and C. Micchelli. "On the approximation order from certain multivariate spline spaces." *J. Australian Math. Soc.,* Ser. B, *26* (1984), 233–246.

[DM3] Dahmen, W., and C. Micchelli. "Translates of multivariate splines." *Linear Alg. Appl. 52* (1983), 217–234.

[DDS] Dahmen, W., R. DeVore, and K. Scherer. "Multi-dimensional spline approximation." *SIAM J. Numer. Anal. 17* (1980), 380–402.

[Dau] Daubechies, I. *Ten Lectures on Wavelets.* SIAM, Philadelphia, 1992.

[DMZ] Davis, G., S. Mallat, and Zhifeng Zhang. "Adaptive time–frequency approximations with matching pursuits." In *Wavelets: Theory, Algorithms, and Applications,* ed. by C. L. Chui, L. Montefusco, and L. Puccio. Academic Press, New York, 1994, pp. 271–293.

[Dav] Davison, M. E. "A singular value decomposition for the Radon transform in n-dimensional Euclidean space." *Numer. Func. Anal. Optim. 3* (1981), 321–340.

[DVP1] de La Vallée Poussin, C. "Sur la convergence des formules d'interpolation entre ordonnées equidistantes." *Bull. Cl. Sci. Acad. Roy. Belg.* no. 4 (1908), 319–403.

[DMV] Della Vecchia, B., G. Mastroianni, and P. Vértesi. "Direct and converse theorems for Shepard rational approximation." *Numer. Func. Anal. Optim. 17* (1996), 537–561.

[DLu] DeVore, R. A., and B. J. Lucier. "Wavelets." *Ann. Numer. Math.* (1991), 1–56.

[DY]   DeVore, R. A., and X. M. Yu. "Degree of adaptive approximation." *Math. Comp. 55* (1990), 625–635.

[Dev]   DeVore, R. A. "Degree of approximation." In *Approximation Theory II,* ed. by G. G. Lorentz, C. K. Chui and L. L. Schumaker. Academic Press, New York, 1976, pp. 117–161.

[Dev2]   DeVore, R. A. *The Approximation of Continuous Functions by Positive Linear Operators.* Lecture Notes in Math, vol. 293. Springer-Verlag, New York, 1972.

[DeS]   Delvos, F.-J., and W. Schempp. *Boolean Methods in Interpolation and Approximation.* Pitman Res. Notes in Math., vol. 230. Longmans/Wiley, New York, 1989.

[DeS2]   Delvos, F.-J., and W. Schempp. "The method of parametric extension applied to right-invertible operators." *Numer. Func. Anal. Optim. 6* (1993), 135–148.

[De1]   Delvos, F.-J. "Convergence of interpolation by translation." Alfred Haar Memorial Volume. *Colloquia Mathematica Societatis Janos Bolya 49* (1985), 273–287.

[De2]   Delvos, F.-J. "Periodic interpolation on uniform meshes." *J. Approx. Theory 51* (1987), 71–80.

[De3]   Delvos, F.-J. "d-Variate Boolean interpolation." *J. Approx. Theory 34* (1982), 99–114.

[De4]   Delvos, F.-J. "Approximation properties of periodic interpolation by translates of one function." *RAIRO Math. Model. Numer. Analysis 28* (1994), 177–188.

[DiS]   Diaconis, P., and M. Shahshahani. "On nonlinear functions of linear combinations." *SIAM J. Sci. Statist. Comput. 5* (1984), 175–191.

[Dier1]   Dierckx, P. *Curve and Surface Fitting with Splines.* Oxford University Press, Oxford, UK, 1993.

[DU]   Diestel, J., and J. J. Uhl, Jr. *Vector Measures.* Math. Surveys, no. 15. Amer. Math. Soc., Providence, RI, 1977.

[Din]   Dingankar, Ajit. *On Applications of Approximation Theory to Identification, Control, and Classification.* Unpublished doctoral dissertation, Dept. of Elect. and Comput. Eng., University of Texas at Austin, 1995.

[Do2]   Donoghue, W. F. *Distributions and Fourier Transforms.* Pure and Applied Mathematics Series, vol. 32. Academic Press, New York, 1969.

[Do1]   Donoghue, W. F. *Monotone Matrix Functions and Analytic Continuation.* Springer-Verlag, New York, 1974.

[Du]   Duchon, J. "Splines minimizing rotation-invariant semi-norms in Sobolev spaces." In *Constructive Theory of Functions of Several Variables,* ed. by W. Schempp and K. Zeller. Lecture Notes in Math, vol. 571. Springer-Verlag, New York, 1976, pp. 85–100.

[Du2]   Duchon, J. "Sur l'erreur d'interpolation des fonctions de plusieurs variables par les D^m-splines." *RAIRO Analyse Numerique 12* (1978), 325–334.

[DS]   Dunford, N., and J. T. Schwartz. *Linear Operators, Part I: General Theory.* Interscience, New York, 1958.

[Dym]   Dym, H. *Contractive Matrix Functions, Reproducing Kernel Hilbert Spaces and Interpolation.* Conference Board of the Mathematical Sciences, Regional Conference Series, vol. 71. Amer. Math. Soc., Providence, RI, 1989.

[DMc]   Dym, H., and H. P. McKean. *Fourier Series and Integrals.* Academic Press, New York, 1972.

[DGM]   Dyn, N., T. Goodman, and C. A. Micchelli. "Positive powers of certain conditionally negative definite matrices." *Proc. Kon. Nederl. Acad. Wetens.* Series A *89* (1986), 163–178.

[DyM]   Dyn, N., and C. A. Micchelli. "Interpolation by sums of radial functions." *Numer. Math.* *58* (1990), 1–9.

[Dy]   Dyn, N. "Interpolation of scattered data by radial functions." In *Topics in Multivariate Approximation,* ed. by C. K. Chui, L. L. Schumaker, and F. I. Utreras. Academic Press, New York, 1987, pp. 47–62.

[Dy2]   Dyn, N. "Interpolation and approximation by radial and related functions." In *Approximation Theory VI,* ed. by C. K. Chui, L. L. Schumaker, and J. D. Ward. Academic Press, New York, 1989, pp. 211–234.

[DLR]   Dyn, N., D. Levin, and S. Rippa. "Numerical procedures for global surface fitting of scattered data by radial functions." *SIAM J. Sci. Statist. Comput. 7* (1986), 639–659.

[DR]   Dyn, N., and A. Ron. "On multivariate interpolation." In *Algorithms for Approximation,* ed. by M. G. Cox and J. C. Mason. Chapman and Hall, London, 1990.

[E1]   Edwards, R. E. *Functional Analysis.* Holt, Rinehart and Winston, New York, 1965.

[E3]   Edwards, R. E. "On functions whose translates are independent." *Ann. Inst. Fourier III* (1951), 31–72.

[Ell]   Ellacott, S. "Aspects of the numerical analysis of neural networks." *Acta Numerica* (1994), 145–202. .

[Er]   Erdős, P. "Problems and results on the theory of interpolation, II." *Acta Math. Acad. Sci. Hungar. 12* (1961), 235–244.

[Er2]   Erdős, P. "Some remarks on polynomials." *Bull. Amer. Math. Soc. 53* (1947), 1169–1176.

[FaH]   Falb, P. L., and U. Haussmann. "Bochner's theorem in infinite dimensions." *Pacific J. Math. 43* (1972), 601–618.

[Fari]   Faridani, A. "A generalized sampling theorem for locally compact Abelian groups." *Math. Comp. 63* (1994), 307–328.

[Far]   Farin, G. *Curves and Surfaces for Computer Aided Design,* 2nd ed. Academic Press, Boston, 1990.

[Fa]   Farwig, R. "Rate of convergence of Shepard's global interpolation formula." *Math. Comp. 46* (1986), 577–590.

[Fass4] Fasshauer, G. E. "Adaptive least squares fitting with radial basis functions on the sphere." In *Mathematical Methods in CAGD III,* ed. by M. Daehlen, T. Lyche, and L. L. Schumaker. Vanderbilt University Press, Nashville, TN, 1995.

[FaSc]   Fasshauer, G. E., and L. L. Schumaker. "Scattered data fitting on the sphere." In *Mathematical Methods for Curves and Surfaces II,* ed. by M. Daehlen, T. Lyche, and L. L. Schumaker. Vanderbilt University Press, Nashville, TN, 1998.

[Fej]   Fejér, L. "Bestimmung derjenigen Abszissen eines Intervalles für welche die Quadratsumme der Grundfunktionen der Lagrangeschen Interpolation im Intervalle ein möglichst kleines Maximum Besitz." *Ann. Scuola Norm. Sup. Pisa 1* (2) (1932), 263–276.

[Fek]   Fekete, M. "Über die Verteilung der Wurzeln bei gewissen algebraischen Gleichungen mit ganzahligen Koeffizienten." *Math. Zeit. 17* (1923), 228–249.

[Fe]   Feller, W. "On Müntz's theorem and completely monotone functions." *Amer. Math. Monthly 75* (1968), 342–349.

[FLE]   Fisher, N. I., T. Lewis, and B. J. J. Embleton. *Statistical Analysis of Spherical Data.* Cambridge University Press, Cambridge, UK, 1987.

[FH]   FitzGerald, C. H., and R. A. Horn. "On fractional Hadamard powers of positive definite matrices." *J. Math. Anal. Appl. 61* (1977), 633–642.

[FS]   Fix, G., and G. Strang. "Fourier analysis of the finite element method in Ritz-Galerkin theory." *Stud. Appl. Math. 48* (1969), 265–273.

[Fol1] Foley, T. A. "Interpolation of scattered data on a spherical domain." In *Algorithms for Approximation II,* ed. by J. C. Mason and M. G. Cox. Chapman and Hall, London, 1990, pp. 303–310.

[Fol2] Foley, T. A. "Three-stage interpolation to scattered data." *Rocky Mt. J. Math. 14* (1984), 141–149.

[Foll] Folland, G. B. *Fourier Analysis and Its Applications.* Wadsworth–Brooks/Cole, Pacific Grove, CA, 1992.

[FJL]   Fontanella, F., K. Jetter, and P. J. Laurent, eds. *Advances in Multivariate Approximation.* World Scientific Publishers, Singapore, 1996.

[FC2]   Franchetti, C., and E. W. Cheney. "Simultaneous approximation and restricted Chebyshev centers in function spaces." In "Approximation Theory and Applications," ed. by Z. Ziegler. Academic Press, New York, 1981, pp. 65–88.

[FC3]   Franchetti, C., and E. W. Cheney. "The embedding of proximinal sets." *J. Approx. Theory 48* (1986), 213–225.

[FC]   Franchetti, C., and E. W. Cheney. "Minimal projections in L_1-spaces." *Duke Math. J. 43* (1976), 501–510.

[Fr]   Franke, R. "Scattered data interpolation: Tests of some methods." *Math. Comp. 38* (1982), 381–200.

[Fred] Frederickson, P. O. "Quasi–interpolation, extrapolation, and approximation in the plane." In *Proceedings of the Manitoba Conference on Numerical Mathematics.* Dept. Comput. Sci., University of Manitoba, 1971, pp. 159–167.

[Free] Freeden, W. "Spherical spline interpolation." *J. Comput. Appl. Math. 11* (1984), 367–375.

[FGS]   Freeden, W., T. Gervens, and M. Schreiner. *Constructive Approximation on the Sphere.* Oxford University Press, Oxford, UK, 1998.

[FSF]   Freeden, W., M. Schreiner, and R. Franke. "A survey on spherical spline approximation." *Berichte der Arbeitsgruppe Technomath., Universität Kaiserslautern 95* (1995).

[Frl]   Friedlander, F. G. *Introduction to the Theory of Distributions.* Cambridge University Press, Cambridge, UK, 1982.

[F]   Friedman, A. *Foundations of Modern Analysis.* Holt, Rinehart and Winston, New York, 1970. Reprint, Dover, New York, 1982.

[Fri]   Friedman, J. H. "Multivariate adaptive regression splines." *Annals of Stat. 19* (1991), 1–141.

[FriS] Friedman, J. H., and W. Stuetzle. "Projection pursuit regression." *J. Amer. Stat. Assoc. 76* (1981), 817–823.

[Fun]   Funahashi, K. "On the approximate realization of continuous mappings by neural networks." *Neural Networks 2* (1989), 183–192.

[GalP]   Galimberti, G., and V. Pereyra. "Numerical differentiation and the solution of multidimensional Vandermonde systems." *Math. Comp. 24* (1970), 357–364.

[Gard]   Gardiner, S. J. *Harmonic Approximation.* Cambridge University Press, Cambridge, UK, 1995. Review: *J. Approx. Theory 86* (1996), 360–361.

[GL]   Gasca, M., and E. Lebron. "Elimination techniques and interpolation." *J. Comput. Appl. Math. 19* (1987), 125–132.

[GM]   Gasca, M., and J. I. Maeztu. "On Lagrange and Hermite interpolation in R^k." *Numer. Math. 39* (1982), 1–14.

[GaMu]   Gasca, M., and G. Mühlbach. "Multivariate polynomial approximation under projectivities, Part I." *Numer. Algorithms 1* (1991), 375–400. Part II, ibid., *2* (1992), 255–278.

[Ga]   Gasca, M. "Multivariate polynomial interpolation." In *Computation of Curves and Surfaces,* ed. by W. Dahmen, M. Gasca, and C. Micchelli. Kluwer Academic, Dordrecht, 1990, pp. 215–235.

[Gasp2]   Gasper, G. "Positivity and special functions." In *Theory and Applications of Special Functions,* ed. by R. A. Askey. Academic Press, New York, 1975, pp. 375–433.

[Gau1]   Gautschi, W. "The condition of Vandermonde-like matrices involving orthogonal polynomials." *Linear Alg. Appl. 52* (1983), 293–300.

[Gau2]   Gautschi, W. "Lower bounds on the condition number of Vandermonde matrices." *Numer. Math. 52* (1988), 241–250.

[GG]   Gelfand, I. M., and S. G. Gindikin (eds.). *Mathematical Problems of Tomography.* Amer. Math. Soc., Providence, RI, 1990.

[GS]   Gelfand, I. M., and G. E. Shilov. *Generalized Functions,* vol. 1. Academic Press, New York, 1964.

[GJ]   Gillman, L., and M. Jerison. *Rings of Continuous Functions.* Springer-Verlag, New York, 1960.

[GP]   Goffman, C., and G. Pedrick. *First Course in Functional Analysis,* 2nd ed., Chelsea, New York. (Now available from Amer. Math. Soc., Providence, RI.)

[Gold]   Goldberg, R. R. *Fourier Transforms.* Cambridge University Press, Cambridge, UK, 1970.

[Golds]   Goldstein, M. (ed.). *Approximation by Solutions of Partial Differential Equations.* NATO-ASI Series C-365. Kluwer, Dordrecht, 1994.

[GoL]   Golitschek, M., and W. Light. "Interpolation by polynomials and radial basis functions on spheres." *Constr. Approx.* (forthcoming).

[Gol2]   Golomb, M. *Lectures on Theory of Approximation.* Argonne National Laboratory, Argonne, IL, 1962.

[GolW]   Golomb, M., and H. F. Weinberger. "Optimal approximation and error bounds." In *On Numerical Approximation,* ed. by R. E. Langer. University of Wisconsin Press, Madison, WI, 1959, pp. 117–190.

[GvL]   Golub, G. H., and C. van Loan. *Matrix Computations.* Johns Hopkins University Press, Baltimore, 1983.

[Goo]   Goodman, T. N. T. "Interpolation in minimum seminorm and multivariate B-splines." *J. Approx. Theory 37* (1982), 286–305.

[GW]   Gordon, W. F., and J. A. Wixom. "Pseudo–harmonic interpolation on convex domains." *SIAM J. Numer. Anal. 11* (1974), 909–933.

[GW2]   Gordon, W. F., and J. A. Wixom. "Shepard's method of metric interpolation to bivariate and multivariate functions." *Math. Comp. 32* (1978), 253–264.

[GC]   Gordon, W. F., and E. W. Cheney. "Bivariate and multivariate interpolation with non-commutative projectors." In *Linear Spaces and Approximation,* ed. by P. L. Butzer and B. Sz.-Nagy. ISNM vol. 40. Birkhäuser, Basel, 1978, pp. 381–387.

[GH]   Gordon, W. J., and C. A. Hall. "Transfinite element methods: Blending function interpolation over arbitrary curved element domains." *Numer. Math. 21* (1973), 109–129.

[Go2]   Gordon, W. J. "Blending-function methods of bivariate and multivariate interpolation and approximation." *SIAM J. Numer. Anal. 8* (1971), 158–177.

[GT]   Groshof, M. S., and G. Taiani. "Vandermonde strikes again." *Amer. Math. Monthly 100* (1993), 575–578.

[Gr]   Gross, Leonard, *Harmonic Analysis on Hilbert Space.* Memoirs Amer. Math. Soc., Amer. Math. Soc., no. 46, Providence, RI, 1963.

[GuRo]   Guenter, R. B., and E. L. Roetman. "Some observations on interpolation in higher dimensions." *Math. Comp. 24* (1970), 517–521.

[GuS]   Guo, K., and Xingping Sun. "Scattered data interpolation by linear combinations of translates of conditionally positive definite functions." *Numer. Func. Anal. Optim. 12* (1991), 137–152.

[GHS]   Guo, K., S. Hu, and Xingping Sun. "Conditionally positive definite functions and Laplace-Stieltjes integrals." *J. Approx. Theory 74* (1993), 249–265.

[GHS2]   Guo, K., S. Hu, and Xingping Sun. "Multivariate interpolation using linear combinations of translates of a conditionally positive definite function." *Numer. Func. Anal. Optim. 14* (1993), 371–381.

[GutI]   Gutzmer, T., and A. Iske. "Detection of discontinuities in scattered data approximation." *Numer. Algorithms 16* (1997), 155–170.

[Hak1]   Hakopian, H. A. "Multivariate divided differences and multivariate interpolation of Lagrange and Hermite type." *J. Approx. Theory 34* (1982), 286–305.

[Hak2]   Hakopian, H. A. "On fundamental polynomials of multivariate interpolation of Lagrange and Hermite type." *Bull. Polish Acad. Sci. Math. 31* (1983), 137–141.

[Hak]   Hakopian, H. A. "Multivariate interpolation of Lagrange and Hermite type." *Studia Math. 80* (1984), 77–88.

[HSSW]   Hamaker, C., K. T. Smith, D. C. Solmon, and S. L. Wagner. "The divergent beam x–ray transform." *Rocky Mt. J. Math. 10* (1980), 253–283.

[HaSo]   Hamaker, C., and D. C. Solmon. "The angles between the null spaces of x-rays." *J. Math. Anal. Appl. 62* (1978), 1–23.

[Ha]   Hardy, R. "Multiquadric equations of topography and other irregular surfaces." *J. Geophys. Res. 76* (1971), 1905–1915.

[Ha2]   Hardy, R. L. "Theory and applications of the multiquadric biharmonic method." *Computers Math. Applic. 19* (1990), 163–208. (This paper has a bibliography of 109 items.)

[HazS]   Hazou, I., and D. C. Solmon. "Approximate inversion formulas and convolution kernels for the exponential x-ray transform." *Zeit. Angew. Math. Mech. 66* (1986), T370–T372.

[HeN]   Hecht-Nelson, R. "Kolmogorov's mapping neural network existence theorem." *Proc. IEEE International Conference on Neural Networks III*. IEEE Press, New York, 1987, pp. 11–13.

[Helg]   Helgason, S. *Groups and Geometric Analysis*. Academic Press, New York, 1984.

[Helg2]   Helgason, S. *The Radon Transform*. Birkhäuser, Basel, 1980.

[Herm]   Herman, G. T. (ed.). *Image Reconstruction from Projections*. Topics in Applied Physics, vol. 32. Springer-Verlag, New York, 1979.

[Herm2]   Herman, G. T. *Image Reconstruction from Projections*. Academic Press, New York, 1980.

[HLN]   Herman, G., A. K. Louis, and F. Natterer. *Mathematical Methods in Tomography*. Lecture Notes in Math., vol. 1497. Springer-Verlag, New York, 1991.

[HN]   Herman, G., and F. Natterer. *Mathematical Aspects of Computerized Tomography*. Springer-Verlag, Berlin, 1981.

[HW]   Hernandez, E., and G. Weiss. *A First Course on Wavelets*. CRC Press, Boca Raton, FL, 1996.

[Her]   Hertle, A. "A characterization of Fourier and Radon transforms on Euclidean space." *Trans. Amer. Math. Soc. 273* (1982), 595–609.

[HewS]   Hewitt, E., and K. Stromberg. *Real and Abstract Analysis*. Springer-Verlag, New York, 1965.

[Hig]   Higgins, J. R. "Five short stories about the cardinal series." *Bull. Amer. Math. Soc. 12* (1985), 45–89.

[HiW]   Hirschman, I. I., and D. V. Widder. *The Convolution Transform*. Princeton University Press, Princeton, NJ, 1955.

[Hof]   Hoff, C. J. "Approximation with kernels of finite oscillation, I." *J. Approx. Theory 3* (1970), 213–228.

[Ho]   Höllig, K. "A remark on multivariate B-splines." *J. Approx. Theory 33* (1982), 119–125.

[Holb]   Holub, J. R. "Daugavet's equation and ideals of operators on $C\,[0, 1]$." *Math. Nachr. 141* (1989), 177–181.

[HJ1]   Horn, R. A., and C. R. Johnson. *Matrix Analysis*. Cambridge University Press, Cambridge, UK, 1985.

[HJ2]   Horn, R. A., and C. R. Johnson. *Topics in Matrix Analysis*. Cambridge University Press, Cambridge, UK, 1991.

[Hu]   Huber, P. J. "Projection pursuit." *Annals of Stat. 13* (1985), 435–475.

[Hu]   Husain, T. "Introduction to Topological Groups." Saunders, Philadelphia, 1966.

[Ji1]   Jackson, I. R. H. "Convergence properties of radial basis functions." *Constr. Approx. 4* (1988), 243–264.

[Ji3]   Jackson, I. R. H. *Radial Basis Function Methods for Multivariable Interpolation*. Unpublished doctoral dissertation, University of Cambridge, 1988.

[Ja]   Jameson, G. J. O. "A lower bound for the projection constant of P_2." *J. Approx. Theory 51* (1987), 163–167.

[JS]   Jawerth, B., and W. Sweldens. "An overview of wavelet based multiresolution analyses." *SIAM Rev. 36* (1994), 377–412.

[Jet]   Jetter, K. "Some contributions to bivariate interpolation and cubature." In *Approxima-tion Theory IV,* ed. by C. K. Chui, L. L. Schumaker, and J. D. Ward. Academic Press, New York, 1983, pp. 533–538.

[Jet2]   Jetter, K. "Multivariate approximation from the cardinal interpolation point of view." In *Approximation Theory VII,* ed. by E. W. Cheney, C. K. Chui, and L. L. Schumaker. Academic Press, New York, 1993, pp. 131–161.

[Jia2]   Jia, R. Q. "Approximation by multivariate splines: An application of Boolean methods." In *Numerical Methods of Approximation Theory,* ed. by D. Braess and L. L. Schumaker. *International Series on Numerical Mathematics,* vol. 105. Birkhäuser, Basel, 1992, pp. 117–134.

[Jia3]   Jia, R. Q. "A counterexample to a result concerning controlled approximation." *Proc. Amer. Math. Soc. 97* (1986), 647–654.

[Jia4]   Jia, R. Q. "Linear independence of translates of a box spline." *J. Approx. Theory 40* (1984), 158–160.

[JL]   Jia, R. Q., and Junjiang Lei. "Approximation by multiinteger translates of functions having global support." *J. Approx. Theory 72* (1993), 2–23.

[JL2]   Jia, R. Q., and Junjiang Lei. "A new version of the Strang-Fix conditions." *J. Approx. Theory 74* (1993), 221–225.

[Jo]   John, F. *Plane Waves and Spherical Means Applied to Partial Differential Equations.* Interscience, New York, 1955.

[Jon]   Jones, L. K. "A simple lemma on greedy approximation in Hilbert space and convergence rates for projection pursuit regression and neural network training." *Annals of Stat. 20* (1992), 608–613.

[Jon2]   Jones, L. K. "Constructive approximations for neural networks by sigmoidal functions." *Proc. of the IEEE 78* (1990), 1586–1589.

[Jon3]   Jones, L. K. "On a conjecture of Huber concerning the convergence of projection pursuit regression." *Annals of Stat. 15* (1987), 880–882.

[KSn]   Kadec, M. I., and M. G. Snowbar. "Some functionals over a compact Minkowski space." *Math. Notes 10* (1971), 694–696.

[K2]   Kansa, E. J. "Multiquadrics—A scattered data approximation scheme with applications to computational fluid dynamics." *Computers Math. Applic. 19* (1990), 127–145.

[KS]   Karlin, S., and W. J. Studden. *Tchebycheff Systems: With Applications in Analysis and Statistics.* Interscience, New York, 1966.

[KSp]   Katsura, H., and D. A. Sprecher. "Computational aspects of Kolmogorov's superposition theorem." *Neural Networks 7* (1994), 455–461.

[Ka]   Katznelson, Y. *Harmonic Analysis.* Wiley, New York, 1968. Reprint, Dover, New York, 1976.

[KN]   Kelley, J. L., I. Namioka, et al. *Linear Topological Spaces.* Van Nostrand, Princeton, NJ, 1963.

[Ke]   Kelley, J. L. *General Topology.* Graduate Texts in Mathematics 27. Springer-Verlag, New York, 1955.

[Kg]   Kergin, P. "A natural interpolation of C^k functions." *J. Approx. Theory 29* (1980), 278–293.

[Khv]   Khavinson, S. Y. *Best Approximation by Linear Superpositions.* Amer. Math. Soc., Providence, RI. *Transl. Math. Monog. 159* (1997).

[Ki2]   Kilgore, T. A. "A characterization of the Lagrange interpolating projection with minimal Tchebycheff norm." *J. Approx. Theory 24* (1978), 273–288.

[Ki1]   Kilgore, T. A. "Optimization of the norm of the Lagrange interpolation operator." *Bull. Amer. Math. Soc. 83* (1977), 1069–1071.

[KC]   Kilgore, T. A., and E. W. Cheney. "A theorem on interpolation in Haar subspaces." *Aequationes Math. 14* (1976), 391–400.

[KLSY]   Kilmer, S. J., W. A. Light, Xingping Sun, and Xiang Ming Yu. "Approximation by translates of positive definite functions." *J. Math. Anal. Appl. 201* (1996), 631–641.

[KinC]   Kincaid, D., and W. Cheney. *Numerical Analysis,* 2nd ed. Brooks/Cole, Pacific Grove, CA, 1996.

[Kl]   Klinger, A. "The Vandermonde matrix." *Amer. Math. Monthly 74* (1967), 571–574.

[KC]   Knowles, R. J., and T. A. Cook. "Some results on Auerbach bases for finite-dimensional normed spaces." *Bull. Soc. Roy. Sci. Liège* (1973), 518–522.

[Kol]   Koldobskii, A. L. "Schoenberg's problem on positive definite functions." *St. Petersburg Math. J. 3* (1992), 563–570.

[Kor]   Körner, T. W. *Fourier Analysis.* Cambridge University Press, Cambridge, UK, 1989.

[Kot]   Kotelnikov, V. A. "On the carrying capacity of the "ether" and wire in telecommunications." *Material for the First All-Union Conference on Questions of Communication, Izd. Red. Upr.* Svyazi RKKA, Moscow, 1933. (Russian)

[KSS]   Kowalski, M. A., K. A. Sikorski, and F. Stenger. *Selected Topics in Approximation and Computation.* Oxford University Press, Oxford, UK, 1995.

[Kro]   Kroó, A. "On approximation by ridge functions." *Constr. Approx. 13* (1997), 447–460.

[Kue]   Kuelbs, J. "Positive definite symmetric functions on linear spaces." *J. Math. Anal. Appl. 42* (1973), 413–426.

[Ky]   Kyriazis, G. C. "Approximation from shift-invariant spaces." *Constr. Approx. 11* (1995), 141–164.

[Lan]   Landau, H. J. (ed.). *Moments in Mathematics.* Amer. Math. Soc., Providence, RI, 1987.

[Law1]   Lawson, C. L. "C^1 surface interpolation for scattered data on a sphere." *Rocky Mt. J. Math. 14* (1984), 177–202.

[LeeS]   Lee, David, and J.-J. H. Shiau. "Thin plate splines with discontinuities and fast algorithms for their computation." *SIAM J. Sci. Comput. 15* (1994), 1311–1330.

[LTT]   Lee, S. L., R. C. Tan, and W. S. Tang. "L^2 approximation by the translates of a function." *Numer. Math. 60* (1992), 549–568.

[LBW]   Lee, Wee Sun, P. L. Bartlett, and R. C. Williamson. "Efficient agnostic learning of neural networks with bounded fan-in." *IEEE Trans. Inform. Theory 42* (1996), 2118–2132.

[Lei1]   Lei, Junjiang. "The multivariate Radon transform." *Approx. Theory Appl. 3* (1987), 30–49.

[Lei2]   Lei, Junjiang. "$L_p(R^d)$-Approximation by certain projection operators." *J. Math. Anal. Appl. 185* (1994), 1–14.

[Lei3]   Lei, Junjiang. "On approximation by translates of globally supported functions." *J. Approx. Theory 77* (1994), 123–138.

[LeC]   Lei, Junjiang, and E. W. Cheney. "Quasi-interpolation on irregular points." *Approximation and Computation,* ed. by R. V. M. Zahar. International Series on Numerical Mathematics vol. 119. Birkhäuser, Basel, 1994, pp. 121–136.

[LJC]   Lei, Junjiang, R. Q. Jia, and E. W. Cheney. "Approximation from shift-invariant spaces by integral operators." *SIAM J. Math. Analysis 28* (1997), 481–498.

[Lev1]   Levesley, J. "Pointwise estimates for multivariate interpolation using conditionally positive definite functions." In *Approximation Theory, Wavelets and Applications,* ed. by S. P. Singh. Kluwer, Dordrecht, 1995, pp. 381–401.

[Lev2]   Levesley, J. "Convolution kernels based on thin-plate splines." *Numer. Algorithms 10* (1995), 401–419.

[LLRS]   Levesley, J., W. Light, D. Ragozin, and Xingping Sun. "Variational theory for interpolation on spheres." In *IDOMAT Conference 1998.* Birkhäuser, Boston, 1998.

[LevS]   Levesley, J., and Xingping Sun. "Scattered Hermite interpolation by ridge functions." *Numer. Func. Anal. Optim. 16* (1995), 989–1001.

[Lew]   Lewicki, W. O. G. *Minimal Projections in Banach Spaces.* Lecture Notes in Math., vol. 1449. Springer-Verlag, Berlin, 1990.

[LC]   Light, W. A., and E. W. Cheney. *Approximation Theory in Tensor Product Spaces.* Lecture Notes in Math., vol. 1169. Springer-Verlag, Berlin, 1985.

[LC1]   Light, W. A., and E. W. Cheney. "Interpolation by periodic radial basis functions." *J. Math. Anal. Appl. 168* (1992), 111–130.

[LC2]   Light, W. A., and E. W. Cheney. "Quasi-interpolation with translates of a function having non-compact support." *Constr. Approx. 8* (1992), 35–48.

[L2]   Light, W. A. "Ridge functions, sigmoidal functions, and neural networks." In *Approximation Theory VII,* ed. by E. W. Cheney, C. K. Chui, and L. L. Schumaker. Academic Press, New York, 1992, pp. 163–206.

[L3]   Light, W. A. "Some aspects of radial basis function approximation." In *Approximation Theory, Spline Functions and Applications,* ed. by S. P. Singh. Kluwer, Dordrecht, 1992, pp. 163–190.

[L4]   Light, W. A. "Using radial functions on compact domains." In *Curves and Surfaces II,* ed. by P. J. Laurent, A. LeMéhauté, and L. L. Schumaker. A. K. Peters, Boston, 1991.

[L5]   Light, W. A. "Techniques for generating approximations via convolution kernels." *Numer. Algorithms 5* (1993), 247–261.

[LW]   Light, W. A., and H. Wayne. "Error estimates for approximation by radial basis functions." In *Approximation Theory, Wavelets and Applications,* ed. by S. P. Singh. Kluwer, Dordrecht, 1994, pp. 215–246.

[LW2]   Light, W., and H. Wayne. "Power functions and error estimates for radial basis function interpolation." *J. Approx. Theory 92* (1998), 245–267.

[LP]   Lin, V. Y., and A. Pinkus. "Fundamentality of ridge functions." *J. Approx. Theory 75* (1993), 295–311.

[LRS]   Linde, U., M. Reimer, and B. Sündermann. "Numerisches Berechnung extremaler Fundamentalsysteme für Polynomräume über der Vollkugel." *Computing 43* (1989), 37–45.

[Lo]   Locher, F. "Interpolation on uniform meshes by translates of one function and related attenuation factors." *Math. Comp. 37* (1981), 403–416.

[Log]   Logan, B. F. "The uncertainty principle in reconstructing functions from projections." *Duke Math. J. 42* (1975), 661–706.

[LoS]   Logan, B. F., and L. A. Shepp. "Optimal reconstruction of a function from its projections." *Duke Math. J. 42* (1975), 645–659.

[LB]   Lund, J., and K. L. Bowers. *SINC Methods for Quadrature and Differential Equations.* Society for Industrial and Applied Mathematics, Philadelphia, 1992.

[Luo]   Luo, Zuhua. "Fundamental and strictly conditionally positive definite kernels on compact groups." *Numer. Func. Anal. Optim.* (in press).

[LR]   Luttman, F. W., and T. J. Rivlin. "Some numerical experiments in the theory of polynomial interpolation." *IBM J. Res. Development 9* (1965), 187–191.

[Lut]   Lutts, J. A. "Topological spaces which admit unisolvent systems." *Trans. Amer. Math. Soc. 111* (1964), 440–448.

[Ma2]   Madych, W. R. "Degree of approximation in computerized tomography." In *Approximation Theory III,* ed. by E. W. Cheney. Academic Press, New York, 1980, pp. 615–621.

[Ma1]   Madych, W. R. "Summability and approximate reconstruction from Radon transform data." *Contemp. Math. 113* (1990), 189–219.

[MN1]   Madych, W. R., and S. A. Nelson. "Multivariate interpolation and conditionally positive definite functions, I." *J. Approx. Theory and Appl. 4* (1988), 77–89.

[MN2]   Madych, W. R., and S. A. Nelson. "Radial sums of ridge functions: A characterization." *Math. Meth. Appl. Sci. 7* (1985), 90–100.

[MN3]   Madych, W. R., and S. A. Nelson. "Multivariate interpolation and conditionally positive definite functions, II." *Math. Comp. 54* (1990), 211–230.

[MN6]   Madych, W. R., and S. A. Nelson. "Bounds on multivariate polynomials and exponential error estimates for multiquadric interpolation." *J. Approx. Theory 70* (1992), 94–114.

[MN7]   Madych, W. R., and S. A. Nelson. "Polynomial based algorithms for computed tomography." *SIAM J. Applied Math. 43* (1983), 157–185. Part II, ibid., *44* (1984), 193–208.

[MN8]   Madych, W. R., and S. A. Nelson. "Approximate inversion formulas for Radon transform data." In *Approximation Theory IV,* ed. by C. K. Chui, L. L. Schumaker, and J. D. Ward. Academic Press, New York, 1983, pp. 599–604.

[MN9]   Madych, W. R., and S. A. Nelson. "Characterization of tomographic reconstructions which commute with rigid motions." *J. Functional Analysis 46* (1982), 258–263.

[Mai]   Mairhuber, J. C. "On Haar's theorem concerning Chebyshev approximation problems having unique solution." *Proc. Amer. Math. Soc. 7* (1956), 609–615.

[Mak]   Makovoz, Y. "Random approximants and neural networks." *J. Approx. Theory 85* (1996), 98–109.

[Mal]   Mallat, S. G. "Multiresolution approximations and wavelet orthonormal bases of $L^2(R)$." *Trans. Amer. Math. Soc. 315* (1989), 69–87.

[MalZ] Mallat, S., and Zhifeng Zhang. "Matching pursuits with time–frequency dictionaries." *IEEE Trans. on Signal Processing 41* (1993), 3397–3415.

[McW]   McCullough, S., and D. E. Wulbert. "The topological spaces that support Haar systems." *Proc. Amer. Math. Soc. 94* (1985), 687–692.

[McS]   McShane, E. J. *Integration.* Princeton University Press, Princeton, NJ, 1944.

[Mein]   Meinguet, J. "Multivariate interpolation at arbitrary points made simple." *Zeit. Angew. Math. Phys. 30* (1979), 292–304.

[Mein2]   Meinguet, J. "An intrinsic approach to multivariate spline interpolation at arbitrary points." In *Polynomial and Spline Approximation,* ed. by B. N. Sahney. Reidel, Dordrecht, 1979, pp. 163–190.

[Men1]   Menegatto, V. A. *Interpolation on Spherical Spaces.* Unpublished doctoral dissertation, University of Texas at Austin, August, 1992.

[Men1]   Menegatto, V. "Interpolation on the complex Hilbert sphere using positive definite and conditionally negative definite kernels." *Acta Math. Hungar. 75* (1997), 215–225.

[Men2]   Menegatto, V. A. "Interpolation on spherical domains." *Analysis 14* (1994), 415–424.

[Men5]   Menegatto, V. A. "Strictly positive definite kernels on the Hilbert sphere." *Applicable Analysis 55* (1994), 91–101.

[Men6]   Menegatto, V. A. "Strictly positive definite kernels on the circle." *Rocky Mt. J. Math. 25* (1995), 1149–1163.

[Men7]   Menegatto, V. A. "Interpolation on the complex Hilbert sphere." *Approx. Theory Appl. 12* (1996), 31–39.

[Men9]   Menegatto, V. A. "Approximation by spherical convolution." *Numer. Func. Anal. Optim. 18* (1997), 995–1012.

[MenPe]   Menegatto, V., and A. P. Peron. "Generalized interpolation on spheres using positive definite and related functions." *Numer. Func. Anal. Optim. 18* (1997), 189–200.

[Mey1]   Meyer, Y. *Wavelets: Algorithms and Applications.* SIAM, Philadelphia, 1993.

[Mh1]   Mhaskar, H. N. "Approximation properties of a multilayered feedforward artificial neural network." *Advances in Comp. Math. 1* (1993), 61–80.

[Mh5]   Mhaskar, H. N. "Approximation of real functions using neural networks." In *Proceedings of International Conference on Computational Mathematics,* ed. by H. P. Dikshit and C. A. Micchelli. World Scientific Publishers, Singapore, 1994.

[MhH]   Mhaskar, H. N., and N. Hahm. "Neural networks for functional approximation and system identification." Preprint December 1995.

[MM1]   Mhaskar, H. N., and C. Micchelli. "Approximation by superpositions of a sigmoidal function." *Advances in Appl. Math. 13* (1992), 350–373.

[MM3]   Mhaskar, H. N., and C. A. Micchelli. "Dimension independent bounds on the degree of approximation by neural networks." *International Business Machines J. Res. 38* (1994), 277–284.

[M2]   Micchelli, C. A. "Algebraic aspects of interpolation." In *Approximation Theory,* ed. by C. de Boor. Proc. Symp. Appl. Math., vol. 36. Amer. Math. Soc., Providence, RI, 1986, pp. 81–102.

[M1]   Micchelli, C. A. "Interpolation of scattered data: Distance matrices and conditionally positive definite functions." *Constr. Approx. 2* (1986), 11–22.

[M5]   Micchelli, C. A. "Optimal estimation of linear operators from inaccurate data: A second look." *Numer. Algorithms 5* (1993), 375–390.

[MR1]   Micchelli, C. A., and T. J. Rivlin. "Lectures on optimal recovery." In *Numerical Analysis Lancaster 1984,* ed. by P. R. Turner. Lecture Notes in Math., vol. 1129. Springer-Verlag, Berlin, 1985, pp. 21–93.

[MR2]   Micchelli, C. A., and T. J. Rivlin. "A survey of optimal recovery." In *Optimal Estimation in Approximation Theory,* ed. by C. A. Micchelli and T. J. Rivlin. Plenum Press, New York, 1977, pp. 1–54.

[Mis]   Misiewicz, J. K. "Positive definite norm dependent functions on ℓ^∞." *Stat. and Probab. Letters 8* (1989), 255–260.

[RLM]   Moore, R. L. "Concerning triods in the plane and the junction points of plane continua." *Proc. Nat. Acad. Science 14* (1928), 85–88.

[MG]   Mühlbach, G., and M. Gasca. "Multivariate interpolation under projectivities, III." *Numer. Algorithms 8* (1994).

[Mul]   Mulcahy, C. "Plotting and scheming with wavelets." *Math. Magazine 69* (1996), 323–343.

[Mu]   Müller, C. *Spherical Harmonics.* Lecture Notes in Math., vol. 17. Springer-Verlag, Berlin, 1966.

[N1]   Narcowich, F. J. "Generalized Hermite interpolation and positive definite kernels on a Riemannian manifold." *J. Math. Anal. Appl. 190* (1995), 165–193.

[NSW]   Narcowich, F. J., N. Sivakumar, and J. Ward. "On condition numbers associated with radial-function interpolation." *J. Math. Anal. Appl. 186* (1994), 457–487.

[NW3]   Narcowich, F. J., and J. D. Ward. "Norm estimates for the inverses of a general class of scattered-data radial-function interpolation matrices." *J. Approx. Theory 69* (1992), 84–110.

[NW4]   Narcowich, F. J., and J. D. Ward. "Generalized Hermite interpolation via matrix-valued conditionally positive definite functions." *Math. Comp. 63* (1994), 661–687.

[NWal] Nashed, M. Z., and G. G. Walter. "General sampling theorems for functions in reproducing kernel Hilbert spaces." *Math. Control Signals Systems 4* (1991), 363–390.

[Na]   Natterer, F. *The Mathematics of Computerized Tomography.* Wiley, New York, 1986.

[Nea]   Neamtu, M. *A Contribution to the Theory and Practice of Multivariate Splines.* Unpublished doctoral dissertation, University of Twente, 1991.

[NS]   von Neumann, J., and I. J. Schoenberg. "Fourier integrals and metric geometry." *Trans. Amer. Math. Soc. 50* (1941), 226–251.

[Neu]   Neumann, G. *Boolesche Interpolatorische Kubatur.* Unpublished doctoral dissertation, Reutlinger, 1982.

[NR]   Newman, D. J., and T. J. Rivlin. "Optimal universally stable interpolation." *Analysis 3* (1983), 355–367.

[NX]   Newman, D. J., and Yuan Xu. "Tchebycheff polynomials on a triangular region." *Constr. Approx. 9* (1993), 543–546.

[NR1]   Nielson, G. M., and R. Ramaraj. "Interpolation on a sphere based upon a minimum norm network." *Computer Aided Geom. Design 4* (1987), 41–57.

[Nur]   Nürnberger, G. *Approximation by Spline Functions.* Springer-Verlag, Berlin, 1989.

[O]   Oberhettinger, F. *Fourier Transforms of Distributions and Their Inverses.* Academic Press, New York, 1973.

[OS]   Oppenheim, A. V., and R. W. Schafer. *Discrete Time Signal Processing.* Prentice Hall, Englewood Cliffs, NJ, 1989.

[ParS]   Park, Joo-Young, and I. W. Sandberg. "Universal approximation using radial basis function networks." *Neural Comp. 3* (1991), 246–257.

[Par]   Partington, J. R. *Interpolation, Identification, and Sampling.* London Math. Soc. Monographs, no. 17. Oxford University Press, Oxford, UK, 1997.

[Per]   Peretto, P. *An Introduction to the Modelling of Neural Networks.* Cambridge University Press, Cambridge, UK, 1992.

[PSS]   Petersen, B. E., K. T. Smith, and D. C. Solmon. "Sums of plane waves, and the range of the Radon transform." *Math. Ann. 243* (1979), 153–161.

[Phe1]   Phelps, R. R. *Lectures on Choquet's Theorem.* Van Nostrand, New York, 1966. Rev. ed., Ergebnisse der Math. Springer-Verlag, Berlin, 1984.

[Phe2]   Phelps, R. R. "Integral representation for elements of convex sets." In *Studies in Functional Analysis,* ed. by R. G. Bartle. Math. Assoc. of America, Washington, DC, 1980, pp. 115–157.

[Ph1]   Phillips, G. M. "Algorithms for piecewise straight line approximations." *Computer J. 11* (1968), 211–212.

[Pin2]   Pinkus, A. "N-Widths and optimal recovery." In *Approximation Theory,* ed. by C. de Boor, Proc. Symp. Appl. Math., vol. 36. Amer. Math. Soc., Providence, RI, 1986, pp. 51–66.

[Pi]   Pisier, G. "Remarques sur un résultat non publié de B. Maury." *Sem. d'Analyse Fonctionelle* 1980–1981, Exp. No. V, 1–12. Ecole Poly., Centre de Math., Palaiseau, 1981.

[Pon]   Pontryagin, L. S. *Foundations of Combinatorial Topology.* Graylock Press, Rochester, NY, 1952.

[PE1]   Pottmann, H., and M. Eck. "Modified multiquadric methods for scattered data interpolation over a sphere." *Computer Aided Geom. Design 7* (1990), 313–321.

[P1]   Powell, M. J. D. "Radial basis functions for multivariable interpolation." In *Algorithms for Approximation,* ed. by J. C. Mason and M. G. Cox. Oxford University Press, Oxford, UK, 1987, pp. 143–167.

[P3]   Powell, M. J. D. "The theory of radial basis function approximation in 1990." In *Advances in Numerical Analysis,* vol. II, ed. by W. A. Light. Oxford University Press, Oxford, UK, 1992, pp. 105–210.

[P6]   Powell, M. J. D. "The uniform convergence of thin plate spline interpolation in two dimensions." *Numer. Math. 68* (1994), 107–128.

[P7]   Powell, M. J. D. "Tabulation of thin plate splines on a very fine two-dimensional grid." In *Numerical Methods of Approximation Theory,* ed. by D. Braess and L. L. Schumaker. Birkhäuser, Basel, 1992, pp. 221–244.

[P8]   Powell, M. J. D. "Truncated Laurent expansions for the fast evaluation of thin plate splines." *Numer. Algorithms 5* (1993), 99–120.

[PSa]   Powell, M. J. D., and M. A. Sabin. "Piecewise quadratic approximation on triangles." *Assoc. for Comp. Mach. Trans. on Math. Software 3* (1977), 316–325.

[Pr]   Prolla, J. B. *Weierstrass-Stone: The Theorem.* Peter Lang, Frankfurt, 1993.

[QSW]   Quak, E., N. Sivakumar, and J. D. Ward. "Least squares approximation by radial functions." *SIAM J. Math. Analysis 24* (1993), 1043–1066.

[Ra]   Rabut, C. "How to build quasi-interpolants with application to polyharmonic B-splines." In *Curves and Surfaces,* ed. by P. J. Laurent, A. Le Méhauté, and L. L. Schumaker. Academic Press, New York, 1991, pp. 391–402.

[Rag2]   Ragozin, D. "Constructive polynomial approximation on spheres and projective spaces." *Trans. Amer. Math. Soc. 162* (1972), 157–170.

[RaSi]   Rassias, T. M., and J. Simsa. *Finite Sum Decompositions in Mathematical Analysis.* Wiley, New York, 1995. (Reviewed in *SIAM Rev. 39* (1997), 157–159. Review has 15 further references.)

[RS]   Reed, M., and B. Simon. *Methods of Modern Mathematical Physics,* vol. 1, *Functional Analysis,* 2nd ed. Academic Press, New York, 1980.

[RSu]   Reid, L., and Xingping Sun. "Distance matrices and ridge function interpolation." *Canad. Math. J. 45* (1993), 1313–1323.

[Rei1]   Reimer, M. "Extremal bases for normed linear spaces." In *Approximation Theory III,* ed. by E. W. Cheney. Academic Press, New York, 1980, pp. 723–728.

[Rei2]   Reimer, M. "Interpolation on the sphere and bounds for the Lagrangian square sums." *Resultate der Mathematik 11* (1987), 43–58.

[Ric]   Richards, D. St. P. "Positive definite symmetric functions on finite dimensional spaces, I." *J. Multivariate Analysis 19* (1986), 280–298. Part II, *Stat. Prob. Letters 3* (1985), 325–329.

[Riv1]   Rivlin, T. J. *Chebyshev Polynomials.* Wiley, New York, 1974. 2nd ed., 1990.

[Riv2]   Rivlin, T. J. "A survey of recent results on optimal recovery." In *Polynomial and Spline Approximation,* ed. by B. N. Sahney. Reidel, Boston, 1979, pp. 225–245.

[RS1]   Ron, A., and Xingping Sun. "Strictly positive definite functions on spheres." *Math. Comp. 65* (1996), 1513–1530.

[Ru1]   Rudin, W. *Fourier Analysis on Groups.* Interscience, New York, 1963.

[Ru2]   Rudin, W. *Functional Analysis.* McGraw-Hill, New York, 1973.

[Ru0]   Rudin, W. *Real and Complex Analysis.* 2nd ed., McGraw-Hill, New York, 1974.

[Rush]   Rushanan, J. J. "On the Vandermonde matrix." *Amer. Math. Monthly 96* (1989), 921–924.

[Rus]   Ruston, A. F. "Auerbach's theorem." *Proc. Camb. Phil. Soc. 58* (1964), 476–480.

[Ry]   Ryan, Patrick J. *Euclidean and Non–Euclidean Geometry.* Cambridge University Press, Cambridge, UK, 1986.

[Sai]   Saitoh, S. *Theory of Reproducing Kernels and Its Applications.* Longmans, London, 1988.

[Sand1]   Sandberg, I. W. "General structures for classification." *IEEE Trans. Circuits and Systems 41* (1994), 372–376.

[Sand2]   Sandberg, I. W. "Approximations for nonlinear functionals." *IEEE Trans. Circuits and Systems 39* (1992), 65–67.

[SD]   Sandberg, I. W., and A. Dingankar. "On approximation of linear functionals on L_p spaces." *IEEE Trans. Circuits Systems I Fund. Theory Appl. 42* (1995), 402–404.

[Sard]   Sard, A. *Linear Approximation,* Amer. Math. Soc., Providence, RI Math. Surveys, No. 9, Providence, RI, 1963.

[Sas]   Sasvári, Z. *Positive Definite and Definitizable Functions.* Akademie Verlag, Berlin, 1994.

[SX]   Sauer, T., and Yuan Xu. "On multivariate Lagrange interpolation." *Math. Comp. 64* (1995), 1147–1170.

[Sck1]   Schaback, R. "Error estimates and condition numbers for radial basis function interpolation." *Adv. Comput. Math. 3* (1995), 251–264.

[Sck2]   Schaback, R. "Comparison of radial basis function interpolants." Preprint, March 1994.

[Sck3]   Schaback, R. "Reproduction of polynomials by radial basis functions." Preprint, March 1994.

[Sck4]   Schaback, R. "Approximation by radial basis functions with finitely many centers." *Constr. Approx. 12* (1996), 331–340.

[Sck5]   Schaback, R. "Lower bounds for norms of inverses of interpolation matrices for radial basis functions." *J. Approx. Theory 79* (1994), 287–306.

[SckW]   Schaback, R., and H. Wendland. "Special cases of compactly supported radial basis functions." Preprint, March 1994.

[Schat]   Schatten, R. *A Theory of Cross-Spaces.* Annals of Math. Studies, vol. 26. Princeton University Press, Princeton, NJ, 1950.

[S1]   Schoenberg, I. J. "Metric spaces and completely monotone functions." *Annals of Math. 39* (1938), 811–841.

[S2]   Schoenberg, I. J. "On certain metric spaces arising from Euclidean spaces." *Annals of Math. 38* (1937), 787–793.

[S3]   Schoenberg, I. J. "Positive definite functions on spheres." *Duke Math. J. 9* (1942), 96–108.

[S4]   Schoenberg, I. J. "Metric spaces and positive definite functions." *Trans. Amer. Math. Soc. 44* (1938), 522–536.

[S5]   Schoenberg, I. J. "Remarks to Maurice Frechet's article." *Annals of Math. 36* (1935), 724–732.

[S7]   Schoenberg, I. J. "Contributions to the problem of approximation of equidistant data by analytic functions." *Quart. Appl. Math 4* (1946), 45–99, 112–141.

[S10]   Schoenberg, I. J. "On the question of unicity in the theory of best approximation." *New York Acad. Sci. Annals 86* (1960), 682–692.

[SY]   Schoenberg, I. J., and C. T. Yang. "On the unicity of solutions of problems of best approximation." *Annali de Matematica Pura ed Applicata 54* (1961), 1–12.

[Scho2]   Schönhage, A. "Fehlerfortpflanzung bei Interpolation." *Numer. Math. 3* (1961), 62–71.

[Schr2]   Schreiner, M. "Locally supported kernels for spherical spline interpolation." *J. Approx. Theory 89* (1997), 172–194.

[Sch2]   Schumaker, L. L. *Spline Functions: Basic Theory.* Wiley, New York, 1981.

[SchW]   Schumaker, L. L., and G. Webb (eds.). *Recent Advances in Wavelet Analysis.* Academic Press, New York, 1994.

[See]   Seeley, R. T. "Spherical harmonics." *Amer. Math. Monthly 73* (Part II) (1966), 115–121.

[Sem]   Semadini, Z. *Banach Spaces of Continuous Functions.* Polish Scientific Publishers, Warsaw, 1971.

[Shan]   Shannon, C. "Communication in the presence of noise." *Proc. IRE 37* (1949), 10–21.

[Shap1]   Shapiro, H. S. *Topics in Approximation Theory.* Lecture Notes in Math., vol. 187. Springer-Verlag, New York, 1971.

[Shap2]   Shapiro, H. S. "Convergence almost everywhere of convolution integrals with a dilation parameter." In *Constructive Theory of Functions of Several Variables.* Lecture Notes in Math., vol. 571. Springer-Verlag, New York, 1977, pp. 250–266.

[Shap3]   Shapiro, H. S. *Smoothing and Approximation of Functions.* Van Nostrand, New York, 1969.

[Shek1] Shekhtman, B. "Interpolating subspaces in R^n." In *Approximation Theory, Wavelets and Applications,* ed. by S. P. Singh. Kluwer, Dordrecht, 1995, pp. 465–471.

[Sh] Shepard, D. "A two-dimensional interpolation function for irregularly spaced data." *Proc. 23rd Nat. Conf. Assoc. Comput. Mach.* (1968), pp. 517–523.

[Shepp] Shepp, L. A. (ed.). *Computed Tomography.* Proc. Symp. in Applied Math., vol. 27. Amer. Math. Soc., Providence, RI, 1983.

[SK] Shepp, L. A., and J. B. Kruskal. "Computerized tomography: The new medical x-ray technology." *Amer. Math. Monthly 85* (1978), 420–439.

[Shi] Shi, Ying Guang. "A minimax problem admitting the equioscillation characterization of Bernstein and Erdős." *J. Approx. Theory 92* (1998), 463–471.

[SG] Shilov, G. E., and B. L. Gurevich. *Integral, Measure and Derivative: A Unified Approach.* Prentice Hall, Englewood Cliffs, NJ, 1966.

[ST] Shohat, J. A., and J. D. Tamarkin. *The Problem of Moments.* Amer. Math. Soc., Providence, RI, 1943.

[SS] Sibson, R., and G. Stone. "Computation of thin-plate splines." *SIAM J. Sci. Statist. Comput. 12* (1992), 1304–1313.

[Si] Sieklucki, K. "Topological properties of sets admitting Tchebycheff systems." *Bull. Acad. Polon. Sci. Ser. Sci. Math. 6* (1958), 603–606.

[Sim] Simmons, G. F. *Topology and Modern Analysis.* McGraw-Hill, New York, 1963.

[Sin] Singer, I. *Best Approximation in Normed Linear Spaces by Elements of Linear Subspaces.* Springer-Verlag, Berlin, 1970.

[Siv] Sivakumar, N. "A note on the Gaussian cardinal interpolation operator." *Proc. Edinburgh Math. Soc. 40* (1997), 137–149.

[SiWa] Sivakumar, N., and J. D. Ward. "On the least squares fit by radial functions to multidimensional scattered data." *Numer. Math. 65* (1993), 219–243.

[Sm] Smith, K. T. "Reconstruction formulas in computed tomography." In *Computed Tomography,* ed. by L. A. Shepp. Proc. Symp. in Applied Math., vol. 27. Amer. Math. Soc., Providence, RI, 1983, pp. 7–24.

[Sm2] Smith, K. T. *Primer of Modern Analysis.* Bogden and Quigley, Boston, 1971. Reprint: Springer-Verlag, New York.

[SSW] Smith, K. T., D. C. Solmon, and S. L. Wagner. "Practical and mathematical aspects of reconstructing objects from radiographs." *Bull. Amer. Math. Soc. 83* (1977), 1227–1270.

[SWa] Smith, P. W., and J. D. Ward. "Quasi-interpolants from spline interpolation operators." *Constr. Approx. 6* (1990), 97–110.

[Sol] Solmon, D. C. "The x-ray transform." *J. Math. Anal. Appl. 71* (1976), 61–83.

[Sp3] Späth, H. *Mathematical Algorithms for Linear Regression.* Academic Press, New York, 1991.

[Spl] Splettstösser, W. "Some extensions of the sampling theorem." In *Linear Spaces and Appproximation,* ed. by P. L. Butzer and B. Sz. Nagy. International Series on Numerical Mathematics, vol. 40. Birkhäuser, Basel, 1978, pp. 615–628.

[Stef] Steffensen, J. F. *Interpolation.* Chelsea, New York, 1950. Amer. Math. Soc., Providence, RI, 1998.

[SW] Stein, E. M., and G. Weiss. *Introduction to Fourier Analysis on Euclidean Spaces.* Princeton University Press, Princeton, NJ, 1971.

[Sten]   Stenger, F. *Numerical Methods Based on Sinc and Analytic Functions.* Springer-Verlag, Berlin, 1993. Review: *Math. Intell. 18* (2), 1996, 71–73.

[St]   Stewart, J. "Positive definite functions and generalizations, an historical survey." *Rocky Mt. J. Math. 6* (1976), 409–434.

[Sto]   Stone, M. H. "A generalized Weierstrass approximation theorem." In *Studies in Modern Analysis,* ed. by R. C. Buck. Math Assoc. Amer., Washington, DC, 1962, pp. 30–87.

[Stra2]   Strang, G. "Wavelets and dilation equations." *SIAM Review 31* (1989), 613–627.

[Stra3]   Strang, G. "Eigenvalues and the convergence of the cascade algorithm." *IEEE Trans. Signal Proc. 44* (1996), 233–238.

[StN]   Strang, G., and T. Nugyen. *Wavelets and Filter Banks.* Wellesley–Cambridge Press, Wellesley, MA, 1996.

[Strau]   Strauss, H. "On interpolation with products of positive definite function." *Numer. Algorithms 15* (1997), 153–165.

[Stri]   Strichartz, R. S. "Radon inversion—variations on a theme." *Amer. Math. Monthly 89* (1982), 377–384.

[Su1]   Sun, Xingping. *Multivariate Interpolation Using Ridge or Related Functions.* Unpublished doctoral dissertation, University of Texas at Austin, August 1990.

[Su2]   Sun, Xingping. "Conditionally positive definite functions and their application to multivariate interpolation." *J. Approx. Theory 74* (1993), 159–180.

[Su3]   Sun, Xingping. "On the solvability of radial function interpolation." In *Approximation Theory VI,* ed. by C. Chui, L. Schumaker, and J. Ward. Academic Press, New York, 1989, pp. 643–646.

[Su5]   Sun, Xingping. "Norm estimates for inverses of Euclidean distance matrices." *J. Approx. Theory 70* (1992), 339–347.

[Su6]   Sun, Xingping. "Cardinal and scattered-cardinal interpolation by functions having non-compact support." *Computers and Mathematics with Application 24* (1997), 195–200.

[Su8]   Sun, Xingping. "Ridge function spaces and their interpolation property." *J. Math. Anal. Appl. 179* (1993), 28–40.

[Su9]   Sun, Xingping. "Cardinal Hermite interpolation using positive definite functions." *Numer. Algorithms 7* (1994), 253–268.

[Su10]   Sun, Xingping. "Scattered Hermite interpolation using radial basis functions." *Linear Alg. Appl. 207* (1994), 135–146.

[Su11]   Sun, Xingping. "The fundamentality of translates of a continuous function on spheres." *Numer. Algorithms 8* (1994), 131–134.

[SC1]   Sun, Xingping, and E. W. Cheney. "The fundamentality of sets of ridge functions." *Aequationes Math. 44* (1992), 226–235.

[SC2]   Sun, Xingping, and E. W. Cheney. "Fundamental sets of continuous functions on spheres." *Constr. Approx. 13* (1997), 245–250.

[SuG]   Sun, Xingping, and K. Guo. "Scattered data interpolation by linear combinations of translates of conditionally positive definite functions." *Numer. Func. Anal. Optim. 12* (1991), 137–152.

[Sund2]   Sündermann, B. "Lebesgue constants in Lagrangian interpolation at the Fekete points." *Mitt. Math. Ges. Hamburg 11* (1983), 204–211.

[Sund3] Sündermann, B. "On projection constants of polynomial spaces on the unit ball in several variables." *Math. Zeit. 188* (1984), 111–117.

[Syn] Synge, J. L. *The Hypercircle in Mathematical Physics.* Cambridge University Press, Cambridge, UK, 1957.

[SzV] Szabados, J., and P. Vértesi. *Interpolation of Functions.* World Scientific Publishers, Singapore, 1990.

[Sz] Szegő, G. *Orthogonal Polynomials.* Colloquium Publ., vol. 23. Amer. Math. Soc., Providence, RI, 1959.

[Tay2] Taylor, Michael E. *Noncommutative Harmonic Analysis.* Mathematical Surveys and Monographs, vol. 22. Amer. Math. Soc., Providence, RI, 1986.

[TD] Thiran, J. P., and C. Detaille. "On real and complex–valued bivariate Chebyshev polynomials." *J. Approx. Theory 59* (1989), 321–337.

[T] Titchmarsh, E. C. *The Theory of Functions,* 2nd ed. Oxford University Press, London, 1939.

[TW] Traub, J. F., and H. Wozniakowski. *A General Theory of Optimal Algorithms.* Academic Press, New York, 1980.

[Ver] Vértesi, P. "On the optimal Lebesgue constants for polynomial interpolation." *Acta Math. Hungar. 47* (1986), 165–178.

[Ver2] Vértesi, P. "Optimal Lebesgue constant for Lagrange interpolation." *SIAM J. Numer. Anal. 27* (1990), 1322–1331.

[Wah1] Wahba, G. "Spline interpolation and smoothing on the sphere." *SIAM J. Sci. Statist. Comput. 2* (1981), 5–16. Ibid. *4* (1982), 385–386.

[Walt] Walter, G. G. *Wavelets and Other Orthogonal Systems with Applications.* CRC Press, Boca Raton, FL, 1994.

[Walz] Walz, G. *Asymptotics and Extrapolation.* Math. Res. 88. Akademie Verlag, Berlin, 1996.

[Wa1] Watkins, D. A. "A generalization of the Bramble-Hilbert lemma and applications to multivariate interpolation." *J. Approx. Theory 26* (1979), 219–231.

[Wa] Watkins, D. S. "Error bounds for polynomial blending function methods." *SIAM J. Numer. Anal. 14* (1977), 721–734.

[Watdf] Watson, D. F. *Contouring: A Guide to the Analysis and Display of Spatial Data.* Pergamon Press, Oxford, 1992.

[WatP] Watson, D. F., and G. M. Phillips. "Neighborhood-based interpolation." *Geobyte 2* (1987), 12–16.

[WeW] Wells, J. H., and L. R. Williams. *Embeddings and Extensions in Analysis.* Springer-Verlag, New York, 1975.

[Wend] Wendland, H. "Error estimates for interpolation by compactly supported radial basis functions of minimal degree." *J. Approx. Theory 93* (1998), 258–272.

[Wer] Werner, D. "The Daugavet equation for operators on function spaces." *J. Functional Analysis 143* (1997), 117–128.

[We] Werner, H. "Remarks on Newton type multivariate interpolation for subsets of grids." *Computing 25* (1980), 181–191.

[Whit1] Whittaker, E. T. "On the functions which are represented by the expansions of the interpolation theory." *Proc. Roy. Soc. Edinburgh 35* (1915), 181–194.

[Whit2] Whittaker, J. M. "The Fourier theory of the cardinal function." *Proc. Math. Soc. Edinburgh 1* (1929), 169–176.

[Whit3] Whittaker, J. M. *Interpolatory Function Theory.* Cambridge University Press, Cambridge, UK, 1935.

[W4] Widder, D. V. *Advanced Calculus,* 2nd ed. Prentice-Hall, Englewood Cliffs, NJ, 1961. Reprint, Dover, New York.

[W1] Widder, D. V. "Necessary and sufficient conditions for the representation of a function as a Laplace integral." *Trans. Amer. Math. Soc. 33* (1932), 851–892.

[W3] Widder, D. V. *An Introduction to Transform Theory.* "Pure and Applied Mathematics Series," vol. 42, ed. by P. A. Smith and S. Eilenberg. Academic Press, New York, 1971.

[W5] Widder, D. V. *The Laplace Transform.* Princeton University Press, Princeton, NJ, 1946.

[Wot] Wojtaszczyk, P. "Some remarks on the Daugavet equation." *Proc. Amer. Math. Soc. 115* (1992), 1047–1052.

[Woj2] Wojtaszczyk, P. *A Mathematical Introduction to Wavelets.* Cambridge University Press, Cambridge, UK, 1997.

[WuS] Wu, Z.-M., and R. Schaback. "Local error estimates for radial basis function interpolation of scattered data." *Inst. Math. Appl. J. Numer. Anal. 13* (1993), 13–27.

[Wul1] Wulbert, D. "Interpolation at a few points." *J. Approx. Theory 96* (1999), 139–148.

[XC1] Xu, Yuan, and E. W. Cheney. "Interpolation by periodic radial functions." *Computers Math. Applic. 24* (1992), 201–215.

[XC2] Xu, Yuan, and E. W. Cheney. "Strictly positive definite functions on spheres." *Proc. Amer. Math. Soc. 116* (1992), 977–981.

[XLC] Xu, Yuan, W. A. Light, and E. W. Cheney. "Constructive methods of approximation by ridge functions and radial functions." *Numer. Algorithms 4* (1993), 205–223.

[XLLC] Xu, Yuan, J. Levesley, W. A. Light, and E. W. Cheney. "Convolution operators for radial basis approximations." *SIAM J. Math. Analysis 27* (1996), 286–304.

[Yos] Yosida, Kôsaku. *Functional Analysis,* 2nd ed. Springer-Verlag, New York, 1968.

[Z1] Zalik, R. A. "Approximation by nonfundamental sequences of translates." *Proc. Amer. Math. Soc. 78* (1980), 261–265.

[Z2] Zalik, R. A. "On approximation by shifts and a theorem of Wiener." *Trans. Amer. Math. Soc. 243* (1978), 299–307.

[Z3] Zalik, R. A. "On fundamental sequences of translates." *Proc. Amer. Math. Soc. 79* (1980), 255–259.

[Zay] Zayed, Ahmed I. *Advances in Shannon's Sampling Theory.* CRC Press, Boca Raton, FL, 1993.

[Zi] Zielke, R. *Discontinuous Čebyšev Systems.* Lecture Notes in Math., vol. 707. Springer-Verlag, New York, 1979.

[Zhao] Zhao, Kang. "Density of dilates of a shift-invariant subspace." *J. Math. Anal. Appl. 184* (1994), 517–532.

[ZhZh] Zhang, Ren Jiang, and Songping Zhou. "Three conjectures on Shepard interpolatory operators." *J. Approx. Theory 93* (1998), 399–414.

[Zw] Zwart, P. B. "Multivariate splines with non-degenerate partitions." *SIAM J. Numer. Anal. 10* (1973), 665–673.

Index

Index of Symbols